Uncertain Dynamical Systems

Stability and Motion Control

PURE AND APPLIED MATHEMATICS

A Program of Monographs, Textbooks, and Lecture Notes

MONOGRAPHS AND TEXTBOOKS IN PURE AND APPLIED MATHEMATICS

Recent Titles

Mircea Sofonea, Weimin Han, and Meir Shillor, Analysis and Approximation of Contact Problems with Adhesion or Damage (2006)

Marwan Moubachir and Jean-Paul Zolésio, Moving Shape Analysis and Control: Applications to Fluid Structure Interactions (2006)

Alfred Geroldinger and Franz Halter-Koch, Non-Unique Factorizations: Algebraic, Combinatorial and Analytic Theory (2006)

Kevin J. Hastings, Introduction to the Mathematics of Operations Research with *Mathematica®*, Second Edition (2006)

Robert Carlson, A Concrete Introduction to Real Analysis (2006)

John Dauns and Yiqiang Zhou, Classes of Modules (2006)

N. K. Govil, H. N. Mhaskar, Ram N. Mohapatra, Zuhair Nashed, and J. Szabados, Frontiers in Interpolation and Approximation (2006)

Luca Lorenzi and Marcello Bertoldi, Analytical Methods for Markov Semigroups (2006)

M. A. Al-Gwaiz and S. A. Elsanousi, Elements of Real Analysis (2006)

Theodore G. Faticoni, Direct Sum Decompositions of Torsion-Free Finite Rank Groups (2007)

R. Sivaramakrishnan, Certain Number-Theoretic Episodes in Algebra (2006)

Aderemi Kuku, Representation Theory and Higher Algebraic K-Theory (2006)

Robert Piziak and P. L. Odell, Matrix Theory: From Generalized Inverses to Jordan Form (2007)

Norman L. Johnson, Vikram Jha, and Mauro Biliotti, Handbook of Finite Translation Planes (2007)

Lieven Le Bruyn, Noncommutative Geometry and Cayley-smooth Orders (2008)

Fritz Schwarz, Algorithmic Lie Theory for Solving Ordinary Differential Equations (2008)

Jane Cronin, Ordinary Differential Equations: Introduction and Qualitative Theory, Third Edition (2008)

Su Gao, Invariant Descriptive Set Theory (2009)

Christopher Apelian and Steve Surace, Real and Complex Analysis (2010)

Norman L. Johnson, Combinatorics of Spreads and Parallelisms (2010)

Lawrence Narici and Edward Beckenstein, Topological Vector Spaces, Second Edition (2010)

Moshe Sniedovich, Dynamic Programming: Foundations and Principles, Second Edition (2010)

Drumi D. Bainov and Snezhana G. Hristova, Differential Equations with Maxima (2011)

Willi Freeden, Metaharmonic Lattice Point Theory (2011)

Murray R. Bremner, Lattice Basis Reduction: An Introduction to the LLL Algorithm and Its Applications (2011)

Clifford Bergman, Universal Algebra: Fundamentals and Selected Topics (2011)

A. A. Martynyuk and Yu. A. Martynyuk-Chernienko, Uncertain Dynamical Systems: Stability and Motion Control (2012)

MONOGRAPHS AND TEXTBOOKS IN
PURE AND APPLIED MATHEMATICS

Recent Titles

Uncertain Dynamical Systems

Stability and Motion Control

Stability and Motion Control

Stability and Motion Control

A. A. Martynyuk
Institute of Mechanics
Kiev, Ukraine

Yu. A. Martynyuk-Chernienko
Institute of Mechanics
Kiev, Ukraine

CRC Press
Taylor & Francis Group
Boca Raton London New York

CRC Press is an imprint of the
Taylor & Francis Group, an **informa** business

A CHAPMAN & HALL BOOK

CRC Press
Taylor & Francis Group
6000 Broken Sound Parkway NW, Suite 300
Boca Raton, FL 33487-2742

First issued in paperback 2019

© 2012 by Taylor & Francis Group, LLC
CRC Press is an imprint of Taylor & Francis Group, an Informa business

No claim to original U.S. Government works

ISBN-13: 978-1-4398-7685-5 (hbk)
ISBN-13: 978-0-367-38207-0 (pbk)

Visit the Taylor & Francis Web site at
http://www.taylorandfrancis.com

and the CRC Press Web site at
http://www.crcpress.com

Contents

Contents

Preface

In this book the methods of analysis of the stability of motion of dynamic systems, whose parameters are specified uncertainly (uncertain systems, for short) are described. The objective of the analysis is to construct sufficient conditions under which the motions of the uncertain systems have a certain type of stability with respect to some moving set or the zero equilibrium state. Lyapunov functions applied in the analysis have the form of scalar, vector, or matrix-valued auxiliary functions.

In Chapter 1 the reader will find a brief review of the results preceding the setting of problems of stability for systems with uncertain parameter values. The setting of the problem of the parametric stability is discussed, as well as the problem of the stability of solutions with respect to a moving invariant set.

In Chapter 2 the stability of the class of uncertain systems is analyzed, which are described by ordinary differential equations. The sufficient conditions for the stability (instability) of different types with respect to a moving invariant (conditionally invariant) set are determined on the basis of the generalized direct Lyapunov method.

In the same chapter the application of the scalar Lyapunov function is considered, and some simple examples are given. The problem of the exponential convergence of motions with respect to a moving invariant set is a certain development of the analyses from the article by Corless and Leitmann.

In Chapter 3 the motions of an uncertain controlled system with respect to a moving conditionally invariant set are analyzed. The problem of the synthesis of controls is solved on the basis of the direct Lyapunov method and the inverse transformation of the initial system, the concept of which goes back to Bendicson.

In Chapter 4 a method of application of the canonical matrix Lyapunov function to the analysis of the stability of motions of a quasilinear time invariant system is proposed. The stability conditions are expressed in terms of the sign definiteness of special matrices.

In Chapter 5 the reader will find the results of the analysis of stability of large-scale uncertain systems. In this case the parameters of functions of connection between the subsystems are uncertain. The vector and the hierarchical Lyapunov functions are the main tools for the analysis of dynamic properties of systems of that kind.

In Chapter 6 the problems of the stability of uncertain systems are analyzed on the basis of new approaches in construction and application of vector Lyapunov functions. Some of the obtained results are illustrated by the example of the choice of parameters of a stable mechanical system and the parametrical stability of a time invariant large-scale system.

In Chapter 7 the reader will find the results of the analysis of stability of an uncertain impulsive system. Those results were obtained under new assumptions on the dynamic properties of the continuous and discrete components of an impulsive system. Under those assumptions it seems to be impossible to apply the known results from the monographs by Lakshmikantham, Bainov and Simeonov, Samoilenko and Perestyuk and others. As an example we consider robust stability of the uncertain impulsive system.

In Chapter 8 uncertain dynamical equations are considered on a time scale. Using the generalized direct Lyapunov method, the sufficient conditions of different types of stability and instability of the zero solution of equations of that time are obtained.

In Chapter 9 the singularly perturbed differential equation is considered. Further the analysis of the stability of the systems with uncertain structure is performed on the basis of matrix-valued Lyapunov's functions.

In Chapter 10 we discuss analysis of the stability of a set of systems of differential equations.

In Chapter 11 we discuss the main ideas of the comparison principle for differential equations with a robust causal operator.

Chapter 12 is devoted to the investigation of stability of a set of impulsive equations via the method of heterogeneous matrix-valued Lyapunov-like functions.

Acknowledgments

First of all, we would like to thank academician A.N. Guz, director of the Institute of Mechanics NAS of Ukraine for the excellent conditions for our scientific work at the Institute during many years.

Many thanks to Professors Pierre Borne (Ecole Centralle de Lille, France), Teodore Burton (Northwest Research Institute, USA) and Efim Galperin (Montreal Polytechnic Institute, Canada) for their careful reading of some parts of the book and their criticism, corrections and suggestions.

Our thanks also go to Professors M. Corless, V. Lakshmikantham, D. Šiljak and A. Vatsala for sending us many copies of their papers in the area of uncertain systems dynamics.

We thank, too, the collaborators of the Department of Processes Stability of Institute of Mechanics NAS of Ukraine for their technical and computer programming support of the book.

Finally, we are grateful to the editors and production staff of Chapman & Hall/CRC Press for their assistance, good ideas, and patience in dealing with publication of this book.

Kiev
A.A.Martynyuk
Yu.A.Martynyuk-Chernienko

Chapter 1

Introduction

The theory of stability of the motion of systems with a finite number of degrees of freedom which was created by A.M. Lyapunov [1935] rests upon the three basic concepts:

(1) deviations of the perturbed motion from the nominal one must be infinitely small;

(2) during the motion there are no perturbing forces, i.e., in the equations of motion all factors of external or internal perturbations should be taken into account;

(3) the functioning interval of the system is unlimited.

There is an implicit assumption that

(4) the parameters of the system, within the accuracy of measurements, are fixed and do not change during the motion.

The wide application of general results of the stability theory started during the World War II, when the first problems were those of the stability of controlled systems, problems of the stability of flat trajectories of shells, and also problems of the stability of motion of hollow-charge shells.

Now the development of the theory of stability of motion is motivated by new engineering problems and by the need for the improvement of the existing methods of investigation of mechanical and other systems.

Within the recent years the theory of stability of motion of mechanical systems with uncertainly defined parameters has been elaborated intensely. This is due to the fact that in many complex engineering systems their multimode operation and multifunctionality can only be ensured under the condition that certain requirements are fulfilled in a fixed, though sufficiently wide range of variation of characteristics of parameters of separate subsystems. In other words, while developing a system design one should provide for long life of its practical operation under the conditions of uncertainties. This work is being done in several directions.

One of them is based on the assumption that parameters of a system (a linear one, as a rule) are changed on a certain interval and there are no controls. Such systems are called interval dynamic systems. Estimates of the stability of their motion are based on the method of characteristic equations

1

or its generalizations. The results obtained in this direction are presented in the review by Šiljak [1978].

Along with the ideas of the method of Lyapunov functions, interval systems (continuous and discrete ones) have been analyzed in many papers.

Another direction is concerned with the investigation of the dynamics of systems under control, i.e., a desired dynamic property of a system with uncertain values of its parameters is achieved through the application of external actions. Here both algebraic and qualitative methods of analysis of the interval stability are used. Here one can mention the H_∞-control method which is closely connected with the solution of the algebraic Riccati equation (see Aliev and Larin [1998] et al.). A generalized view on the role of Riccati equations and many results obtained in this direction are presented in the joint monograph edited by Bittanti, Laub, Willems [1991].

Some time ago Boyd et al. [1994] noted that many problems arising in the system theory, including the H_∞-control of uncertain linear systems, can be reduced to the analysis of linear matrix inequalities.

This approach is closely connected with the works of Yakubovich [1963] and others. For linear differential inclusions it was developed and presented in the above mentioned monograph by Boyd et al. [1994], where

(a) uncertainties parameters in the linear system

$$\frac{dx}{dt} = A(t)x + B_w(t)w$$

are determined by the polytope

$$[A(t),\, B_w(t)] \in \mathrm{co}\big\{[A_1 B_{w1}],\ldots,[A_L B_{wL}]\big\},$$

where $\mathrm{co}S$ is a convex envelope of the set $S \subset R^n$, determined by the formula

$$\mathrm{co}S \triangleq \bigg\{ \sum_{i=1}^{p} \lambda_i x_i \,\big|\, x_i \in S,\ p \geq 0 \bigg\};$$

(b) uncertainties parameters in the system

$$\frac{dx}{dt} = Ax + B_p p + B_w w, \quad q = C_q x, \quad p = \Delta(t)q$$

are bounded in norm

$$\|\Delta(t)\| \leq 1$$

and finally,

(c) uncertainties parameters in the linear system

$$\frac{dx}{dt} = Ax + B_p p + B_w w, \quad q = C_q x, \quad p_i = \delta_i(t)q_i$$

are bounded in absolute value, i.e. $|\delta_i(t)| \leq 1$, $i = 1, 2, \ldots, n_q$.

Common in those analyzes is the search of sufficient (and sometimes also necessary) conditions for the stability or the asymptotic stability of the zero solution of an uncertain system of equations of perturbed motion.

One of the effective approaches applied in the analysis of such kind of systems is the method of Lyapunov functions. Both scalar Lyapunov functions (see Corless and Leitmann [1986] and [1999], Gutman [1979], Corless [1990], Leitmann [1990] and [1993], Rotea and Khargonekar [1989], and others) and vector Lyapunov functions are used (see Chen [1996] and the list of literature to it).

Thus, the construction of the theory of stability of systems with uncertain values of their parameters (both controlled and free ones) is an actual problem of current investigations in nonlinear dynamics and the system theory.

Let us deal with two general concepts of stability for systems with uncertain (changing) parameters.

1.1   Parametric Stability

The presence of physical parameters in a real system, which can change their values in the process of functioning of the latter, presupposes the adequate reflection of this phenomenon in the mathematical model of the process and/or the system. As against the situation when the parameters of a system are fixed, changing parameters (or one parameter) causes the occurrence of new equilibrium states of the system. This circumstance does not allow the direct application of the generally used technique of analysis of stability, which was developed for the unique equilibrium state of the system.

The concept of the parametric stability which takes in account this phenomenon, as per Ikeda, Ohta and Šiljak [1991], is formulated as follows.

Consider the time invariant system

$$\frac{dx}{dt} = f(x, \alpha), \tag{1.1}$$

where $x(t) \in R^n$ is the state of the system (1.1) at the point $t \in R$, $\alpha \in R^d$ is a vector physical parameter, $f \in C(R^n \times R^d, R^n)$ is a sufficiently smooth function such that at any $\alpha \in R^d$ and the initial state x_0 at a point $t_0 = 0$ the system (1.1) has the unique solution $x(t; x_0, \alpha) = x(t, \alpha)$.

Suppose that for some nominal value α^* of the parameter α there exists an equilibrium state x^* of the system (1.1), i. e.

$$f(x^*, \alpha^*) = 0,$$

and x^* is stable.

Let the parameter α change its value from α^* to some other value. Here some questions of qualitative analysis of the system (1.1) arise, including the following ones:

* do any new equilibrium states x^e of the system (1.1) exist, and how "far" are they from the state x^*?

* if the equilibrium state x^e exists and is stable, is this stability of the same type as the stability of the equilibrium state x^*?

* is the change of equilibrium states of the system (1.1) from x^* to x^e followed by the loss of stability when the parameter is changed from α^* to α?

The answers to those questions can be found within the framework of the concept of the parametric stability of the system (1.1).

Consider the equilibrium state $x^e \colon R^d \to R^n$ as a function of vector parameter $x^e = x^e(\alpha)$ at $\alpha \in R^d$, and note the following definitions.

Definition 1.1 The system (1.1) is *parametrically stable* at the value of parameter $\alpha^* \in R^d$, if there exists an open neighbourhood $N(\alpha^*)$ of the value of parameter α^* such that for any $\alpha \in N(\alpha^*)$ the following conditions are satisfied:

(a) there exists an equilibrium state $x^e(\alpha) \in R^n$;

(b) for an arbitrary $\varepsilon > 0$ there exists $\delta = \delta(\varepsilon, \alpha) > 0$ such that the condition

$$\|x_0 - x^e(\alpha)\| < \delta$$

implies the estimate

$$\|x(t; x_0, \alpha) - x^e(\alpha)\| < \varepsilon$$

at all $t \in R_+$.

Definition 1.2 The system (1.1) is *asymptotically parametrically stable* at the values of parameter $\alpha^* \in R^d$, if it:

(a) is parametrically stable at the value α^* of the parameter α and;

(b) at all $\alpha \in N(\alpha^*)$ there exists a number $\mu(\alpha) > 0$ such that the condition

$$\|x_0 - x^e(\alpha)\| < \mu$$

implies the limit relation

$$\lim_{t \to \infty} x(t; x_0, \alpha) = x^e(\alpha). \tag{1.2}$$

If (1.2) is true at any initial values $x_0 \in R^n$ and μ, $\mu \to \infty$, then the system (1.1) is asymptotically parametrically stable as a whole.

By slight modification of the Definitions 1.1 and 1.2 one can take into account the estimate of the parameter domain $P \subset R^d$, under which a certain type of parametric stability of motion is obtained. In Appendix reader will find the general results of the estimate of the parameter domain and the phase space domain, under which the conclusions about the parametric stability of a dynamic system are correct.

If the system (1.1) does not have the type of stability as per the Definition 1.1 (or 1.2), then it is parametrically unstable (or asymptotically parametrically unstable).

Thus, to analyze the system (1.1) in the context with the Definitions 1.1 or 1.2, first it is necessary to determine the conditions for the existence of $x^e(\alpha) \in \bar{B}(x^*, r)$ at any $\alpha \in \bar{B}(\alpha^*, q)$, where $\bar{B}(x^*, r) = \{x \in R^n \colon \|x - x^*\| \le r, \ r > 0\}$ and

$$\bar{B}(\alpha^*, q) = \{\alpha \in R^d \colon \|\alpha - \alpha^*\| \le q, \ q > 0\},$$

and then apply an appropriate method of the analysis of stability, the modified Lyapunov's direct method in particular.

The noted concept of the parametric stability has been developed and applied to a certain extent. In particular, the problem of the parametric stability of the system (1.1) by Lyapunov's direct method has been analyzed on the basis of the scalar and the vector Lyapunov functions. The conditions of quadratic stabilizability of an uncertain quasilinear system (see Ohta and Šiljak [1994]) were found. The "variability" of the system parameters is simulated by some matrices with unknown parameters which take on values from compact sets. In the work by Wada, Ikeda, Ohta and Šiljak [1998] the parametric absolute stability of a Lur'e system with many inputs/outputs is analyzed on the basis of the Popov criterion. The parametric absolute stability of one class of singularly perturbed systems is analyzed in the work by Silva and Dzul [1998] on the basis of the comparison principle for slow and fast variables.

Close to this direction are the results of the analysis of stability of systems under nonclassic parametric perturbations, made on the basis of matrix-valued Lyapunov functions (see Martynyuk and Miladzhanov [2009]).

1.2   Stability with Respect to Moving Invariant Sets

Along with the concept of the parametric stability, the problem of *stability of motion with respect to a moving invariant set* (see Lakshmikantham and Vatsala [1997]) is also of interest. The "uncertainties" of the system parameters in this statement are reflected in the "mobility" of some set with respect to which the behavior of solutions of the following uncertain system is analyzed:

$$\frac{dx}{dt} = f(t, x, \alpha), \quad x(t_0) = x_0, \tag{1.3}$$

where $x \in R^n$, $t \in T_0 = [t_0, +\infty)$, $t_0 \in T_i$, $T_i \subseteq R$, $f \in C(T_0 \times R^n \times R^d, R^n)$. The parameter $\alpha \in S \subseteq R^d$, where $d \geq 1$, S is a compact set, represents the "uncertainties" of the considered system, $x(t, \alpha) = x(t; t_0, x_0, \alpha)$ is a solution of the system (1.3) with the initial conditions (t_0, x_0) at any $\alpha \in S \subseteq R^d$.

The stability of solutions of the system (1.3) is analyzed with respect to the moving set

$$A(r) = \{x \in R^n : \|x\| = r(\alpha)\},$$

where $r(\alpha) > 0$, $r(\alpha) \to r_0$ ($r_0 = \text{const} > 0$) at $\|\alpha\| \to 0$ and $r(\alpha) \to +\infty$ at $\|\alpha\| \to +\infty$.

Note that the representation of the function $r(\alpha)$ determining the set $A(r)$ is connected with the quality of motions of the system being synthesized. In particular cases this function can be determined by the expressions

(a) $r(\alpha) = \exp(\|\alpha\|)$, $0 \leq \|\alpha\| < +\infty$;

(b) $r(\alpha) = \ln\left[\exp(\|\alpha\|)\right]$, $0 \leq \|\alpha\| < +\infty$,

implying that in the case (a) $r(\alpha) \to 1$ at $\|\alpha\| \to 0$, and in the case (b) $r(\alpha) \to 0$ at $\|\alpha\| \to 0$.

From the physical point of view, the case (a) corresponds to the dynamic properties of the system (1.3), at which in a system with uncertainly specified parameters a "limit cycle" must exist.

The case (b) corresponds to a different situation: the "disappearance" of the uncertainties parameter in a real system reduces the considered stability problem to the classical one of the equilibrium state $x = 0$.

Note that the investigation of motions of the system (1.3) with respect to the set

$$\bar{A}(r) = \{x \in R^n : \|x\| \leq r(\alpha)\},$$

including the case $r(\alpha) \to r_0$ ($r_0 = \text{const}$), is the problem of the estimation of the domain of attraction of solutions of the system (1.3).

Unlike the work by Lakshmikantham and Vatsala [1997], our definition of the set $A(r)$ makes it possible to consider the systems (1.3) with uncertain values of parameters, whose linear approximation may not contain uncertainties parameters. The results of the above mentioned work by Lakshmikantham and Vatsala [1997] have actually been formulated for nonlinear and nonlinearizable systems with uncertain values of parameters (see Skowronski [1984]) or for such systems whose linear approximation "vanishes" when the uncertainties parameter disappears.

Later in this monograph the results of the analysis of stability of solutions of some types of simultaneous equations with uncertain values of parameters are made on the basis of both traditional and new approaches.

Chapter 2

Lyapunov's Direct Method for Uncertain Systems

Passing over to the generalization of Lyapunov's direct method, in the first place it is necessary to note that among its directions there are the following ones:

(a) the application of the method to simultaneous equations different from ordinary linear ones;

(b) the reduction of requirements to classical Lyapunov functions and their derivatives in view of equations of motion;

(c) the search of new classes of auxiliary Lyapunov functions.

The objects for analysis in this monograph are some classes of systems of nonlinear and linear equations with uncertain values of parameters (in short, uncertain systems).

The application of matrix-valued, vector, and scalar Lyapunov functions in combination with the comparison principle brings our generalization into correspondence with the directions (a)–(c) above, which are being intensely developed today.

The accepted definitions of the stability of motion are given below; they are made specific in separate chapters of this monograph as applied to the considered systems.

2.1 Problem Setting and Auxiliary Results

Consider a mechanical system whose motion is described by the simultaneous differential equations

$$\frac{dx}{dt} = f(t, x, \alpha), \quad x(t_0) = x_0, \tag{2.1}$$

where $x(t) \in R^n$, $t \in \mathcal{T}_0 = [t_0, +\infty)$, $t_0 \in \mathcal{T}_i$, $\mathcal{T}_i \subseteq R$ and $f \in C(\mathcal{T}_0 \times R^n \times R^d, R^n)$. The parameter $\alpha \in \mathcal{S} \subseteq R^d$, where $\mathcal{S}$ is a compact set, is the "uncertainties" parameter of the system (2.1).

In accordance with Leitmann [1990] and Chen [1996], the parameter α:

(a) can present an uncertain value of some physical parameter of the system or the estimate of an external perturbation;

(b) can be a function mapping R into R^d and representing an uncertainly measurable value of input action of one of subsystems upon another;

(c) can be a function mapping $T_0 \times R^n$ into R^d and reflecting nonlinear elements of the considered mechanical system, which are difficult to measure accurately;

(d) can be simply an index denoting the existence of some uncertainties in the system;

(e) can be a combination of the characteristics (a) – (c).

Let a function $r = r(\alpha) > 0$ be given, such that $r(\alpha) \to r_0$ $(r_0 = \text{const})$ at $\|\alpha\| \to 0$ and $r(\alpha) \to +\infty$ at $\|\alpha\| \to +\infty$. In the Euclidian space $(R^n, \|\cdot\|)$ define a moving set

$$A(r) = \{x \in R^n \colon \|x\| = r(\alpha)\} \qquad (2.2)$$

and suppose that at any $(\alpha \neq 0) \in S \subseteq R^d$ the set $A(r)$ is not empty.

Definition 2.1 The set $A(r)$ is called a *moving invariant set* of the system (2.1), if for each $x_0 \in A(r)$ and for all unextensible solutions $x(t, \alpha) = x(t; t_0, x_0, \alpha)$ of the system (2.1) defined on some interval $J \subset T_0$ and such that $x(t_0; t_0, x_0, \alpha) = x_0$, at all $(\alpha \neq 0) \in S \subseteq R^d$ the inclusion $x(t, \alpha) \in A(r)$ holds at every $t \in J$.

Remark 2.1 If the system (2.1) is essentially nonlinear (does not have a linear approximation), then $r(\alpha) \to 0$ is admissible at $\|\alpha\| \to 0$, and the set $A(r)$ collapses into a point 0.

Remark 2.2 If $r(\alpha) \to +\infty$ at $\|\alpha\| \to +\infty$, then the set $A(r)$ unlimitedly increases.

We will give some definitions of the stability of motions with respect to a movable invariant set.

Definition 2.2 *The movable set $A(r)$ for motions of the system (2.1)*

(a) *is stable with respect to the set $T_i \subset R$ if, and only if, for specified $r(\alpha)$, $\varepsilon > 0$ and $t_0 \in T_i$ there exists $\delta = \delta(t_0, \varepsilon) > 0$ such that under the initial conditions*

$$r(\alpha) - \delta < \|x_0\| < r(\alpha) + \delta$$

motions of the system (2.1) satisfy the estimate

$$r(\alpha) - \varepsilon < \|x(t, \alpha)\| < r(\alpha) + \varepsilon$$

at all $t \in T_0$ and $\alpha \in S \subseteq R^d$;

(b) *is uniformly stable with respect to the set* T_i if, and only if, the conditions (a) of the Definition 2.2 are satisfied and at any $\varepsilon > 0$ the corresponding maximal value δ_M satisfying the conditions (a) of this definition, is such that

$$\inf\left[\delta_M(t,\varepsilon)\colon \ t \in T_i\right] > 0;$$

(c) *is stable in large with respect to* T_i if, and only if, the conditions of the Remark 2.2 are satisfied, as well as the conditions (a) of the Definition 2.2 with the function

$$\delta_M(t,\varepsilon) \to +\infty \quad \text{at} \quad \varepsilon \to +\infty \quad \forall t \in T_i;$$

(d) *is uniformly stable in large with respect to* T_i if, and only if, the conditions (b) and (c) of the Definition 2.2 are satisfied.

The expression "with respect to T_i" in the Definition 2.2 is omitted if, and only if, $T_i = R$.

Definition 2.3 *The moving set* $A(r)$ *for motions of the system (2.1) is:*

(a) *attractive with respect to* T_i if, and only if, for a given function $r(\alpha)$ and $t_0 \in T_i$ there exists $\delta(t_0) > 0$ and for any $\zeta > 0$ there exists $\tau(t_0, x_0, \zeta) \in [0, \infty)$ such that the condition

$$r(\alpha) - \delta < \|x_0\| < r(\alpha) + \delta$$

implies the estimate

$$r(\alpha) - \zeta < \|x(t,\alpha)\| < r(\alpha) + \zeta$$

at all $t \in (t_0 + \tau(t_0, x_0, \zeta), +\infty)$ and $\alpha \in S \subseteq R^d$;

(b) x_0-*uniformly attractive with respect to* T_i if, and only if, the conditions (a) of the Definition 2.3 are satisfied, for any $t_0 \in T_i$ there exists $\delta(t_0) > 0$ and for any $\zeta \in (0, +\infty)$ there exists $\tau_u(t_0, \Delta(t_0), \zeta) \in [0, \infty)$ such that

$$\sup\left\{\tau_m(t_0, x_0, \zeta)\colon \ r(\alpha) - \Delta \le \|x_0\| < r(\alpha) + \Delta\right\} = \tau_u(t_0, \Delta(t_0), \zeta);$$

(c) t_0- *uniformly attractive with respect to* T_i if, and only if, the conditions (a) of the Definition 2.3 are satisfied, there exists $\Delta^* > 0$ and for any

$$(x_0, \zeta) \in \{r(\alpha) - \Delta^* \le \|x_0\| < r(\alpha) + \Delta^*\} \times (0, +\infty)$$

there exists $\tau_u(T_i, x_0, \zeta) \in [0, +\infty)$ such that

$$\sup\left\{\tau_m(t_0, x_0, \zeta)\colon \ t_0 \in T_i\right\} = \tau_u(T_i, x_0, \zeta);$$

(d) *uniformly attractive with respect to* T_i, if the conditions (b) and (c) of the Definition 2.3 are satisfied or, which is the same, the conditions (a) of the Definition 2.3, there exists $\delta > 0$ and for any $\zeta \in (0, +\infty)$ there exists $\tau_u(T_i, \Delta, \zeta) \in [0, \infty)$ such that

$$\sup\left[\tau_m(t_0, x_0, \zeta): (t_0, x_0) \in T_i \times \{r(\alpha) - \Delta < \|x_0\| < r(\alpha) + \Delta\}\right]$$
$$= \tau_u(T_i, \Delta, \zeta);$$

(e) the properties of attraction 2.3 (a)–(d) *are true in large*, if the conditions (a) of the Definition 2.3 are satisfied for any $\Delta(t_0) \in (0, +\infty)$ and any $t_0 \in T_i$, if only $r(\alpha) \to +\infty$ at $\|\alpha\| \to +\infty$.

The expression "with respect to T_i" in the Definition 2.3 is omitted if, and only if, $T_i = R$.

Definition 2.4 *Motion of the system* (2.1) *with respect to the moving set* $A(r)$ *is:*

(a) *asymptotically stable with respect to* $A(r)$ if, and only if, it is stable with respect to $A(r)$ and attractive with respect to $A(r)$;

(b) *equiasymptotically stable with respect to* $A(r)$ if it is stable with respect to $A(r)$ and x_0 is uniformly attractive with respect to $A(r)$;

(c) *quasiuniformly asymptotically stable with respect to* $A(r)$, if it is uniformly stable with respect to $A(r)$ and t_0 is uniformly attractive with respect to $A(r)$;

(d) *uniformly asymptotically stable with respect to the sets* $A(r)$ *and* T_i, if it is uniformly stable with respect to the sets $A(r)$ and T_i and uniformly attractive with respect to the sets $A(r)$ and T_i;

(e) *uniformly exponentially stable with respect to* $A(r)$, if for the given function $r(\alpha)$ and constant β_1, β_2, λ there exists $\delta > 0$ such that the condition
$$r(\alpha) - \delta < \|x_0\| < r(\alpha) + \delta$$
implies the estimate
$$r(\alpha) - \beta_1 \|x_0\| \exp[-\lambda(t - t_0)] \leq \|x(t, \alpha)\|$$
$$\leq r(\alpha) + \beta_2 \|x_0\| \exp[-\lambda(t - t_0)] \quad \forall t \in T_0 \quad \forall t_0 \in T_i;$$

(f) *exponentially stable in large with respect to* $A(r)$, if the conditions of the Definition 2.4 (e) are satisfied at $r(\alpha) \to \infty$, $\|\alpha\| \to +\infty$, $\delta \to \infty$.

The expression "with respect to T_i" in the Definition 2.4 is omitted if, and only if, $T_i = R$.

2.2 Classes of Lyapunov Functions

Later in this monograph three classes of Lyapunov functions are applied and their description is given in this section. As required, the general definitions given here will be modified in conformity with the considered class of systems. It is assumed that auxiliary functions can be constructed for nominal systems corresponding to the considered uncertain systems.

2.2.1 Matrix-valued Lyapunov functions

For the nominal system corresponding to the uncertain system (2.1) we will consider the matrix-valued function

$$U(t,x) = [v_{ij}(t,x)], \quad i,j = 1,2,\ldots,m, \tag{2.3}$$

where $v_{ij} \in C(\mathcal{T}_\tau \times R^n, R)$ at all $i,j = 1,2,\ldots,m$.

It is supposed that the following conditions are satisfied:

(i) $v_{ij}(t,x)$, $i,j = 1,2,\ldots,m$, are locally Lipshitz with respect to x;

(ii) $v_{ij}(t,0) = 0$ at all $t \in R_+$ $(t \in \mathcal{T}_\tau)$, $i,j = 1,2,\ldots,m$;

(iii) $v_{ij}(t,x) = v_{ji}(t,x)$ in any open connected neighborhood $\mathcal{N}$ of the point $x = 0$ at all $t \in R_+$ $(t \in \mathcal{T}_\tau)$.

Definition 2.5 All functions of the form

$$V(t,x,a) = a^{\mathrm{T}} U(t,x) a, \quad a \in R^m, \tag{2.4}$$

where $U \in C(\mathcal{T}_\tau \times \mathcal{N}, R^{m \times m})$, will be called functions of the *SL class*.

In the expression (2.4) the vector a can be chosen in the form:

(i) $a = y \in R^m$, $y \neq 0$;

(ii) $a = \xi \in C(R^n, R_+^m)$, $\xi(0) = 0$;

(iii) $a = \psi \in C(\mathcal{T}_\tau \times R^n, R_+^m)$, $\psi(t,0) = 0$;

(iv) $a = \eta \in R_+^m$, $\eta > 0$.

Note that the choice of the vector a has an impact on the property of sign definiteness of the function (2.4) and its full derivative along the solutions of the system (2.1).

2.2.2   Comparison functions

Comparison functions are applied at upper and lower estimates of functions of the form (2.4) and their full derivative along the solutions of the considered system is usually denoted by φ, $\varphi\colon R_+ \to R_+$. The research of these functions and their application in the stability theory was initiated by Hahn [1967]. Following this monograph, we will describe some properties of comparison functions.

Definition 2.6 The *function* φ, $\varphi\colon R_+ \to R_+$, belongs to:

(i) *the class* $K_{[0,b)}$, $0 < b \le +\infty$ if, and only if, it is defined, continuous, strictly increasing over $[0,b)$ and $\varphi(0) = 0$;

(ii) *the class* K if, and only if, the condition (i) is satisfied at $b = +\infty$, $K = K_{[0,+\infty)}$;

(iii) *the class* KR if, and only if, it belongs to the class K and $\varphi(\zeta) \to +\infty$ at $\zeta \to +\infty$;

(iv) *the class* $L_{[0,b)}$ if, and only if, it is defined, continued, strictly decreasing over $[0,b)$ and $\lim [\varphi(\zeta)\colon \zeta \to +\infty] = 0$;

(v) *the class* L if, and only if, the condition (iv) is satisfied at $b = +\infty$, $L = L_{[0,+\infty)}$.

Let φ^{-1} denote the *inverse function* of the function φ, $\varphi^{-1}[\varphi(\zeta)] \equiv \zeta$. The following results were obtained by Hahn [1967].

Proposition 2.1 For the functions $\varphi \in K$ and $\psi \in K$ there are following statements:

(1) if $\varphi \in K$ and $\psi \in K$, then $\varphi(\psi) \in K$;

(2) if $\varphi \in K$ and $\sigma \in L$, then $\varphi(\sigma) \in L$;

(3) if $\varphi \in K_{[0,b)}$ and $\varphi(b) = \xi$, then $\varphi^{-1} \in K_{[0,\xi)}$;

(4) if $\varphi \in K$ and $\lim [\varphi(\zeta)\colon \zeta \to +\infty] = \xi$, then φ^{-1} is not defined on $(\xi, +\infty]$;

(5) if $\varphi \in K_{[0,b)}$, $\psi \in K_{[0,b)}$ and $\varphi(\zeta) > \psi(\zeta)$ on $[0,b)$ then $\varphi^{-1}(\zeta) < \psi^{-1}(\zeta)$ on $[0, \beta]$, where $\beta = \psi(b)$.

Definition 2.7 Two *functions* φ_1, $\varphi_2 \in K$ (or φ_1, $\varphi_2 \in KR$) have *the same order of growth*, if there exist positive constants α, β such that

$$\alpha\varphi_1(\zeta) \le \varphi_2(\zeta) \le \beta\varphi_1(\zeta) \quad \text{at all} \quad \zeta \in [0, \zeta_1]$$
$$(\text{or at all} \quad \zeta \in [0, \infty)).$$

2.2.3 Properties of matrix-valued functions

Properties of sign definiteness of a matrix-valued function are defined on the basis of a function of class SL. Below readers will find some definitions applied in this monograph in the analysis of the dynamic behavior of uncertain systems.

Definition 2.8 *The matrix-valued function* $U: \mathcal{T}_\tau \times R^n \to R^{m \times m}$ *is called:*

(i) *positive-semidefinite on* $\mathcal{T}_\tau = [\tau, +\infty)$, $\tau \in R$, *if there exists a time-invariant neighbourhood* $\mathcal{N}$ *of the point* $x = 0$, $\mathcal{N} \subseteq R^n$, *and a vector* $y \in R^m$, $y \neq 0$, *such that:*

    (a) $V(t, x, y)$ *is continuous on* $(t, x) \in \mathcal{T}_\tau \times \mathcal{N} \times R^m$,

    (b) $V(t, x, y)$ *is nonnegative on* $\mathcal{N}$, $V(t, x, y) \geq 0$ *at all* $(t, x, y \neq 0) \in \mathcal{T}_\tau \times \mathcal{N} \times R^m$,

    (c) *vanishes in the origin of coordinates:* $V(t, 0, y) = 0$ *at all* $t \in \mathcal{T}_\tau \times R^m$,

    (d) *if the conditions* (a) – (c) *are satisfied at any* $t \in \mathcal{T}_\tau$ *and there exists* $w \in \mathcal{N}$ *such that* $V(t, w, y) > 0$, *then* V *is a function strictly semidefinite on* $\mathcal{T}_\tau$;

(ii) *a function positive-semidefinite on* $\mathcal{T}_\tau \times \mathcal{G}$, *if the condition* (i) *is satisfied at* $\mathcal{N} = \mathcal{G}$;

(iii) *a function positive-semidefinite in large on* $\mathcal{T}_\tau$ *if the condition* (i) *is satisfied at* $\mathcal{N} = R^n$;

(iv) *a function negative-semidefinite (in large) on* $\mathcal{T}_\tau$ *(on* $\mathcal{T}_\tau \times \mathcal{N}$), *if* $(-V)$ *is positively semidefinite (in large) on* $\mathcal{T}_\tau$ *(on* $\mathcal{T}_\tau \times \mathcal{N}$).

The expression "*on* $\mathcal{T}_\tau$" is omitted if all conditions of the definition are satisfied at any $\tau \in R$.

Definition 2.9 *The matrix-valued function* $U: \mathcal{T}_\tau \times R^n \to R^{m \times m}$ *is called:*

(i) *positive-definite on* $\mathcal{T}_\tau$, $\tau \in R$, *if there exists a time-invariant neighbourhood* $\mathcal{N}$ *of the point* $x = 0$, $\mathcal{N} \subseteq R^n$ *and a vector* $y \in R^m$, $y \neq 0$, *such that the function* (2.4) *is positive-semidefinite on* $\mathcal{T}_\tau \times \mathcal{N}$ *and for the positive-definite function* w *on* $\mathcal{N}$, $w: R^n \to R_+$, *the inequality* $w(x) \leq V(t, x, y)$ *holds at all* $(t, x, y) \in \mathcal{T}_\tau \times \mathcal{N} \times R^m$;

(ii) *positive-definite on* $\mathcal{T}_\tau \times \mathcal{G}$ *if the conditions of the Definition* (i) *are satisfied at* $\mathcal{N} = \mathcal{G}$;

(iii) *positive-definite in large on* $\mathcal{T}_\tau$ *if the conditions of the Definition* (i) *are satisfied at* $\mathcal{N} = R^n$;

(iv) *negative-definite (in large) on* $\mathcal{T}_\tau$ *(on* $\mathcal{T}_\tau \times \mathcal{N} \times R^m$*) if* $(-V)$ *is positive-definite (in large) on* $\mathcal{T}_\tau$ *(on* $\mathcal{T}_\tau \times \mathcal{N} \times R^m$*);*

(v) *weakly decreasing, if there exist a constant* $\Delta_1 > 0$ *and a function* $a \in CK$ *such that* $V(t,x,y) \le a(t, \|x\|)$ *at* $\|x\| < \Delta_1$;

(vi) *asymptotically decreasing, if there exist a constant* $\Delta_2 > 0$ *and a function* $b \in KL$ *such that* $V(t,x,y) \le b(t, \|x\|)$ *at* $\|x\| < \Delta_2$.

The expression "*on* $\mathcal{T}_\tau$" is omitted if all conditions of the Definition 2.9 are satisfied for every $\tau \in R$.

Proposition 2.2 For the matrix-valued function $U \colon R \times R^n \to R^{m \times m}$ to be positive-definite on $\mathcal{T}_\tau$, $\tau \in R$, it is necessary and sufficient that the following relation should be true:

$$y^{\mathrm{T}} U(t,x)y = y^{\mathrm{T}} U_{+}(t,x)y + a(\|x\|),$$

where $U_{+}(t,x)$ is a positive-semidefinite matrix-values function and $a \in K$.

Definition 2.10 *The matrix-valued function* $U \colon R \times R^n \to R^{s \times s}$ *is called:*

(i) *decreasing on* $\mathcal{T}_\tau$, $\tau \in R$, *if there exist a time-invariant neighbourhood* $\mathcal{N}$ *of the point* $x = 0$ *and a positive-definite function* w *on* $\mathcal{N}$, $w \colon R^n \to R_{+}$, *such that* $y^{\mathrm{T}} U(t,x)y \le w(x)$ *at all* $(t,x) \in \mathcal{T}_\tau \times \mathcal{N}$;

(ii) *decreasing on* $\mathcal{T}_\tau \times \mathcal{G}$, *if the conditions of the Definition (i) are satisfied at* $\mathcal{N} = \mathcal{G}$;

(iii) *decreasing in large in large on* $\mathcal{T}_\tau$, *if the conditions of the Definition (i) are satisfied at* $\mathcal{N} = R^n$.

The expression "*on* $\mathcal{T}_\tau$" is omitted if all respective conditions are satisfied for any $\tau \in R$.

Proposition 2.3 For the matrix-valued function $U \colon R \times R^n \to R^{m \times m}$ to be decreasing on $\mathcal{T}_\tau$, $\tau \in R$, it is necessary and sufficient that the following relation should be true

$$y^{\mathrm{T}} U(t,x)y = y^{\mathrm{T}} U_{-}(t,x)y + b(\|x\|), \quad (y \ne 0) \in R^m,$$

where $U_{-}(t,x)$ is a negative-semidefinite matrix-valued function and $b \in K$.

Definition 2.11 *The matrix-valued function* $U \colon R \times R^n \to R^{m \times m}$ *is called:*

(i) *radially unbounded on* $\mathcal{T}_\tau$, $\tau \in R$, *if from the condition* $\|x\| \to \infty$ *it follows that* $y^{\mathrm{T}} U(t,x)y \to +\infty$ *at all* $t \in \mathcal{T}_\tau$, $y \in R^m$, $y \ne 0$;

(ii) *radially unbounded, if from the condition* $\|x\| \to \infty$ *it follows that* $y^{\mathrm{T}} U(t,x)y \to +\infty$ *at all* $t \in \mathcal{T}_\tau$ *and all* $\tau \in R$, $y \in R^m$, $y \ne 0$.

Proposition 2.4 For the matrix-valued function $U\colon \mathcal{T}_\tau \times R^n \to R^{m\times m}$ to be radially unbounded in large (on $\mathcal{T}_\tau$), it is necessary and sufficient that the following relation should be true

$$y^{\mathrm{T}} U(t,x) y = y^{\mathrm{T}} U_+(t,x) y + a(\|x\|) \quad \text{at all} \quad x \in R^n,$$

where $U_+(t,x)$ is a positive-semidefinite in large matrix-valued function (on $\mathcal{T}_\tau$) and $a \in KR$.

The matrix-valued function (2.3) and/or the scalar function (2.4) are applied below in the analysis of the dynamics of the uncertain system (2.1) together with its full derivative along the solutions of that system. It is supposed that every element $v_{ij}(t,x)$ of the matrix-valued function (2.3) is defined on the open set $\mathcal{T}_\tau \times \mathcal{N}$, $\mathcal{N} \subset R^n$, i.e. $v_{ij}(t,x) \in C(\mathcal{T}_\tau \times \mathcal{N}, R)$ and is locally Lipshitz with respect to x. For any point of the set $\mathcal{N}$ there exists a neighbourhood Δ and a positive number $L = L(\Delta)$ such that

$$|v_{ij}(t,x) - v_{ij}(t,y)| \le L\|x - y\|, \quad i,j = 1,2,\ldots,m,$$

for any $(t,x) \in \mathcal{T}_\tau \times \Delta$, $(t,y) \in \mathcal{T}_\tau \times \Delta$. In this case the full derivative of the elements of the matrix-valued function is calculated as follows:

$$
\begin{aligned}
D^+ v_{ij}(t,x) = \limsup\{[v_{ij}(t+\sigma,\, x+\sigma f(t,x)) \\
- v_{ij}(t,x)]\sigma^{-1}\colon \sigma \to 0^+\}, \quad i,j = 1,2,\ldots,m.
\end{aligned}
\tag{2.5}
$$

If the matrix-valued function $U(t,x)$ belongs to $C^{(1,1)}(\mathcal{T}_\tau \times \mathcal{N}, R^{m\times m})$, i.e. all its elements $v_{ij}(t,x)$ are continuously differentiable functions with respect to t and x, then

$$Dv_{ij}(t,x) = \frac{\partial v_{ij}}{\partial t}(t,x) + \sum_{s=1}^{n} \frac{\partial v_{ij}}{\partial x_s}(t,x)\, f_s(t,x),$$

where $f_s(t,x)$ are components of the vector function $f(t,x) = (f_1(t,x),\ldots, f_n(t,x))^{\mathrm{T}}$.

2.2.4 Vector Lyapunov functions

The vector function

$$V(t,x) = (v_1(t,x), v_2(t,x),\ldots, v_m(t,x))^{\mathrm{T}}$$

can be obtained on the basis of the matrix function (2.3) by several methods.

Definition 2.12 All vector functions of the form

$$L(t,x,b) = AU(t,x)b,$$

where $U \in C(\mathcal{T}_\tau \times R^n, R^{s\times s})$, A is a constant $s \times s$-matrix and the vector b is determined like the vector a from the Definition 2.5, form the *class VL*.

If in the matrix-valued function (2.3) at all $i \neq j$ the elements $v_{ij}(t,x) = 0$, then

$$L(t,x) = \mathrm{diag}\,(v_{11}(t,x),\, \ldots,\, v_{mm}(t,x)),$$

where $v_{ii} \in C(\mathcal{T}_\tau \times R^n,\, R)$, $i = 1, 2, \ldots, m$, is a vector function. In the general case, components of the vector function $L(t,x,b)$ are the functions

$$L_k(t,x,b) = \sum_{i=1}^{m} a_{ki} b_i v_{ii}(t,x), \quad k = 1, 2, \ldots, m.$$

The application of vector Lyapunov functions is connected with the construction of linear or nonlinear comparison systems and the analysis of their stability.

2.2.5 Scalar Lyapunov functions

As a result of the generalization of the ideas of Lagrange and Poincaré, A.M. Lyapunov discovered a class of auxiliary functions which make it possible to investigate the stability of solutions of simultaneous equations of perturbed motion of quite a general form without its direct integration. Those functions have a number of special properties described in the Definitions 2.8 – 2.11.

The functions

$$V(t,x) \in C(\mathcal{T}_0 \times R^n,\, R_+),$$

for which

(a) $V(t,0) = 0$ at all $t \in \mathcal{T}_\tau$;

(b) $V(t,x)$ is locally Lipshitz with respect to x belong to the simplest type of auxiliary functions of the SL class.

The sign definiteness of the Lyapunov function and its full derivative, in view of the system (2.1), allows us to solve the problem of stability of the equilibrium state $x = 0$ of the system (2.1) or the problem of stability of its solutions with respect to a moving invariant set.

Classical auxiliary functions are the basis for the direct Lyapunov method. We will give examples of some types of Lyapunov functions.

Example 2.1 (i) The function

$$V(t,x) = (1 + \sin^2 t)\, x_1^2 + (1 + \cos^2 t)\, x_2^2$$

is positive-definite and decreasing, while the function

$$V(t,x) = (x_1^2 + x_2^2)\, \sin^2 t$$

is decreasing and positive-semidefinite. (ii) The function

$$V(t,x) = x_1^2 + (1+t)\, x_2^2$$

is positive-definite but not decreasing, whereas the function

$$V(t, x) = x_1^2 + \frac{x_2^2}{1 + t}$$

is decreasing but not positive-semidefinite. (iii) The function

$$v(t, x) = (1 + t)(x_1 - x_2)^2$$

is positive-semidefinite but not decreasing.

Further on the auxiliary functions (matrix-valued, vector, or scalar) solving the problem of stability (instability) of the equilibrium state $x = 0$ of the system (2.1) will be called Lyapunov functions.

2.3 Theorems on Stability and Uniform Stability

In the context of the direct Lyapunov method we will further use the function (2.4) and its upper right Dini derivative (2.5) along the solutions of the system (2.1). The sufficient conditions for stability of the solutions of the uncertain system (2.1) with respect to the moving set $A(r)$ are contained in the following statement.

Theorem 2.1 *Suppose that $f(t, x, \alpha)$ in the system (2.1) is continuous on $T_0 \times R^n \times R^d$ and the following conditions are satisfied:*

(1) *for every $\alpha \in S \subseteq R^d$ there exists a function $r = r(\alpha) > 0$ such that the set $A(r)$ is not empty at all $\alpha \in S$;*

(2) *there exists a matrix-valued function $U \in C(T_0 \times R^n, R^{s \times s})$, $U(t, x)$ locally Lipshitz with respect to x, a vector $y \in R^s$ and $(s \times s)$ matrices $\theta_1(r)$, $\theta_2(r)$ such that:*

(a) $a(\|x\|) \leq V(t, x, y)$ *at* $\|x\| > r(\alpha)$,

(b) $V(t, x, y) \leq b(\|x\|)$ *at* $\|x\| < r(\alpha)$, *where a, b belong to the K-class,*

(c) $D^+ V(t, x, y)|_{(2.1)} < \varphi^T(\|x\|)\theta_1(r)\varphi(\|x\|)$, *if $\|x\| > r(\alpha)$ at all $\alpha \in S$, where $\varphi^T(\|x\|) = (\varphi_1^{1/2}(\|x_1\|), \dots, \varphi_s^{1/2}(\|x_s\|))$, $\varphi_i \in K$,*

(d) $D^+ V(t, x, y)|_{(2.1)} = 0$, *if $\|x\| = r(\alpha)$ at all $\alpha \in S$,*

(e) $D^+ V(t, x, y)|_{(2.1)} > \psi^T(\|x\|)\theta_2(r)\psi(\|x\|)$, *if $\|x\| < r(\alpha)$ at all $\alpha \in S$, where $\psi^T(\|x\|) = (\psi_1^{1/2}(\|x_1\|), \dots, \psi_s^{1/2}(\|x_s\|))$, $\psi_i \in K$;*

(3) *there exist constant $(s \times s)$–matrices $\bar{\theta}_1$, $\bar{\theta}_2$ such that:*

(a) $\frac{1}{2}\left(\theta_1(r) + \theta_1^{\mathsf{T}}(r)\right) \leq \bar{\theta}_1$ *at all* $\alpha \in \mathcal{S}$,

(b) $\frac{1}{2}\left(\theta_2(r) + \theta_2^{\mathsf{T}}(r)\right) \geq \bar{\theta}_2$ *at all* $\alpha \in \mathcal{S}$,

and $\bar{\theta}_1$ *is negative-semidefinite, and* $\bar{\theta}_2$ *is positive-semidefinite;*

(4) *for any* $r(\alpha) > 0$ *and functions* $a(r)$, $b(r)$ *the following relation holds:*

$$a(r) = b(r).$$

Then the set $A(r)$ is invariant with respect to motions of the system (2.1), *and motions of the system* (2.1) *are stable with respect to the set $A(r)$.*

Proof Let all the conditions of the Theorem 2.1 be satisfied for some given function $r(\alpha) > 0$ at any $\alpha \in \mathcal{S}$. Let us prove that the set $A(r)$ is invariant with respect to the solutions of the uncertain system (2.1). Suppose this is not so. Then for the solution $x(t, \alpha)$ of the system (2.1) with the initial conditions $\|x_0\| = r(\alpha)$ one can find points of time $t_2 > t_1 \geq t_0$ such that

$$\begin{aligned} & \|x(t_1, \alpha)\| = r(\alpha), \\ \text{(A)} \quad & \|x(t_2, \alpha)\| > r(\alpha), \\ & \|x(t, \alpha)\| \geq r(\alpha) \quad \text{at any} \quad t \in [t_1, t_2], \end{aligned}$$

or

$$\begin{aligned} & \|x(t_1, \alpha)\| = r(\alpha), \\ \text{(B)} \quad & \|x(t_2, \alpha)\| < r(\alpha), \\ & \|x(t, \alpha)\| \leq r(\alpha) \quad \text{at any} \quad t \in [t_1, t_2]. \end{aligned}$$

From the conditions (2) (c)–(e) and (3) of the Theorem 2.1 obtain at $t_1 \geq t \geq t_2$

$$V(t, x(t, \alpha), y) < V(t_1, x(t_1, \alpha), y),$$

or

$$V(t, x(t, \alpha), y) > V(t_1, x(t_1, \alpha), y).$$

Hence, taking into account the conditions (2) (a), (b), in the case (A) we obtain the inequalities

$$\begin{aligned} a(r(\alpha)) &< a(\|x(t_2, \alpha)\|) \leq V(t_2, x(t_2, \alpha), y) \\ &< V(t_1, x(t_1, \alpha), y) \leq b(\|x(t_1, \alpha)\|) = b(r(\alpha)), \end{aligned} \tag{2.6}$$

which result in the contradiction with the condition (4) of the Theorem 2.1. In the case (B) we have the sequence of inequalities

$$\begin{aligned} b(r(\alpha)) &> b(\|x(t_2, \alpha)\|) \geq V(t_2, x(t_2, \alpha), y) \\ &> V(t_1, x(t_1, \alpha), y) \geq a(\|x(t_1, \alpha)\|) = a(r(\alpha)), \end{aligned} \tag{2.7}$$

which also result in the contradiction with the condition (4) of the Theorem 2.1.

The inequalities (2.6), (2.7) along with the condition (2)(d) of the Theorem 2.1 prove the invariance of the set $A(r)$ with respect to the solutions of the system (2.1).

Note that the condition (2)(d) sometimes allows to determine the function $r(\alpha) > 0$ which participates in the setting of $A(r)$.

Now prove the stability of the solutions of the system (2.1) with respect to the set $A(r)$. Let $t_0 \in \mathcal{T}_i$, $\varepsilon > 0$ and $r(\alpha) > 0$ be given. Suppose that $A(r)$ is invariant with respect to the solutions of the system (2.1). For the given (t_0, ε) choose $\delta = \delta(t_0, \varepsilon) > 0$ so that at all $\alpha \in \mathcal{S} \subseteq R^d$

$$a(r(\alpha) + \varepsilon) > b(r(\alpha) + \delta), \tag{2.8}$$

and

$$b(r(\alpha) - \varepsilon) < a(r(\alpha) - \delta). \tag{2.9}$$

Such choice of δ is possible, since the functions a, b belong to the K-class. Let us prove that with the chosen value of δ the solutions of the system (2.1) will be stable with respect to the moving set $A(r)$ in the sense of the Definition 2.1(a).

Suppose that this is not so. Then for the solution $x(t, \alpha)$ of the system (2.1) with initial conditions satisfying the condition

$$r(\alpha) - \delta < \|x_0\| < r(\alpha) + \delta, \tag{2.10}$$

one can find values of time $t_2 > t_1 > t_0$ such that at all $t \in [t_1, t_2]$

$$(A) \qquad \begin{aligned} \|x(t_2, \alpha)\| &= r(\alpha) + \varepsilon, \\ \|x(t_1, \alpha)\| &= r(\alpha) + \delta, \\ \|x(t, \alpha)\| &> r(\alpha) \end{aligned}$$

or

$$(B) \qquad \begin{aligned} \|x(t_2, \alpha)\| &= r(\alpha) - \varepsilon, \\ \|x(t_1, \alpha)\| &= r(\alpha) - \delta, \\ \|x(t, \alpha)\| &< r(\alpha). \end{aligned}$$

Consider the case (A). According to the conditions (2)(a) and (3)(a), obtain

$$\begin{aligned} a(r(\alpha) + \varepsilon) = a(\|x(t_2, \alpha)\|) &\le V(t_2, x(t_2, \alpha), y) \\ &< V(t_1, x(t_1, \alpha), y) \le b(\|x(t_1, \alpha)\|) = b(r(\alpha) + \delta) \end{aligned}$$

at all $\alpha \in \mathcal{S}$. This contradicts the inequality (2.8).

Similarly, in the case (B), according to the conditions (2)(b) and (3)(b), we obtain the inequalities

$$\begin{aligned} b(r(\alpha) - \varepsilon) = b(\|x(t_2, \alpha)\|) &\ge V(t_2, x(t_2, \alpha), y) \\ &> V(t_1, x(t_1, \alpha), y) \ge a(\|x(t_1, \alpha)\|) = a(r(\alpha) - \delta) \end{aligned}$$

at all $\alpha \in S$, which contradicts the condition (2.9). Hence, for the solutions $x(t, \alpha)$ of the system (2.1) with the initial conditions (2.10) the following estimate is true

$$r(\alpha) - \varepsilon < \|x(t, \alpha)\| < r(\alpha) + \varepsilon$$

at all $t \geq t_0$ and $\alpha \in S \subseteq R^d$. The Theorem 2.1 is proved.

For the cases when for the system (2.1) it is possible to construct a scalar Lyapunov function with relevant properties, the stability of solutions can be analyzed on the basis of the following statement.

Theorem 2.2 *Let the following conditions be satisfied:*

(1) *for each $\alpha \in S \subseteq R^d$ there exists a function $r = r(\alpha) > 0$ such that $r(\alpha) \to r_0$ at $\|\alpha\| \to 0$ and $r(\alpha) \to +\infty$ at $\|\alpha\| \to +\infty$;*

(2) *there exist scalar functions $V \in C^{(1.1)}(\mathcal{T}_0 \times R^n, R_+)$, $W_1 \colon R^n \times S \to R$ and $W_2 \colon R^n \times S \to R$ such that:*

    (a) $a(\|x\|) \leq V(t, x)$ *at* $\|x\| > r(\alpha)$,

    (b) $V(t, x) \leq b(\|x\|)$ *at* $\|x\| < r(\alpha)$, *where a, b belongs to the K-class,*

    (c) $DV(t, x)\big|_{(2.1)} \leq W_1(x, \alpha)$ *at* $\|x\| > r(\alpha)$, $\alpha \in S \subseteq R^d$,

    (d) $DV(t, x)\big|_{(2.1)} = 0$ *at* $\|x\| = r(\alpha)$, $\alpha \in S$,

    (e) $DV(t, x)\big|_{(2.1)} \geq W_1(x, \alpha)$ *at* $\|x\| < r(\alpha)$, $\alpha \in S$;

(3) *there exist functions $\overline{W}_1(x)$ and $\underline{W}_2(x)$ such that:*

    (a) $W_1(x, \alpha) \leq \overline{W}_1(x) < 0$ *at all* $\alpha \in S$,

    (b) $W_2(x, \alpha) \geq \underline{W}_2(x) > 0$ *at all* $\alpha \in S$;

(4) *for any $r(\alpha) > 0$ and functions $a(r)$, $b(r)$ the following relation is true:*

$$a(r) = b(r).$$

Then the set $A(r)$ is invariant with respect to motions of the system (2.1), and the solutions of the system (2.1) are stable with respect to the set $A(r)$.

The statement of the Theorem 2.2 follows from the Theorem 2.1.

Example 2.2 A non-conservative Hamilton system with uncertain values of parameters was analyzed in the article by Skowronski [1984]. Apply the Theorem 2.1 for the investigation of a system of the form

$$\dot{x}_1 = \frac{\partial H}{\partial x_2} + \Phi_1(t, x_1, x_2, \alpha_1),$$

$$\dot{x}_2 = -\frac{\partial H}{\partial x_1} + \Phi_1(t, x_1, x_2, \alpha_2), \tag{2.11}$$

where $H\colon R^2 \to R$, $\Phi_i\colon R \times R^2 \times S \to R$, $i = 1, 2, \ldots$, $H(x_1, x_2)$ is a positive-definite convex C^1-function, $H(0,0) = 0$, $\alpha_1(t), \alpha_2(t) \in [0,1]$ are unknown scalar parameters, Φ_1, Φ_2 are nonpotential dissipative forces.

Let E_0 be the constant energy of the system (2.11) at $\Phi_1 = \Phi_2 = 0$. For the given function $r(\alpha) > 0$ consider the set

$$A(r) = \{x \in R^2 \colon \|x\| = r(\alpha)\}, \quad \alpha \in S \subseteq R^d, \tag{2.12}$$

where $\lim r(\alpha) = E_0$ at $\|\alpha\| \to 0$.

It is known that

$$\dot{H}(x_1(t), x_2(t)) = \sum_{i=1}^{2} \frac{\partial H}{\partial x_i} \Phi_i(t, x_1, x_2, \alpha).$$

Let functions $W_i(x_1, x_2)$, $i = 1, 2$ exist, such that at $\alpha \in S$ the following conditions are satisfied:

(a) $\displaystyle\sum_{i=1}^{2} \frac{\partial H}{\partial x_i} \Phi_i(t, x_1, x_2, \alpha) < W_1(x_1, x_2)$ at $\|x\| > r(\alpha)$;

(b) $\displaystyle\sum_{i=1}^{2} \frac{\partial H}{\partial x_i} \Phi_i(t, x_1, x_2, \alpha) = 0$ at $\|x\| = r(\alpha)$,

(c) $\displaystyle\sum_{i=1}^{2} \frac{\partial H}{\partial x_i} \Phi_i(t, x_1, x_2, \alpha) > W_2(x_1, x_2)$ at $\|x\| < r(\alpha)$.

If along with the conditions (a) – (c) the following inequalities hold

(a)′ $W_1(x_1, x_2) < 0$ at $\|x\| > r(\alpha)$, $\alpha \in S$,
(b)′ $W_2(x_1, x_2) > 0$ at $\|x\| < r(\alpha)$, $\alpha \in S$,

then the solutions of the system (2.11) converge to the moving surface (2.12).

The conditions (a), (b) indicate that outside the moving surface $A(r)$ the system (2.11) is dissipative, while inside $A(r)$ energy is pumped.

Remark 2.3   Many physical systems are simulated by first-order differential equations. Among them are: fall of a point body with the mass m in a viscous medium with resistance proportionate to velocity; a power grid consisting of one linear constant resistance, a linear constant energy store, and a single external source of energy; processes including chemical changes or compounding processes. In all those cases some parameters are known and/or can be determined uncertainly. Here is an example of such a system.

Example 2.3   Consider a first-order uncertain system

$$\frac{dx}{dt} = x - f^2(\alpha)x^3, \quad x(0) \neq 0, \tag{2.13}$$

where $f(\alpha)$ is a nondifferentiable function of the uncertainties parameter $\alpha \in S \subseteq R^d$, $f(\alpha) \to f_0$ at $\|\alpha\| \to 0$ and $f(\alpha) \to 0$ at $\|\alpha\| \to \infty$.

The zero solution $x = 0$ of this system is uncertain by Lyapunov because its first approximation

$$\frac{dx}{dt} = x, \quad x(0) \neq 0,$$

has the eigenvalue $\lambda = 1 > 0$.

Let $r(\alpha) = \dfrac{1}{f(\alpha)} > 0$. It is clear that $r(\alpha) \to r_0$ at $\|\alpha\| \to 0$ and $r(\alpha) \to \infty$ at $\|\alpha\| \to \infty$.

The set $A(r)$ has the form

$$A(r) = \left\{ x: \ |x| = \frac{1}{f(\alpha)} \right\}.$$

Take $V = x^2$ and calculate

$$\frac{dV}{dt} = 2x \frac{dx}{dt} = 2x^2 \left(1 - f^2(\alpha)x^2 \right).$$

Hence it is clear that

$$\frac{dV}{dt} < 0 \quad \text{at} \quad |x| > \frac{1}{f(\alpha)};$$

$$\frac{dV}{dt} = 0 \quad \text{at} \quad |x| = \frac{1}{f(\alpha)};$$

$$\frac{dV}{dt} > 0 \quad \text{at} \quad |x| < \frac{1}{f(\alpha)}, \quad t \geq 0.$$

Therefore, if $f(\alpha)$ satisfies the conditions $\lim\limits_{\|\alpha\| \to 0} f(\alpha) = f_0$, $\lim\limits_{\|\alpha\| \to \infty} f(\alpha) = \infty$, then according to the Theorem 2.2 the set $A(r)$ is invariant with respect to solutions of the equation (2.14), and all solutions of this equation are stable with respect to the set $A(r)$ in the sense of the Definition 2.2(a).

Remark 2.4 At $f^2(\alpha) = \beta^2$ (β is a control parameter) the equation (2.13) was considered in the work by Galperin and Skowronsky [1985]. Note that in the case in point the equation (2.13) cannot be reduced to the normal form, like the Abel equation, due to the nondifferentiability of the function $f(\alpha)$, $\alpha \in S \subseteq R^d$.

Remark 2.5 Examples of physical systems simulated by second-order equations or two systems of first-order equations are: the motion of a body with the mass m, suspended on a spring and subject to viscous depreciation; an electric circuit with two sources of energy; a branched RLC-circuit; the oscillation of a material point with the mass m, suspended on a thread of the length l, and others. As in the previous case, some parameters in the relevant systems of differential equations may be known uncertainly. We will give several examples of second-order systems.

Example 2.4 Consider the class of second-order systems

$$\frac{dx}{dt} = n(t)y + x\left[1 - m(\alpha)(x^2 + y^2)\right], \qquad x(t_0) = x_0,$$

$$\frac{dy}{dt} = -n(t)x + y\left[1 - m(\alpha)(x^2 + y^2)\right], \qquad y(t_0) = y_0,$$

(2.14)

where $t \in T_0$, $x(t) \in R$, $y(t) \in R$, $n(t)$ is a continuous function for any $t \in T_0$, $m(\alpha)$ is an unknown function of parameter $\alpha \in S \subseteq R^d$, increasing at $\|\alpha\| \to \infty$; $\lim m(\alpha) \neq +\infty$ at $\|\alpha\| \to +\infty$ and $\lim \dfrac{1}{m(\alpha)} = m_0$ at $\|\alpha\| \to 0$.

The set $A(r)$ will be determined as follows:

$$A(r) = \left\{ x, y : \ x^2 + y^2 = \frac{1}{m(\alpha)} \right\}, \tag{2.15}$$

i. e., $r(\alpha) = \dfrac{1}{m(\alpha)} > 0$ at all $\alpha \in S \subseteq R^d$.

For the analysis of the behavior of the solutions $((x(t, \alpha), y(t, \alpha))^T$ of the system (2.15) with respect to the set (2.15), take a function V of the form

$$V = x^2 + y^2, \tag{2.16}$$

whose derivative, in view of the given equations, has the form

$$\frac{dV}{dt} = 2(x^2 + y^2)\left[1 - m(\alpha)(x^2 + y^2)\right]. \tag{2.17}$$

The function (2.16) satisfies all the conditions of the Theorem 2.2. From (2.17) it follows that at $\alpha \in S \subseteq R^d$

$$\left.\frac{dV}{dt}\right|_{(2.14)} < 0 \quad \text{at} \quad x^2 + y^2 > r(\alpha),$$

$$\left.\frac{dV}{dt}\right|_{(2.14)} = 0 \quad \text{at} \quad x^2 + y^2 = r(\alpha),$$

$$\left.\frac{dV}{dt}\right|_{(2.14)} > 0 \quad \text{at} \quad x^2 + y^2 < r(\alpha).$$

Thus, for the function (2.17) the conditions of the Theorem 2.2 are satisfied, and any solutions of the system (2.14) with the initial conditions x_0, y_0:

$$r(\alpha) - \delta < x_0^2 + y_0^2 < r(\alpha) + \delta, \qquad \alpha \in S,$$

will be stable with respect to the set (2.15) in the sense of the Definition 2.2(a).

Theorem 2.3 *Assume that $f(t, x, \alpha)$ in the system (2.1) is continuous on $T_0 \times R^n \times R^d$ and the following conditions are satisfied:*

(1) *for any $\alpha \in S \subseteq R^d$ there exists a function $r = r(\alpha) > 0$ such that the set $A(r)$ is not empty at all $\alpha \in S$;*

(2) *there exists a matrix-valued function $U \in C(\mathcal{T}_0 \times R^n, R^{s \times s})$, $U(t,x)$ locally Lipshitz with respect to x, vector $y \in R^s$, $(s \times s)$-matrices $\Phi_1(r)$ and $\Phi_2(r)$ such that:*

  (a) $a(\|x\|) \leq V(t,x,y)$ *at* $\|x\| > r(\alpha)$,

  (b) $0 < V(t,x,y) \leq b(\|x\|)$ *at* $\|x\| < r(\alpha)$, *where a, b belong to the K-class,*

  (c) $D^+V(t,x,y)|_{(2.1)} < \varphi^{\mathrm{T}}(\|x\|)\Phi_1(r)\varphi(\|x\|)$ *at* $\|x\| > r(\alpha)$, $\alpha \in S \subseteq R^d$,

  (d) $D^+V(t,x,y)|_{(2.1)} = 0$ *at* $\|x\| = r(\alpha)$, $\alpha \in S \subseteq R^d$,

  (e) $D^+V(t,x,y)|_{(2.1)} > \psi^{\mathrm{T}}(\|x\|)\Phi_2(r)\psi(\|x\|)$ *at* $\|x\| < r(\alpha)$, $\alpha \in S \subseteq R^d$,

where

$$\varphi^{\mathrm{T}}(\|x\|) = \left(\varphi_1^{1/2}(\|x_1\|), \dots, \varphi_s^{1/2}(\|x_s\|)\right), \quad \varphi_i \in K,$$

$$\psi^{\mathrm{T}}(\|x\|) = \left(\psi_1^{1/2}(\|x_1\|), \dots, \psi_s^{1/2}(\|x_s\|)\right), \quad \psi_i \in K,$$

$$x_s \in R^{n_s}, \quad n_1 + n_2 + \dots n_s = n;$$

(3) *there exist constant $(s \times s)$-matrices $\overline{\Phi}_1$, $\overline{\Phi}_2$ such that*

  (a) $\dfrac{1}{2}\left(\Phi_1(r) + \Phi_1^{\mathrm{T}}(r)\right) \leq \overline{\Phi}_1$ *at all* $\alpha \in S$,

  (b) $\dfrac{1}{2}\left(\Phi_2(r) + \Phi_2^{\mathrm{T}}(r)\right) \geq \overline{\Phi}_2$ *at all* $\alpha \in S$,

here $\overline{\Phi}_1$ is negative-definite, and $\overline{\Phi}_2$ is positive-definite;

(4) *for any $r(\alpha) > 0$ and functions $a(r)$, $b(r)$ the following relation is true:*

$$a(r) = b(r).$$

    Then the set $A(r)$ is invariant with respect to motions of the system (2.1), and the solutions of the system (2.1) are uniformly asymptotically stable with respect to the set $A(r)$.

Proof The invariance of the set $A(r)$ with respect to the system (2.1) follows from the Theorem 2.1. All its conditions are satisfied when the conditions of the Theorem 2.3 are satisfied.

    The uniform asymptotic stability of solutions of the system (2.1) with respect to the set $A(r)$ will be proved if, with all conditions of the Theorem 2.3

satisfied, $T = T(\varepsilon) > 0$ (ε is the same as in the Theorem 2.1) such that the condition

$$r(\alpha) - \delta_0 < \|x(t_0, \alpha)\| < r(\alpha) + \delta_0,$$

where $\delta_0 = \delta_0(\varepsilon)$, implies the existence of $t^* \in [t_0, t_0 + T]$, for which

$$r(\alpha) - \delta < \|x(t^*, \alpha)\| < r(\alpha) + \delta. \qquad (2.18)$$

Here $\delta = \delta(\varepsilon)$.

Let, with all conditions of the Theorem 2.3 satisfied, there does not exist $t^* \in [t_0, t_0 + T(\varepsilon)]$, for which the inequality (2.18) would hold true. Here two cases are possible:

$$\text{(a)} \quad \|x(t, \alpha)\| \geq r + \delta \quad \text{at all} \quad t \in [t_0, t_0 + T]$$

or

$$\text{(b)} \quad \|x(t, \alpha)\| \leq r - \delta \quad \text{at all} \quad t \in [t_0, t_0 + T].$$

Consider the case (a). Since in view of the conditions (2)(c) and (3)(a) of the Theorem 2.3

$$D^+V(t, x(t), y)\big|_{(2.1)} < 0 \quad \text{at} \quad \|x\| > r(\alpha), \quad t_0 \leq t < \infty, \qquad (2.19)$$

the function $V(t, x(t), y)$ monotonely decreases, i. e.,

$$\lim_{t \to \infty} V(t, x(t, \alpha), y) = \inf_t V(t, x(t, \alpha), y) = \Delta(r).$$

Since the matrix $\overline{\Phi}_1$ is negative-definite, i. e. Re $\lambda_M(\overline{\Phi}_1) < 0$, then

$$D^+V(t, x(t, \alpha), y)\big|_{(2.1)} < -\lambda_M(\overline{\Phi}_1)\Pi(\|x\|) \qquad (2.20)$$

at $\|x\| > r(\alpha)$, where $\Pi(\|x\|) \geq \varphi^T(\|x\|)\varphi(\|x\|)$, $\Pi(0) = 0$ and Π belongs to the K-class.

According to the conditions of the Theorem 2.1, solutions of the system (2.1) are uniformly stable, therefore it can be assumed that $\|x(t, \alpha)\| \leq H < +\infty$ at all $\alpha \in S \subseteq R^d$.

On the set $r(\alpha) + \delta \leq \|x\| \leq H$ calculate

$$\gamma(r) = \inf_{r+\delta \leq \|x\| \leq H} \Pi(\|x\|).$$

Obviously, $\gamma(r) \geq 0$. From the inequality (2.20) obtain

$$V(t_0 + T(\varepsilon), x(t, \alpha), y) < V(t_0, x(t_0, \alpha), y) - \lambda_M(\overline{\Phi}_1) \int_{t_0}^{t_0 + T(\varepsilon)} \Pi(\|x(\tau)\|)\, d\tau,$$

or, since

$$-\Pi(\|x\|) \leq -\gamma(r) \quad \text{at} \quad r + \delta \leq \|x\| \leq H,$$

then

$$V(t_0 + T(\varepsilon), x(t, \alpha), y) < V(t_0, x(t_0, \alpha), y) - \lambda_M(\overline{\Phi}_1) \int\limits_{t_0}^{t_0 + T(\varepsilon)} \gamma(r)\, d\tau$$

$$< V(t_0, x(t_0, \alpha), y) - \lambda_M(\overline{\Phi}_1)\gamma(r)T(\varepsilon).$$

Hence follows that at a sufficiently large $T = T(\varepsilon)$

$$V(T, x(t, \alpha), y) < 0 \quad \text{at} \quad \|x\| > r(\alpha).$$

But this contradicts the condition (2)(a) of the Theorem 2.3. Therefore the case (a) is impossible, i.e., there should exist a value $t^* \in [t_0, t_0 + T]$, at which

$$\|x(t^*, \alpha)\| \le r(\alpha) + \delta. \tag{2.21}$$

Now show that $\|x(t, \alpha)\| \to r(\alpha)$ at $t \to +\infty$. Indeed, let $\varepsilon > 0$ be an arbitrarily small number, and

$$l(r) = \inf \lambda_M(\overline{\Phi}_1)\Pi(\|x\|) > 0 \quad \text{at} \quad r(\alpha) + \varepsilon \le \|x\| \le H.$$

From (2.20) it follows that there exists $T(\varepsilon) > t_0$ such that

$$V(T(\varepsilon), x(T(\varepsilon), \alpha), y) > l(r). \tag{2.22}$$

But, in view of the conditions (2.19), the function $V(t, x(t), y)$ is monotonically decreasing, so

$$V(t, x(t, \alpha), y) < l(r) \quad \text{at all} \quad t > T(\varepsilon) \quad \text{and} \quad \|x\| > r(\alpha). \tag{2.23}$$

Hence, at $t > T(\varepsilon)$ obtain $\|x(t, \alpha)\| < r(\alpha) + \varepsilon$.

Assume that this be not so, and let $t_1 > T(\varepsilon)$ exist, for which

$$\|x(t_1, \alpha)\| > r(\alpha) + \varepsilon.$$

Then from (2.22) and (2.23) it follows that

$$l(r) > V(t_1, x(t_1, \alpha), y) > \lambda_M(\overline{\Phi}_1)\Pi(\|x\|) \ge l(r),$$

which is a contradiction. Therefore

$$\lim_{t \to \infty} \|x(t, \alpha)\| = r(\alpha).$$

Now consider the case (b). Since the matrix $\overline{\Phi}_2$ is positive-definite, for the function $D^+V(t, x, y)$ one can easily obtain the estimate

$$D^+V(t, x, y)\big|_{(2.1)} > \lambda_m(\overline{\Phi}_2)\chi(\|x\|) \quad \text{at} \quad \|x\| < r(\alpha),$$

where $\chi(\|x\|) \leq \psi^{\mathrm{T}}(\|x\|)\psi(\|x\|)$, $\chi(0) = 0$, χ belongs to the K-class, $\lambda_m(\overline{\Phi}_2)$ is the minimum eigenvalue of the matrix $\overline{\Phi}_2$. From the condition (2)(b) of the Theorem 2.3 we obtain

$$V(t, x, y) \leq b(\|x\|) \quad \text{at} \quad \|x\| < r(\alpha).$$

Hence $V(t, x, y)$ is bounded, i.e., there exists $M > 0$ such that

$$|V(t, x, y)| \leq M$$

at $t_0 \leq t < \infty$ and $\|x\| < r(\alpha) < H$, where M and H are some positive numbers.

Let $\delta > 0$ ($\delta < r(\alpha) \; \forall \alpha \in \mathcal{S} \subseteq R^d$) be an arbitrarily small number. In view of the conditions (2)(a) of the Theorem 2.3 there exists a point (t_0, x_0), $t_0 \in R_+$ and $0 < \|x_0\| < r(\alpha)$, such that

$$V(t_0, x_0, y) = \sigma > 0.$$

Let the solution $x(t, \alpha)$ of the uncertain system (2.1) be determined by the initial conditions $x(t_0) = x_0$, and

$$0 < \|x(t, \alpha)\| < r(\alpha) - \delta \quad \text{at} \quad t \in [t_0, t_0 + T].$$

Due to the conditions (2)(d) and (3)(b), the function $V(t, x(t, \alpha), y)$ at $\|x\| \leq r(\alpha)$ is monotonely increasing with t, therefore at $t \geq t_0$ we obtain

$$V(t, x(t, \alpha), y) > V(t_0, x(t_0), y) = \sigma > 0.$$

At some $t^* \in [t_0, t_0 + T]$ the following inequality will be true:

$$\|x(t^*, \alpha)\| > r(\alpha) - \delta. \tag{2.24}$$

Indeed, assume that this is not so, and

$$\|x(t, \alpha)\| < r(\alpha) - \delta \quad \text{at all} \quad t \geq t_0.$$

The solution $x(t, \alpha)$ is infinitely extendable to the right, and one can find $\beta < r(\alpha) - \delta$ at all $\alpha \in \mathcal{S} \subseteq R^d$ such that

$$0 < \beta \leq \|x(t, \alpha)\| < r(\alpha) - \delta \quad \text{at} \quad t_0 \leq t < \infty.$$

Calculate

$$\tilde{\gamma}(r) = \inf_{\beta \leq \|x\| \leq r(\alpha) - \delta} \chi(\|x\|).$$

It is clear that $\tilde{\gamma}(r) > 0$, and in the domain $\|x\| < r(\alpha)$ we obtain

$$D^+ V(t, x(t, \alpha), y)\big|_{(2.1)} > \lambda_m(\overline{\Phi}_2)\tilde{\gamma}(r) \quad \text{at} \quad t_0 \leq t < \infty.$$

Hence, at $t_0 \leq t < \infty$ we obtain

$$V(t, x(t, \alpha), y) > V(t_0, x_0, y) + \lambda_m(\overline{\Phi}_2)\tilde{\gamma}(r)(t - t_0),$$

which contradicts the boundedness of the function $V(t, x, y)$ in the domain $\|x\| < r(\alpha)$, $t_0 \leq t < \infty$. Thus, it has been proved that there exists a value $t^* \in [t_0, t_0 + T]$, for which the inequality (2.24) holds true. Along with the inequality (2.21), the following estimate holds:

$$r(\alpha) - \delta < \|x(t^*, \alpha)\| < r(\alpha) + \delta$$

at $t^* \in [t_0, t_0 + T]$. Hereby we have determined the uniform asymptotic stability of solutions of the uncertain system (2.1) in the sense of the Definition 2.4 (d).

Remark 2.6 For the class of systems (2.1) which do not have linear approximation, the Theorem 2.3 is the basis of the following two statements.

Proposition 2.5 Let the vector-function $f(t, x, \alpha)$ in the system (2.1) have no linear approximation, let all the conditions of the Theorem 2.3 be satisfied, and, in addition, $r(\alpha) \to 0$ at $\|\alpha\| \to 0$, $f(t, x, \alpha) = 0$, if, and only if, $x = 0$, $T_i = R$. Then the set $A(r)$ tends to zero, and the equilibrium state $x = 0$ of the uncertain system (2.1) is uniformly asymptotically stable.

Proposition 2.6 Let the vector-function $f(t, x, \alpha)$ in the system (2.1) have no linear approximation, let all the conditions of the Theorem 2.3 be satisfied, and, in addition, $r(\alpha) \to \infty$ at $\|\alpha\| \to \infty$, $f(t, x, \alpha) = 0$, if, and only if, $x = 0$, $T_i = R$, φ_i, and ψ_i belong to the KR-class. Then the set $A(r)$ is unlimitedly increasing, and the equilibrium state $x = 0$ of the uncertain system (2.1) is uniformly asymptotically stable.

Like the Theorem 2.1, the Theorem 2.3 has a corollary based on the scalar Lyapunov function.

Theorem 2.4 *Let the following conditions be satisfied:*

(1) *for each $\alpha \in S \subseteq R^d$ there exists a function $r = r(\alpha) > 0$ such that $r(\alpha) \to r_0$ at $\|\alpha\| \to 0$ and $r(\alpha) \to +\infty$ at $\|\alpha\| \to +\infty$;*

(2) *there exist scalar functions $V \in C^{(1,1)}(T_0 \times R^n, R_+)$, $W_1 \colon R^n \times R^d \to R$ and $W_2 \colon R^n \times R^d \to R$ such that:*

    (a) $a(\|x\|) \leq V(t, x)$ *at* $\|x\| > r(\alpha)$,

    (b) $V(t, x) \leq b(\|x\|)$ *at* $\|x\| < r(\alpha)$,

    where a, b belong to the K-class;

    (c) $DV(t, x)|_{(2.1)} < W_1(x, \alpha)$ *at* $\|x\| > r(\alpha)$, $\alpha \in S$,

    (d) $DV(t, x)|_{(2.1)} = 0$ *at* $\|x\| = r(\alpha)$, $\alpha \in S$,

    (e) $DV(t, x)|_{(2.1)} > W_1(x, \alpha)$ *at* $\|x\| < r(\alpha)$, $\alpha \in S$;

(3) *there exist functions sign definite in the Lyapunov sense* $\overline{W}_1(x)$ *and* $\underline{W}_2(x)$ *such that:*

   (a) $W_1(x, \alpha) \leq \overline{W}_1(x) < 0$ *at all* $\alpha \in \mathcal{S}$,

   (b) $W_2(x, \alpha) \geq \underline{W}_2(x) > 0$ *at all* $\alpha \in \mathcal{S}$;

(4) *for any* $r(\alpha) > 0$ *and functions* $a(r)$, $b(r)$ *the following relations are true:*

$$a(r) = b(r).$$

Then the set $A(r)$ *is invariant with respect to the motions of the system* (2.1), *and the solutions of the systems* (2.1) *are uniformly asymptotically stable with respect to the set* $A(r)$.

Example 2.5 Let the following equations be given:

$$\frac{dx}{dt} = n(t)y + \left[1 - \frac{1}{a^2}\tilde{m}^2(\alpha)(x^2 + y^2)\right] x(x^2 + y^2),$$

$$\frac{dy}{dt} = -n(t)x + \left[1 - \frac{1}{a^2}\tilde{m}^2(\alpha)(x^2 + y^2)\right] y(x^2 + y^2), \tag{2.25}$$

where $n \in C(R, R)$, $\tilde{m}(\alpha)$ is an uncertainties function which has the same properties as the function $m(\alpha)$ in the Example 2.4.

Let $r(\alpha) = \dfrac{a}{\tilde{m}(\alpha)} > 0$, $\alpha \in \mathcal{S} \subseteq R^d$. The set $A(r)$ is determined as follows:

$$A(r) = \left\{x, y: \ (x^2 + y^2)^{1/2} = r(\alpha)\right\}.$$

Take a function V in the form

$$V = x^2 + y^2,$$

whose derivative, in view of the equations (2.25), is

$$\left.\frac{dV}{dt}\right|_{(2.25)} = 2\left[1 - \frac{1}{a^2}\tilde{m}^2(\alpha)(x^2 + y^2)\right](x^2 + y^2)^2.$$

Hence for $t \geq t_0$

$$\left.\frac{dV}{dt}\right|_{(2.25)} < 0 \quad \text{at} \quad (x^2 + y^2)^{1/2} > r(\alpha),$$

$$\left.\frac{dV}{dt}\right|_{(2.25)} = 0 \quad \text{at} \quad (x^2 + 1y^2)^{1/2} = r(\alpha),$$

$$\left.\frac{dV}{dt}\right|_{(2.25)} > 0 \quad \text{at} \quad (x^2 + y^2)^{1/2} < r(\alpha).$$

Thus all conditions of the Theorem 2.3 are satisfied, and the set $A(r)$ is invariant for the system (2.25), and all its solutions are uniformly asymptotically stable with respect to the set $A(r)$.

The asymptotic approximation of the solutions $x(t, \alpha)$ of the system (2.1) to the bound of the moving set can be determined by using two auxiliary functions, with the conditions of the Theorem 2.4 duly modified.

Theorem 2.5 *Assume that $f(t, x, \alpha)$ in the system (2.1) is continuous on $T_0 \times R^n \times R^d$ and the following conditions are satisfied:*

(1) *for any $\alpha \in S \subseteq R^d$ there exists a function $r = r(\alpha) > 0$ such that the set $A(r)$ is not empty at all $\alpha \in S$;*

(2) *There exist functions $V(t, x)$ and $W(t, x)$ and comparison functions a, b belonging to the K-class, such that:*

    (a) $V \in C^{(1,1)}(T_0 \times R^n, R^n)$, $V(t, x)$ *is bounded with respect to x and $V(t, 0) = 0$ at all $t \in T_0$;*

    (b) $W \in C(T_0 \times R^n, T_0)$ *and $b(\|x\|) \leq W(t, x) \leq a(\|x\|)$ at all $(t, x) \in T_0 \times R^n$;*

(3) *There exists a comparison function c belonging to the K-class, a locally integrable function $k(t) \geq 0$ on $T_\tau = [\tau, \infty)$, $k(t) \notin L^1(I)$, and a bounded function $\Phi(t)$ such that:*

    (a) $DV(t, x)\big|_{(2.1)} \leq \Phi(t) - [c(W(t, x)) - r(\alpha)]k(t)$ *at $\|x\| > r(\alpha)$,*

    (b) $DV(t, x)\big|_{(2.1)} = 0$ *at $\|x\| = r(\alpha)$,*

    (c) $DV(t, x)\big|_{(2.1)} \geq \Phi(t) - [c(W(t, x)) - r(\alpha)]k(t)$ *at $\|x\| < r(\alpha)$*

and for any $\tau \geq 0$ the following estimates hold true:

    (d) $\liminf\limits_{t \to \infty} \left(\int\limits_\tau^t \Phi(s)\, ds \Big/ \int\limits_\tau^t k(s)\, ds \right) \leq 0$,

    (e) $\limsup\limits_{t \to \infty} \left(\int\limits_\tau^t \Phi(s)\, ds \Big/ \int\limits_\tau^t k(s)\, ds \right) \geq 0$;

(4) *For any $r(\alpha) \geq 0$ there exists $\varkappa = \varkappa(r) \geq 0$ such that $a(r) = \varkappa = b(r)$ and $c(b(r)) = r = c(a(r))$.*

Then any bounded solution $x(t, \alpha)$ of the system (2.1), for which there exists $\lim\limits_{t \to \infty} W(t, x(t, \alpha))$, asymptotically approximates the bound of the domain $A(r)$ at $t \to \infty$.

For the proof we will need the following statement.

Lemma 2.1 *If the absolutely continuous and bounded below function Φ: $[\tau, \infty) \to R$ satisfies the inequality $\Phi'(t) \leq -p(t)$ almost all over $I = [\tau, \infty)$, where $p(t)$ is a function locally integrable on I, then*

$$\limsup_{t \to \infty} \int_\tau^t p(s)\, ds < +\infty.$$

Proof From the absolute continuity of the function $\Phi(t)$ it follows that at all $t \in [\tau, \infty)$

$$\Phi(t) - \Phi(\tau) = \int_{\tau}^{t} \Phi'(s) \, ds.$$

Let $m = \inf_{t \in [\tau, +\infty)} \Phi(t)$, then, integrating the inequality $\Phi'(t) \leq -p(t)$ within the bounds from τ to t,

$$\Phi(t) - \Phi(\tau) \leq - \int_{\tau}^{t} p(s) \, ds,$$

$$\int_{\tau}^{t} p(s) \, ds \leq \Phi(\tau) - \Phi(t) < \Phi(\tau) - m,$$

i. e.,

$$\limsup_{t \to \infty} \int_{\tau}^{t} p(s) \, ds < +\infty.$$

If $\lim_{t \to \infty} W(t, x(t, \alpha)) = \varkappa$, then the statement of the Theorem 2.5 follows from the conditions (2)(b) and (4). If any bounded solution $x(t, \alpha)$ of the system (2.1) does not satisfy the condition $\lim_{t \to \infty} W(t, x(t, \alpha)) = \varkappa$, then the two cases are possible:

(a) there exists a solution $x(t, \alpha)$ for which $W(t, x(t, \alpha)) \geq \varkappa + \eta$ at $t \geq \tau$,

or

(b) there exists a solution $x(t, \alpha)$ for which $W(t, x(t, \alpha)) \leq \varkappa - \eta$ at $t \geq \tau$.

Here τ is sufficiently large, and $\eta > 0$.

According to the condition (4) of the Theorem 2.5 for the given $0 < \eta_1, \eta_2 < \eta$ one can find $\varepsilon_1 > 0$, $\varepsilon_2 > 0$ such that

$$b(r(\alpha) + \varepsilon_1) = \varkappa + \eta_1, \qquad c(b(r(\alpha) + \varepsilon_1)) = r(\alpha) + \varepsilon_1,$$
$$b(r(\alpha) - \varepsilon_2) = \varkappa - \eta_2, \qquad c(b(r(\alpha) - \varepsilon_2)) = r(\alpha) - \varepsilon_2.$$

Consider the case (a). According to the condition (3)(a) of the Theorem 2.5,

$$DV(t, x)\big|_{(2.1)} \leq \Phi(t) - [c(b(r(\alpha) + \varepsilon_1)) - r(\alpha)]k(t)$$

at $\|x\| > r(\alpha)$.

Denote $p(t) = -\Phi(t) + [c(b(r(\alpha) + \varepsilon_1)) - r(\alpha)]$, then $p(t) = -\Phi(t) + [r(\alpha) + \varepsilon_1 - r(\alpha)]k(t) = -\Phi(t) + \varepsilon_1 k(t)$ at all $t \in \mathcal{T}_0$. Obviously, $p(t)$ is a locally integrable function on $\mathcal{T}_0$. According to the condition (2)(a) of the

Theorem 2.5, on any bounded solution $x(t, \alpha)$ the function $V(t, x(t, \alpha))$ is limited, therefore the function $\psi(t) = V(t, x(t, \alpha))$ is bounded below.

From the condition $\psi'(t) \leq -p(t)$ and the Lemma 2.1 it follows that

$$\limsup_{t \to \infty} \int_{\tau}^{t} p(s)\, ds < +\infty. \tag{2.26}$$

From the condition (3)(d) of the Theorem 2.5 it follows that there exists a sequence $t_i \to \infty$ such that

$$\int_{\tau}^{t_i} \Phi(s)\, ds \leq \frac{1}{2}\, \varepsilon_1 \int_{\tau}^{t_i} k(s)\, ds.$$

Hence

$$\int_{\tau}^{t_i} p(s)\, ds \geq \frac{1}{2}\, \varepsilon_1 \int_{\tau}^{t_i} k(s)\, ds$$

and at $i \to \infty$ the integral diverges to $+\infty$, since $k(t) \notin L^1(I)$. This contradicts the inequality (2.26).

Now consider the case (b). From the condition (3)(c) of the Theorem 2.5 obtain

$$DV(t, x)\big|_{(2.1)} \geq \Phi(t) - [c(a(r(\alpha) - \varepsilon_2)) - r(\alpha)]k(t)$$

at $\|x\| < r(\alpha)$.

Denote $q(t) = \Phi(t) - [c(a(r(\alpha) - \varepsilon_2)) - r(\alpha)]k(t)$, then $q(t) = \Phi(t) - [r(\alpha) - \varepsilon_2 - r(\alpha)]k(t) = \Phi(t) + \varepsilon_2 k(t)$ at all $t \in \mathcal{T}_0$.

Obviously, $q(t)$ is a locally integrable function on $\mathcal{T}_0$. Like above, $V(t, x(t, \alpha))$ is a bounded function, therefore $\chi(t) = -V(t, x(t, \alpha))$ is bounded below. From the condition $\chi'(t) \leq -q(t)$ and the Lemma 2.1 it follows that

$$\limsup_{t \to \infty} \int_{\tau}^{t} q(s)\, ds < +\infty. \tag{2.27}$$

On the other hand, according to the condition (3)(e) of the Theorem 2.5, there exists a sequence $t_i \to \infty$ such that at all i the following estimate holds true:

$$\int_{\tau}^{t} \Phi(s)\, ds \geq \frac{1}{2}\, \varepsilon_2 \int_{\tau}^{t_i} k(s)\, ds.$$

Hence

$$\int_{\tau}^{t} q(s)\, ds \geq \frac{1}{2}\, \varepsilon_2 \int_{\tau}^{t_i} k(s)\, ds,$$

and at $i \to \infty$ divergence occurs, since $k(t) \notin L^1(I)$. This contradicts the condition (2.27) and proves the statement of the Theorem 2.5, i. e., $\|x(t, x)\| \to r(\alpha)$ uniformly with respect to t at $t \to +\infty$.

2.4 Exponential Convergence of Motions to a Moving Invariant Set

Continue the analysis of solutions of the uncertain system (2.1) with respect to the moving invariant set (2.2). Taking into account some results of the article by Corless and Leitmann [1996], one can give the following definition.

Definition 2.13 The motion $x(t, \alpha)$ of the uncertain system (2.1) is *exponentially stable with respect to the moving invariant set $A(r)$ with the index γ*, if there exist constants $\beta_1 \geq 0$, $\beta_2 \geq 0$ such that:

(1) for each $t_0 \in R$, $\alpha \in S \subseteq R^d$ there exists $\delta > 0$ such that the solution $x(t, \alpha)$: $[t_0, t_1) \to R^n$ of the system (2.1) is determined at all $t_0 < t_1$ and x_0: $r(\alpha) - \delta \leq \|x_0\| \leq r(\alpha) + \delta$;

(2) any solution $x(t, \alpha)$: $[t_0, t_1) \to R^n$ of the system (2.1) with the initial value x_0 has the extension $\bar{x}(t, \alpha)$: $[t_0, \infty) \to R^n$, i.e., $\bar{x}(t, \alpha) = x(t, \alpha)$ at all $t \in [t_0, t_1)$ and $\bar{x}(t, \alpha)$ is a solution of the system (2.1);

(3) any solution $x(t, \alpha)$: $[t_0, t_1) \to R^n$ of the system (2.1) with the initial value x_0 satisfies the estimate

$$r(\alpha) - \beta_1 \|x_0\| \exp[-\gamma(t - t_0)] \leq \|x(t, \alpha)\| \leq$$
$$\leq r(\alpha) + \beta_2 \|x_0\| \exp[-\gamma(t - t_0)] \quad \text{at all} \quad t \geq t_0.$$

Thus, the purpose of our analysis is to find the conditions under which the solutions of the system (2.1) will be exponentially stable with respect to the moving invariant set $A(r)$.

The next theorem contains the conditions which guarantee the exponential convergence of motions of the uncertain system (2.1) to the moving invariant set $A(r)$.

Theorem 2.6 *Assume that $f(t, x, \alpha)$ in the system (2.1) is continuous on $R_+ \times R^n \times R^d$ and the following conditions are satisfied:*

(1) *for any $\alpha \in S \subseteq R^d$ there exists a function $r = r(\alpha) > 0$ such that the set $A(r)$ is not empty at all $\alpha \in S$;*

(2) *there exists a matrix-valued function $U \in C(R_+ \times R^n, R^{s \times s})$, $U(t, x)$ continuously differentiable with respect to t, x, a vector $\eta \in R_+^s$, $\eta > 0$, and constants $q \geq 0$, $a_1(\alpha) > 0$, $a_2(\alpha) > 0 \; \forall \alpha \in S \subseteq R^d$ such that the function*

$$V(t, x, \eta) = \eta^{\mathrm{T}} U(t, x) \eta$$

satisfies the estimates:

(a) $a_1(\alpha)\|x\|^q \leq V(t,x,\eta) \leq a_2(\alpha)\|x\|^q$ *at all* $(t,x) \in T_0 \times R^n$,

(b) *for any* $\alpha \in S \subseteq R^d$ *there exist constants* $\delta_1(\alpha)$ *and* $\delta_2(\alpha)$, $0 < \delta_1(\alpha) < \delta_2(\alpha) < +\infty$, *such that*

$$\delta_1(\alpha) \leq V(t,x,\eta) \leq \delta_2(\alpha) \quad \forall \alpha \in S \subseteq R^d,$$

$$\left(\frac{\delta_1(\alpha)}{w_1}\right)^{1/q} = \left(\frac{\delta_2(\alpha)}{w_2}\right)^{1/q}, \quad w_1 = \min_{\alpha \in S} a_1(\alpha), \quad w_2 = \min_{\alpha \in S} a_2(\alpha);$$

(3) *for the full derivative* $DV(t,x,\eta)$ *of the function* $V(t,x,\eta)$ *along the solutions of the uncertain system* (2.1) *the following inequalities are true:*

(a) $DV(t,x,\eta)|_{(2.1)} < -q\gamma[V(t,x,\eta) - \delta_1(\alpha)]$ *at* $\|x\| > r(\alpha)$, $\alpha \in S \subseteq R^d$,

(b) $DV(t,x,\eta)|_{(2.1)} = 0$ *at* $\|x\| = r(\alpha)$, $\alpha \in S \subseteq R^d$,

(c) $DV(t,x,\eta)|_{(2.1)} > -q\gamma[V(t,x,\eta) - \delta_2(\alpha)]$ *at* $\|x\| < r(\alpha)$, $\alpha \in S \subseteq R^d$;

(4) *for any* $r(\alpha) > 0$ *the following relation holds true:*

$$a_1(\alpha)[r(\alpha)]^q = a_2(\alpha)[r(\alpha)]^q.$$

Then:

(a) *the moving set*

$$A(r) \subset \{x \in R^n: \ V(t,x,\eta) \leq \delta_2(\alpha)\}$$

is invariant with respect to the solutions of the system (2.1);

(b) *motions of the system* (2.1) *exponentially converge to the moving invariant set* $A(r)$ *with the index* γ.

Proof First, prove the invariance of the moving set $A(r)$. Let $t_0 \in R$ and let $x_0 \in R^n$ satisfy the condition $\|x_0\| = r(\alpha)$. According to the condition (2) (a), (b) of the Theorem 2.6

$$A(r) \subset \{x \in R^n: \ V(t,x,\eta) \leq \delta_2(\alpha)\},$$

where $A(r) = \{x \in R^n: \|x\| = r(\alpha)\}$, $r(\alpha)$ is determined below.

Assume that the set $A(r)$ is not invariant. Then one can find time values $t_2 > t_1 \geq t_0$ such that at all $t \in [t_1, t_2]$

$$\begin{aligned} &\|x(t_1,\alpha)\| = r(\alpha),\\ \text{(A)} \quad &\|x(t_2,\alpha)\| > r(\alpha),\\ &\|x(t,\alpha)\| \geq r(\alpha),, \end{aligned}$$

or

$$\|x(t_1, \alpha)\| = r(\alpha),$$
(B) $$\|x(t_2, \alpha)\| < r(\alpha),$$
$$\|x(t, \alpha)\| \le r(\alpha).$$

From the conditions (2) (b) and (3) (a) it follows that if

$$\delta_1(\alpha) \le V(t, x, \eta) \le \delta_2(\alpha) \quad \forall \alpha \in \mathcal{S} \subseteq R^d,$$

then

$$DV(t, x, \eta)|_{(1.1)} < -q\gamma[V(t, x, \eta) - \delta_1(\alpha)]$$

at $\|x\| > r(\alpha)$ and therefore

$$V(t, x(t, \alpha), \eta) < V(t_1, x(t_1, \alpha), \eta) \quad \text{at} \quad t_1 \le t \le t_2 \quad \text{and} \quad \|x\| > r(\alpha).$$

Similarly,

$$DV(t, x, \eta)|_{(1.1)} > -q\gamma[V(t, x, \eta) - \delta_2(\alpha)] \ge 0 \quad \text{at} \quad \|x\| < r(\alpha).$$

and therefore

$$V(t, x(t, \alpha), \eta) > V(t_1, x(t_1, \alpha), \eta) \quad \text{at} \quad t_1 \le t \le t_2.$$

According to the condition (2)(a) the inequality

$$a_1(\alpha)[r(\alpha)]^q < a_1(\alpha)\|x(t_2, \alpha)\|^q \le V(t_2, x(t_2, \alpha), \eta)$$
$$< V(t_1, x(t_1, \alpha), \eta) \le a_2(\alpha)\|x(t_1, \alpha)\|^q = a_2(\alpha)[r(\alpha)]^q, \tag{2.28}$$

contradicts the condition (4) of the Theorem 2.6.

In the case (B) obtain the sequence of inequalities

$$a_2(\alpha)[r(\alpha)]^q > a_2(\alpha)\|x(t_2, \alpha)\|^q \ge V(t_2, x(t_2, \alpha), \eta)$$
$$> V(t_1, x(t_1, \alpha), \eta) \ge a_1(\alpha)\|x(t_1, \alpha)\|^q = a_1(\alpha)[r(\alpha)]^q, \tag{2.29}$$

which also contradicts the condition (4) of the Theorem 2.6. The inequalities (2.28), (2.29) together with the condition (3)(a) of the Theorem 2.6 prove its statement.

Now prove the exponential stability of the solutions $x(t, \alpha)$ of the system (2.1) with respect to the moving invariant set $A(r)$. Let x_0 be such that

$$r(\alpha) - \delta < \|x_0\| < r(\alpha) + \delta,$$

where $\delta > 0$ is some constant.

Consider the case (A), where at $\|x\| > r(\alpha)$ the condition (3)(a) of the Theorem 2.6 is satisfied. Denote

$$\varphi(t) = V(t, x, \eta) - \delta_1(\alpha), \quad \alpha \in \mathcal{S} \subseteq R^d,$$
$$\tilde{\varphi} = \delta_2^* - \delta_1^*, \tag{2.30}$$
$$\delta_1^* = \min_{\alpha \in \mathcal{S}} \delta_1(\alpha), \quad \delta_2^* = \max_{\alpha \in \mathcal{S}} \delta_2(\alpha).$$

It is clear that $\varphi(t_0) \geq 0$ and $\varphi(t) < \tilde{\varphi}$ at all $t \geq t_0$.
Since

$$\frac{d\varphi}{dt} = DV(t, x, \eta),$$

taking into account the inequalities $\varphi(t) < \tilde{\varphi}$ and the condition (3)(a) of the Theorem 2.6,

$$\frac{d\varphi(t)}{dt} < -q\gamma\varphi(t) \quad \text{at all} \quad t \geq t_0.$$

Hence

$$\varphi(t) < \varphi(t_0) \exp[-q\gamma(t - t_0)] \quad \text{at all} \quad t \geq t_0.$$

Taking into account the notation (2.30),

$$V(t, x(t, \alpha), \eta) < \delta_1(\alpha) + [V(t_0, x_0, \eta) - \delta_1(\alpha)] \exp[-q\gamma(t - t_0)]. \qquad (2.31)$$

If the condition (2)(b) of the Theorem 2.6 is satisfied, $V(t_0, x_0, \eta) - \delta_1 > 0$ and therefore there exists a constant $k_1 > 0$ such that

$$k_1 V(t_0, x_0, \eta) = V(t_0, x_0, \eta) - \delta_1. \qquad (2.32)$$

Taking into account the inequality (2.32), the inequality (2.31) transforms to the form

$$V(t, x(t, \alpha), \eta) < \delta_1(\alpha) + k_1 V(t_0, x_0, \eta) \exp[-q\gamma(t - t_0)]. \qquad (2.33)$$

In conformity with the condition (2)(a),

$$\|x(t, \alpha)\| < \left[V(t, x(t, \alpha), \eta)/a_1(\alpha)\right]^{1/q} < \left[V(t, x(t, \alpha), \eta)/w_1\right]^{1/q},$$

where $w_1 = \min_{\alpha \in \mathcal{S}} a_1(\alpha)$. Hence, taking into account (2.33),

$$\|x(t, \alpha)\| < [\delta_1(\alpha)/w_1 + (k_1 V(t_0, x_0, \eta)/w_1) \exp[-q\gamma(t - t_0)]]^{1/q}$$
$$< (\delta_1(\alpha)/w_1)^{1/q} + (k_1 V(t_0, x_0, \eta)/w_1)^{1/q} \exp[-\gamma(t - t_0)] \qquad (2.34)$$
$$< r_1(\alpha) + \beta_1 \|x_0\| \exp[-\gamma(t - t_0)],$$

where $\beta_1 = k_1^{1/q} \left(\dfrac{w_2}{w_1}\right)^{1/q}$, $r_1(\alpha) = \left(\dfrac{\delta_1(\alpha)}{w_1}\right)^{1/q}$.

Consider the case (B) where at $\|x\| < r(\alpha)$ the condition (3)(c) in the Theorem 2.6 is satisfied. Denote

$$\psi(t) = V(t, x, \eta) - \delta_2(\alpha), \quad \alpha \in \mathcal{S} \subseteq R^d. \qquad (2.35)$$

From the condition (2)(b) of the Theorem 2.6 it follows that $\psi(t_0) \leq 0$ and $\psi(t) < \tilde{\varphi}$ at all $t \geq t_0$. From

$$\frac{d\psi}{dt} = DV(t, x, \eta)$$

and $\widetilde{\varphi} > \psi(t) \geq 0$ at all $t \geq t_0$

$$\frac{d\psi(t)}{dt} > -q\gamma\psi(t) \quad \text{at all} \quad t \geq t_0.$$

Hence

$$\psi(t) > \psi(t_0) \exp[-q\gamma(t - t_0)] \quad \text{at all} \quad t \geq t_0. \tag{2.36}$$

From (2.36), taking into account (2.35), we obtain

$$V(t, x(t, \alpha), \eta) > \delta_2(\alpha) + [V(t_0, x_0, \eta) - \delta_2(\alpha)] \exp[-q\gamma(t - t_0)]. \tag{2.37}$$

Since $V(t, x, \eta) - \delta_2(\alpha) \leq 0$ at all $t \geq t_0$ and $\alpha \in \mathcal{S} \subseteq R^d$, there exists a constant $k_2 < 0$ such that

$$-k_2 V(t_0, x_0, \eta) = V(t_0, x_0, \eta) - \delta_2^*. \tag{2.38}$$

Taking into account the equality (2.38), represent the inequality (2.37) in the form

$$V(t, x(t, \alpha), \eta) > \delta_2(\alpha) - k_2 V(t_0, x_0, \eta) \exp[-q\gamma(t - t_0)]. \tag{2.39}$$

According to the condition (2)(a) we obtain

$$\|x(t, \alpha)\| > [V(t, x(t, \alpha), \eta)/a_2(\alpha)]^{1/q} > [V(t, x(t, \alpha), \eta)/w_2]^{1/q}, \tag{2.40}$$

where $w_2 = \max\limits_{\alpha \in \mathcal{S}} a_2(\alpha)$. From the inequality (2.40), taking into account (2.39),

$$\begin{aligned}
\|x(t, \alpha)\| &> [\delta_2(\alpha)/w_2 - (k_2 V(t_0, x_0, \eta)/w_2) \exp[-q\gamma(t - t_0)]]^{1/q} \\
&> (\delta_2(\alpha)/w_2)^{1/q} - (k_2 V(t_0, x_0, \eta)/w_2)^{1/q} \exp[-\gamma(t - t_0)] \\
&> r_2(\alpha) - \beta_2 \|x_0\| \exp[-\gamma(t - t_0)],
\end{aligned} \tag{2.41}$$

where $r_2(\alpha) = \left(\dfrac{\delta_2(\alpha)}{w_2}\right)^{1/q}$ and $\beta_2 = k_2^{1/q} \left(\dfrac{w_1}{w_2}\right)^{1/q}$.

Taking into account the condition (2)(b) of the Theorem 2.6, the estimates (2.34), (2.41) imply the statement (b) of the Theorem 2.6 with the function $r(\alpha) = \min\{r_1(\alpha), r_2(\alpha)\}$.

Remark 2.7 If in the system (2.1) $f(t, 0, \alpha) = 0$ and $r(\alpha) \to 0$ at $\|\alpha\| \to 0$ or $r(\alpha) \to \infty$ at $\|\alpha\| \to +\infty$, then the Theorem 2.6 contains the conditions for the exponential stability of the zero solution of the uncertain system (2.1) or the exponential stability in large of its zero solution.

Remark 2.8 If $f(t, x, \alpha) \equiv P(t)x$ at all $\alpha \in \mathcal{S} \subseteq R^d$, where $P(t)$ is an $(n \times n)$-matrix with continuous real elements on R, then the Definition 2.13 implies the definition of the exponential stability of the condition $x = 0$ of the system $\dot{x} = P(t)x$, based on the estimate (Zubov [1994])

$$a_1 e^{-b_1(t - t_0)} \|x_0\| \leq \|x(t; t_0, x_0)\| \leq a_2 e^{-b_2(t - t_0)} \|x_0\|,$$

where a_1, b_1, a_2, b_2 are some positive constants.

The Theorem 2.6 implies the result of the article by Corless and Leitmann [1996], which is given below.

Corollary 2.1 *Assume that $f(t, x, \alpha)$ in the system (2.1) is continuous on $R_+ \times R^n \times R^d$ and the following conditions are satisfied:*

(1) *there exists a scalar function $V : R_+ \times R^n \to R_+$ and constants $q \geq 1$, $\omega_1 > 0$, $\omega_2 > 0$ such that*

$$\omega_1 \|x\|^q \leq V(t, x) \leq \omega_2 \|x\|^q$$

at all $x \in R^n$;

(2) *for any $\alpha \in S \subseteq R^d$ there exist constants $\delta_1(\alpha)$ and $\delta_2(\alpha)$, $0 < \delta_1(\alpha) < \delta_2(\alpha) < +\infty$, such that $\delta_1(\alpha) < V(t, x) < \delta_2(\alpha)$ at all $\alpha \in S \subseteq R^d$;*

(3) *there exists a constant $\gamma > 0$ such that*

$$D^+ V(t, x)\big|_{(2.1)} \leq -q\gamma[V(t, x) - \delta_1(\alpha)]$$

at all $t \in R_+$.

Then:

(a) *the set $A = \{x \in R^n : V(t, x) < \delta_2(\alpha)\}$ is invariant for the solutions of the system (2.1);*

(b) *the solutions of the system (2.1) uniformly with respect to t exponentially converge to the surface of the sphere $B(r) = \{x \in R^n : \|x\| = r\}$, where $r = (\delta_1(\alpha)/\omega)^{1/q}$, with the estimate*

$$\|x(t, \alpha)\| \leq r + \beta \|x(t_0)\| \exp[-\gamma(t - t_0)],$$

where $\beta = (\omega_2/\omega_1)^{1/q}$.

Example 2.6 Consider the scalar equation

$$\frac{dx}{dt} = -x + \alpha(t),$$

where $\alpha(t)$ is an unknown external perturbation with the known boundary e^ρ, $0 \leq \rho \leq 1$, $|\alpha(t)| \leq e^\rho$ at all $t \geq t_0$.
    Let $V(x) = |x|$, then

$$D^+ V(x)[-x + \alpha(t)] \leq -(|x| - e^\rho).$$

We now study the behavior of solutions of the scalar equation with respect to the moving set

$$A(\rho) = \{x \in R : |x| = e^\rho\}.$$

Note that at $\rho \to 0$ the set $A(\rho)$ becomes a two-point set $A(\rho) = \{x \colon |x| = 1\}$ on the line $x \in R$.

It is obvious that

$$D^+ V(x)[-x + \alpha(t)] < 0 \qquad \text{at} \qquad |x| > e^\rho,$$
$$D^+ V(x)[-x + \alpha(t)] = 0 \qquad \text{at} \qquad |x| = e^\rho,$$
$$D^+ V(x)[-x + \alpha(t)] > 0 \qquad \text{at} \qquad |x| < e^\rho.$$

Hence follows the convergence of solutions to the invariant moving set $|x| = e^\rho$, $0 \le \rho \le 1$, with the estimate

$$|x(t)| \le e^\rho + |x_0| \exp(-(t - t_0))$$

at all $t \ge t_0$.

2.5 Instability of Solutions with Respect to a Given Moving Set

In this section we will consider the problem of the instability of solutions of the uncertain system (2.1) with respect to the given moving set $A(r)$.

Definition 2.14 The motions $x(t, \alpha)$ of the system (2.1) are *unstable with respect to the moving set* $A(r)$, if with the given function $r(\alpha) > 0$ there exist $\varepsilon > 0$, $t_0 \in \mathcal{T}_i$ and with any $\delta > 0$ there exist x_0 and $t^* \in \mathcal{T}_i$ such that if

$$r(\alpha) - \delta < \|x_0\| < r(\alpha) + \delta,$$

the norm of the solution $x(t, \alpha) = x(t; t_0, x_0, \alpha)$ satisfies one of the inequalities

$$\|x(t^*, \alpha)\| > r(\alpha) + \varepsilon \tag{2.42}$$

or

$$\|x(t^*, \alpha)\| < r(\alpha) - \varepsilon. \tag{2.43}$$

The sufficient conditions for the instability of solutions of the system (2.1) in the sense of the Definition 2.14 are contained in the following statement.

Theorem 2.7 *Assume that $f(t, x, \alpha)$ in the system (2.1) is continuous on $\mathcal{T}_0 \times R^n \times R^d$, and the following conditions are satisfied:*

(1) *for any $\alpha \in \mathcal{S} \subseteq R^d$ there exists a function $r = r(\alpha) > 0$ such that the set $A(r)$ is not empty at all $\alpha \in \mathcal{S}$;*

(2) *there exists a matrix-valued function $U \in C(\mathcal{T}_0 \times R^n, R^{s \times s})$, $U(t, x)$ locally Lipschitz with respect to x, a vector $y \in R^s$ and $(s \times s)$-matrices $Q_1(r)$ and $Q_2(r)$ such that:*

(a) $0 < V(t, x, y) \leq b(\|x\|)$ *at* $\|x\| > r(\alpha)$ *for some function b belonging to the KR-class,*

(b) $a(\|x\|) \leq V(t, x, y) \leq c(\|x\|)$ *at* $\|x\| \leq r(\alpha)$, *where a and c belong to the K-class,*

(c) $D^+V(t, x, y)\big|_{(2.1)} > \varphi^{\mathrm{T}}(\|x\|)Q_1(r)\varphi(\|x\|)$ *at* $\|x\| > r(\alpha)$, $\alpha \in \mathcal{S}$, *where*

$$\varphi^{\mathrm{T}}(\|x\|) = \left(\varphi_1^{1/2}(\|x_1\|), \ldots, \varphi_s^{1/2}(\|x_s\|)\right), \quad \varphi_i \in K,$$

(d) $D^+V(t, x, y)\big|_{(2.1)} = 0$ *at* $\|x\| = r(\alpha)$, $\alpha \in \mathcal{S}$,

(e) $D^+V(t, x, y)\big|_{(2.1)} < \psi^{\mathrm{T}}(\|x\|)Q_2(r)\psi(\|x\|)$ *at* $\|x\| < r(\alpha)$, $\alpha \in \mathcal{S}$, *where*

$$\psi^{\mathrm{T}}(\|x\|) = \left(\psi_1^{1/2}(\|x_1\|), \ldots, \psi_s^{1/2}(\|x_s\|)\right), \quad \psi_i \in K;$$

(3) *there exist symmetrical constant* $(s \times s)$*-matrices* $\overline{Q}_1, \overline{Q}_2$ *such that:*

(a) $\dfrac{1}{2}\left(Q_1(r) + Q_1^{\mathrm{T}}(r)\right) \geq \overline{Q}_1 \quad \forall \alpha \in \mathcal{S} \subseteq R^d,$

(b) $\dfrac{1}{2}\left(Q_2(r) + Q_2^{\mathrm{T}}(r)\right) \leq \overline{Q}_2 \quad \forall \alpha \in \mathcal{S} \subseteq R^d,$

$\overline{Q}_1$ *is positive-semidefinite, and* $\overline{Q}_2$ *is negative-definite.*

Then the solutions of the uncertain system (2.1) are unstable with respect to the given moving set A(r).

Proof At first consider the case of the instability of the solution $x(t, \alpha)$ of the system (2.1), which is realized on the basis of the inequality (2.42). From the conditions (1) and (2)(a) of the Theorem 2.7 it follows that for each $\delta > 0$ one can find x_0 such that $\|x_0\| < r(\alpha) + \delta$ and $V(t_0, x_0, y) > 0$. While the solution $x(t, \alpha) \in \{x \colon r(\alpha) - \varepsilon \leq \|x\| < r(\alpha) + \varepsilon\}$ from the conditions (2)(a), (c) for any $t \in J \subseteq R_+$ one can find

$$V(t, x(t), y) = V(t_0, x_0, y) + \int_{t_0}^{t} D^+V(\tau, x(\tau), y)\, d\tau \tag{2.44}$$

$$> V(t_0, x_0, y) + \xi(\|x_0\|)\lambda_m(\overline{Q}_1)(t - t_0),$$

where $\lambda_m(\overline{Q}_1) \geq 0$ is the minimum eigenvalue of the matrix $\overline{Q}_1$ and $\xi(\|x\|) \geq \varphi^{\mathrm{T}}(\|x\|)\varphi(\|x\|)$, $\varphi \in K$-class. From the condition (2)(a) of the Theorem 2.7 it follows that the function $V(t, x, y)$ is bounded, i. e., $|V(t, x, y)| \leq M$ at $t \geq t_0$ and $\|x\| > r(\alpha)$, $\alpha \in \mathcal{S} \subseteq R^d$, M is some positive constant.

From the estimate (2.44) it follows that the function $V(t, x, y)$ in the domain $t \geq t_0$ and $\|x\| > r(\alpha)$, $\alpha \in \mathcal{S} \subseteq R^d$, is unbounded. Therefore the solution $x(t, \alpha)$ must leave the domain $\|x\| \geq r(\alpha) + \varepsilon$ upon some point of time $t_0 \leq t^* < \infty$.

If $r(\alpha) \to r_0$ at $\|\alpha\| \to 0$, then the instability of solutions of the system (2.1) with respect to the set $A(r)$ is proved.

Now consider the case when the function $r(\alpha)$ is given and does not vanish at any $\alpha \in \mathcal{S} \subseteq R^d$. The instability of the solutions of the system (2.1) with respect to the set $A(r)$ can occur when the inequality (2.43) is true.

With the given function $r(\alpha)$, denote

$$V^{-1}(t, r(\alpha)) = \{x \colon V(t, x, y) \leq a(r(\alpha))\},$$

where $a \in K$ is chosen so that $V^{-1}(t, r(\alpha)) \subset \overline{A}(r)$ for each $t \in R_+$ at any $\alpha \in \mathcal{S}$. Let $t_0 \in R_+$ and $x_0 \colon \|x_0\| < r(\alpha) + \delta$, $x_0 \in V^{-1}(t, r(\alpha))$. Then according to the conditions (2)(c), (e) and (3)(b) find that $x(t, \alpha) \in V^{-1}(t_0, \alpha)$ and, since

$$V^{-1}(t, r(\alpha)) \subset \overline{A}(r) \subset R^n,$$

the motion $x(t, \alpha)$ cannot reach the boundary of the set $A(r)$. Then for any $\varepsilon > 0$ choose $\eta > 0$ so that

$$c(r(\alpha) - \eta) < a(r(\alpha) - \varepsilon) \quad \text{at all} \quad \alpha \in \mathcal{S},$$

and take $T(\varepsilon)$ larger than $c(r(\alpha) + \delta)/\lambda_M(\overline{Q}_2)\psi(r(\alpha) - \eta)$. Now $\|x(t, \alpha)\|$ cannot exceed the value $r(\alpha) - \eta$ for all $t \in [t_0, t_0 + T(\varepsilon)]$ because in that case we would obtain for $t = t_0 + T(\varepsilon)$

$$V(t, x(t), y) < V(t_0, x_0, y) - \lambda_M(\overline{Q}_2) \int_{t_0}^{t_0+T(\varepsilon)} \Psi(\|x(s, \alpha)\|) \, ds$$

$$< c(r(\alpha) + \delta) - \lambda_M(\overline{Q}_2)\psi(r(\alpha) - \eta)T(\varepsilon) < 0,$$

which contradicts the condition (2)(b).

Therefore one can find $t_1 \in [t_0, t_0 + T(\varepsilon)]$ such that $c(\|x(t_1, \alpha)\|) < c(r(\alpha) - \eta) < a(r(\alpha) - \varepsilon)$ and since $V(t, x(t, \alpha), y)$ is decreasing, for $t \geq t_0 + T(\varepsilon)$ we obtain

$$a(\|x(t, \alpha)\|) \leq V(t, x(t, \alpha), y) < V(t_1, x(t_1, \alpha), y)$$
$$< c(\|x(t_1, \alpha)\|) < c(r(\alpha) - \eta) < a(r(\alpha) - \varepsilon).$$

Hence, for $t \geq t_0 + T(\varepsilon)$ $\|x(t, \alpha)\| < r(\alpha) - \varepsilon$. According to the Definition 2.14, this means the instability of the solutions of the system (2.1) with respect to the given moving set $A(r)$.

The Theorem 2.7 is proved.

Remark 2.9 If $x = 0$ is the only equilibrium state of the system (2.1) at all $\alpha \in \mathcal{S} \subseteq R^d$ and $f(t, 0, \alpha) = 0$, then the conditions (1), (2)(b), (d) and (3)(b) of the Theorem 2.7 guarantee the uniform asymptotic stability of solutions of the system (2.1) with respect to $B_\alpha \subset V^{-1}(t, r(\alpha))$, and the set $A(r)$ is an unstable limit cycle for such solutions, from which they tend to zero at $t \to +\infty$.

Below is a variant of the Theorem 2.7, which is based on the scalar Lyapunov function.

Theorem 2.8 *Assume that the following conditions are satisfied:*

(1) *for any $\alpha \in S \subseteq R^d$ there exists a function $r = r(\alpha) > 0$ such that the set $A(r)$ is not empty at all $\alpha \in S$;*

(2) *there exists a function $V \in C^{(1,1)}(T_0 \times R^n, R_+)$ and functions $H_1 \colon R^n \times R^d \to R$ and $H_2 \colon R^n \times R^d \to R$ such that:*

> (a) $0 < V(t, x) \le b(\|x\|)$ *at* $\|x\| > r(\alpha)$ *for some function b belonging to the KR-class,*

> (b) $a(\|x\|) \le V(t, x) \le c(\|x\|)$ *at* $\|x\| \le r(\alpha)$, *where a and c belong to the K-class,*

> (c) $D^+V(t, x)\big|_{(2.1)} \ge H_1(x, \alpha)$ *at* $\|x\| > r(\alpha)$, $\alpha \in S$,

> (d) $D^+V(t, x)\big|_{(2.1)} = 0$ *at* $\|x\| = r(\alpha)$, $\alpha \in S$,

> (e) $D^+V(t, x)\big|_{(2.1)} \le H_2(x, \alpha)$ *at* $\|x\| < r(\alpha)$, $\alpha \in S \subseteq R^d$;

(3) *there exist sign-definite functions $\underline{H}_1(x)$ and $\overline{H}_2(x)$ such that:*

> (a) $H_1(x, \alpha) \ge \underline{H}_1(x) > 0$ *at all* $\alpha \in S$,

> (b) $H_2(x, \alpha) \le \overline{H}_2(x) < 0$ *at all* $\alpha \in S$.

Then the motions $x(t, \alpha)$ of the system (2.1) are unstable with respect to the given moving set $A(r)$,

Example 2.7 Let the equations of the motion of the uncertain system have the form

$$
\begin{aligned}
\frac{dx}{dt} &= -p(t)y - \left[1 - \frac{1}{a^2}m^2(\alpha)(x^2 + y^2)\right]x(x^2 + y^2), \\
\frac{dy}{dt} &= p(t)x - \left[1 - \frac{1}{a^2}m^2(\alpha)(x^2 + y^2)\right]y(x^2 + y^2),
\end{aligned}
\tag{2.45}
$$

where $p \in C(R_+, R_+)$ is a single-valued function $t \ge t_0 \ge 0$, $m(\alpha)$ is a function which characterizes the uncertainties, with the same properties as in the Example 2.2.

The function $r(\alpha)$ will be chosen as follows: $r(\alpha) = \dfrac{a}{m(\alpha)}$, $\alpha \in S \subseteq R$, $r(\alpha) > 0$, and therefore

$$
A(r) = \{x, y \colon \|z\| = r(\alpha)\},
\tag{2.46}
$$

where $\|z\| = (x^2 + y^2)^{1/2}$.

To investigate the instability of the solutions $(x(t, \alpha), y(t, \alpha))$ of the system (2.45) with respect to the given moving set (2.46), apply the function V of the form

$$V = x^2 + y^2.$$

This function is positive-definite and satisfies the conditions (2)(a), (b) of the Theorem 2.8. Its derivative, in view of the equations (2.45), is determined by the formula

$$\left.\frac{dV}{dt}\right|_{((2.45))} = -2\left[1 - \frac{1}{a^2}m^2(\alpha)(x^2 + y^2)\right](x^2 + y^2)^2.$$

It is easy to prove that for $\alpha \in \mathcal{S} \subseteq R^d$

$$\left.\frac{dV}{dt}\right|_{(2.45)} > 0 \quad \text{at} \quad \|z\| > \frac{a}{m(\alpha)}, \tag{2.47}$$

$$\left.\frac{dV}{dt}\right|_{(2.45)} = 0 \quad \text{at} \quad \|z\| = \frac{a}{m(\alpha)},$$

$$\left.\frac{dV}{dt}\right|_{(2.45)} < 0 \quad \text{at} \quad \|z\| < \frac{a}{m(\alpha)}. \tag{2.48}$$

Thus, if the function $m(\alpha)$ satisfies the above conditions, then the conditions (2)(c), (d) of the Theorem 2.8 are satisfied. So, according to the Theorem 2.8, any motion $(x(t, \alpha), y(t, \alpha))$ of the system (2.45), which begins in the domain

$$r(\alpha) - \delta < (x_0^2 + y_0^2)^{1/2} < r(\alpha) + \delta,$$

at arbitrarily small $\delta > 0$ will leave the domain

$$r(\alpha) - \varepsilon > (x^2 + y^2)^{1/2} > r(\alpha) + \varepsilon \tag{2.49}$$

at some point $t^* > t_0$.

The motion $(x(t, \alpha), y(t, \alpha))$ can leave the domain (2.49) in two ways: either under the condition (2.47) the solutions $(x(t, \alpha), y(t, \alpha))$ leave the ε-neighbourhood of the set (2.46) through the external boundary of the ε-neighbourhood, or under the condition (2.48), they leave it through its internal boundary.

Example 2.8 Let the equations of the motion of the uncertain system have the form

$$\frac{dx}{dt} = -(x - \beta y)\left(1 - m^2(\alpha)(x^2 + y^2)\right),$$
$$\frac{dy}{dt} = -(y + \gamma x)\left(1 - m^2(\alpha)(x^2 + y^2)\right), \tag{2.50}$$

where γ, β, $\gamma < \beta$, are positive constants, the function of the uncertainties parameter $m(\alpha)$ satisfies the same conditions as in the Example 2.5.

Determine the moving set $A(r)$ by the formula

$$A(r) = \{x, y \colon \|z\| = r(\alpha)\}, \quad r(\alpha) = \frac{1}{m(\alpha)} > 0.$$

For the system (2.50) the positive-definite function

$$V = \gamma x^2 + \beta y^2$$

satisfies the conditions (2)(a), (b) of the Theorem 2.8. Its derivative

$$\left.\frac{dV}{dt}\right|_{(2.50)} = -2\left(\gamma x^2 + \beta y^2\right)\left(1 - m^2(\alpha)(x^2 + y^2)\right)$$

satisfies the conditions (2)(c), (d):

$$\left.\frac{dV}{dt}\right|_{(2.50)} < 0 \quad \text{at} \quad \|z\| < r(\alpha),$$

$$\left.\frac{dV}{dt}\right|_{(2.50)} = 0 \quad \text{at} \quad \|z\| = r(\alpha),$$

$$\left.\frac{dV}{dt}\right|_{(2.50)} > 0 \quad \text{at} \quad \|z\| > r(\alpha),$$

where $\|z\| = (x^2 + y^2)^{1/2}$, $\alpha \in \mathcal{S} \subseteq R^d$.

Thus, the solutions $(x(t, \alpha), y(t, \alpha))$ of the system (2.50), outgoing from an arbitrary point x_0, y_0 located in the moving ring

$$r(\alpha) - \delta \le (x_0^2 + y_0^2)^{1/2} \le r(\alpha) + \delta,$$

for an arbitrarily small $\delta > 0$, will be unstable with respect to the given moving set $A(r)$.

2.6   Stability with Respect to a Conditionally Invariant Moving Set

In this section we will describe the approach which gives the opportunity to reduce the analysis of the stability of moving conditionally invariant sets to the study of the external part of the phase space with respect to the boundary of the moving set. This is attained by the inverse transformation of the initial dynamic system, which is used in the next chapter for the synthesis of controls of motions of an uncertain system.

Along with the concept of stability of the equilibrium state $x = 0$ of the system (2.1) in the sense of the Definitions 2.1–2.4, it is interesting to consider other concepts of stability of solutions of that system.

Let $0 < \rho_0(\alpha) \leq r_0(\alpha) \leq \rho(\alpha)$ be some known functions of parameter α, determining the moving sets

$$B^+ = \{x \in R^n : \|x\| \geq \rho_0(\alpha)\},$$
$$B^- = \{x \in R^n : \|x\| \leq \rho_0(\alpha)\},$$
$$A^+ = \{x \in R^n : \|x\| \geq r_0(\alpha)\},$$
$$A^- = \{x \in R^n : \|x\| \leq r_0(\alpha)\},$$
$$B = B^+ \cap B^-, \quad A = A^+ \cap A^-.$$

Let M denote one of the sets B^+, B^- or B, and let N denote one of the sets A^+, A^- or A.

Definition 2.15 The *pair of moving sets* (M, N) is called *conditionally invariant and uniformly asymptotically stable* with respect to the solutions of the system (2.1) if the following conditions are satisfied:

(a) at any $(t_0, x_0) \in R_+ \times N$ the motion of the system (2.1) $x(t, \alpha)$ belongs to M at all $t \geq t_0$;

(b) at any $\varepsilon > 0$ and $t_0 \in R_+$ there exists $\delta = \delta(\varepsilon) > 0$ such that motions of the system (2.1) beginning in the δ-neighborhood of the set N at all $t \geq t_0$ stay in the ε-neighborhood of the set M, i.e., for any x_0 such that $d(N, x_0) < \delta$, the inequality $d(M, x(t, \alpha)) < \varepsilon$ holds at all $t \geq t_0$;

(c) there exist $\delta_0 > 0$ and $T = T(\varepsilon) > 0$ such that at all x_0 such that $d(N, x_0) < \delta_0$ the estimate $d(M, x(t, \alpha)) < \varepsilon$ holds at all $t \geq t_0 + T(\varepsilon)$.

Here $d(M, x) = \inf_{y \in M} \|x - y\|$.

Later along with the system (2.1) we will consider the scalar comparison equation

$$\frac{du}{dt} = g(t, x, \mu), \quad u(t_0) = u_0 \geq 0, \tag{2.51}$$

where $u \in R$, $g \in C(R^3, R)$ and $\mu = \mu(\alpha) \geq 0$ is some function of uncertainties parameter $\alpha \in S$.

Assume that solutions of the equation (2.51) exist at all $t \geq t_0$ and they are unique for the given initial conditions $(t_0, u_0) \in R_+ \times R_+$.

We introduce the sets

$$\Omega^+(H) = \{u \in R_+ : u \leq H\},$$
$$\Omega^-(H) = \{u \in R_+ : u \geq H_0\},$$
$$\Omega(H, H_0) = \Omega^+(H) \cap \Omega^-(H_0), \quad 0 \leq H_0 \leq H < +\infty,$$

and formulate the following definition.

Definition 2.16 The set $\Omega^+(H)$ is *invariant and uniformly asymptotically stable* with respect to solutions of the equation (2.51) if the following conditions are satisfied:

(a) from the condition $u_0 \leq H$ it follows that $u(t, t_0, u_0) \leq H$ at all $t \geq t_0$;

(b) for the given $\varepsilon > 0$ and $t_0 \in R_+$ there exists $\delta = \delta(\varepsilon) > 0$ such that from the condition $u_0 < H + \delta$ it follows that $u(t, t_0, u_0) < H + \varepsilon$ at all $t \geq t_0$;

(c) there exist values $\delta_0 > 0$ and $T = T(\varepsilon) > 0$ such that from $u_0 < H + \delta_0$ it follows that $u(t, t_0, u_0) < H + \varepsilon$ at all $t \geq t_0 + T(\varepsilon)$.

We now determine the conditions under which the motions of the system (2.1) are conditionally invariant and uniformly asymptotically stable with respect to the pairs of sets (A, B), (A^+, B^+), (A^-, B^-).

The correctness of the following statement can easily be verified.

Proposition 2.7 The pair of moving sets (A, B) is invariant and uniformly asymptotically stable with respect to solutions of the system (2.1) if, and only if, the pairs of moving sets (A^+, B^+) and (A^-, B) are conditionally invariant and uniformly asymptotically stable with respect to the system (2.1).

In the terms of the existence of a relevant Lyapunov function, the sign of the conditional invariance and the uniform asymptotic stability of the pair of moving sets (A^+, B^+) is formulated as follows.

Theorem 2.9 *Assume that the equations of perturbed motion (2.1) are such that:*

(1) *for any $\alpha \in S \subseteq R^d$ there exist functions $r(\alpha)$ and $\rho(\alpha)$, $\rho(\alpha) \geq r(\alpha)$ at all $\alpha \in S$, such that $r(\alpha) \to q - \text{const} > 0$ at $\|\alpha\| \to 0$ and $\rho(\alpha) \to +\infty$ at $\|\alpha\| \to +\infty$;*

(2) *there exists a function $V \in C(R_+ \times R^n, R_+)$, $V(t, x)$ locally Liptshitz with respect to x at each $t \in R_+$, and the functions a, b belonging to the KR-class, such that:*

    (a) $b(\|x\|) \leq V(t, x) \leq a(\|x\|)$ *at* $\|x\| \geq r(\alpha)$,

    (b) $D^+V(t, x)|_{(2.1)} \leq g(t, V(t, x), r(\alpha))$ *at* $\|x\| \geq r(\alpha)$,

    where $g \in C(R^3, R)$, $g(t, 0, r(\alpha)) = 0$ at all $t \in R_+$;

(3) *the set $\Omega^+(H)$, where $H = a(r) = b(\rho)$, is invariant and uniformly asymptotically stable with respect to solutions of the comparison equation (2.51).*

Then the pair of moving sets (A^+, B^+) is conditionally invariant and uniformly asymptotically stable with respect to solutions of the system (2.1).

Proof First, prove the conditional invariance of the pair of moving sets (A^+, B^+). Assume that under the conditions of the Theorem 2.9 the pair of sets (A^+, B^+) is not conditionally invariant. In this case there exists a solution $x(t, \alpha)$ of the system (2.1) with the initial conditions $(t_0, x_0) \in R_+ \times A^+$ and points of time $t_2 > t_1 \geq t_0$ such that

$$\|x(t_1, \alpha)\| = r(\alpha), \quad \|x(t_2, \alpha)\| > \rho(\alpha) \quad \text{and}$$
$$\|x(t, \alpha)\| \geq r(\alpha) \quad \text{at} \quad t \in [t_1, t_2]. \tag{2.52}$$

In consequence of the condition 2 (c) and the comparison principle,

$$V(t\, x(t, \alpha)) \leq u^+(t, t_1, V(t_1, x(t_1, \alpha))) \quad \text{at} \quad t \in [t_1, t_2], \tag{2.53}$$

where $u^+(t, \cdot)$ is the upper solution of the comparison equation (2.51) with the initial conditions $(t_1, V(t_1, x(t_1, \alpha)))$.

Taking into account the condition 2 (a) of the theorem and the relation (2.52), from the inequality (2.53) we obtain

$$b(\rho) < b(\|x(t_2, \alpha)\|) \leq V(t_2, x(t_2, \alpha)) \leq u^+(t_2, t_1, a(\|x(t_1, \alpha)\|))$$
$$= u^+(t_2, t_1, a(r)) \leq H = a(r) = b(\rho).$$

The obtained contradiction proves that under the conditions of the Theorem 2.9 the relations (2.52) are impossible, i. e., the pair of sets (A^+, B^+) is conditionally invariant with respect to solutions of the system (2.1).

Now we prove that solutions of the system (2.1) are asymptotically stable with respect to the pair of sets (A^+, B^+).

According to the condition (3) of the Theorem 2.9, for arbitrary $\tilde{\varepsilon} > 0$ and $t_0 \in R_+$ there exists $\tilde{\delta} = \tilde{\delta}(\tilde{\varepsilon}) > 0$ such that the condition $u_0 < H + \tilde{\delta}$ implies $u(t) < H + \tilde{\varepsilon}$ at all $t \geq t_0$. Now for an arbitrary $\varepsilon > 0$ determine $\tilde{\varepsilon} = b(\rho + \varepsilon) - b(\rho) > 0$ and choose $\delta(\varepsilon) = a^{-1}(H + \tilde{\delta}) - r > 0$. Let $\|x_0\| < r + \delta$, then $\|x_0\| < a^{-1}(H + \tilde{\delta})$. According to the condition (2) of the Theorem 2.9,

$$u_0 = V(t_0, x_0) < a(a^{-1}(H + \tilde{\delta})) = H + \tilde{\delta},$$

and for $u(t)$ the estimate $u(t) < H + \tilde{\varepsilon}$ holds at all $t \geq t_0$. From the condition 2(a) obtain $b(\|x(t, \alpha)\|) \leq V(t, x(t)) \leq u^+(t, u_0) < H + \tilde{\varepsilon} = b(\rho + \varepsilon)$ at all $t \geq t_0$. Therefore $\|x(t, \alpha)\| < b^{-1}(b(\rho + \varepsilon)) = \rho + \varepsilon$ at all $t \geq t_0$. The stability of the pair of moving sets (A^+, B^+) is proved.

Show that the stability will be asymptotic. From the condition (3) of the Theorem 2.9 it follows that for any $\tilde{\varepsilon} > 0$ there exist $\tilde{\delta}_0 > 0$ and $\tilde{T}(\tilde{\varepsilon}) > 0$ such that at all $u_0 < H + \tilde{\delta}_0$ the estimate $u(t) < H + \tilde{\varepsilon}$ holds at all $t \geq t_0 + \tilde{T}(\tilde{\varepsilon})$. Assume $\delta_0 = a^{-1}(H + \tilde{\delta}_0) - r > 0$ and for an arbitrary $\varepsilon > 0$ choose $T(\varepsilon) = \tilde{T}(\tilde{\varepsilon})$, where $\tilde{\varepsilon} = b(\rho + \varepsilon) - b(\rho) > 0$. In this case, under the condition $\|x_0\| < H + \delta_0$ we obtain $u_0 = v(t_0, x_0) \leq a(\|x_0\|) < a(H + \delta_0) = r = \tilde{\delta}_0$ and $u(t) < H + \tilde{\varepsilon}$ at $t \geq t_0 + \tilde{T}(\tilde{\varepsilon}) = t_0 + T(\varepsilon)$.

From the condition (2)(a) we obtain

$$b(\|x(t,\alpha)\|) \leq V(t,x(t)) \leq u^+(t,t_0,u_0) < H + \tilde{\varepsilon} = b(\rho + \varepsilon).$$

Hence, at all $t \geq t_0 + T(\varepsilon)$ $\|x(t,\alpha)\| < \rho + \varepsilon$, which proves the Theorem 2.9.

Definition 2.17 A *function* $f(t,x)$, given in $R_+ \times R^n$, is called *homogeneous of order* $\mu = \frac{p}{q}$, (p and q are natural numbers, q is odd), if the following equality is true

$$f(t, \lambda x_1, \ldots, \lambda x_n) = \lambda^\mu f(t, x_1, \ldots, x_n)$$

at all $(t,x) \in R_+ \times R^n$.

Now make the following assumption with respect to the system (2.1)
The vector-function $f \in C(R_+ \times R^n \times R^d, R^n)$ in the system (2.1) can be represented in the form

$$f(t, x, \alpha) = \sum_{i=1}^{s} f_i(t, x_1, \ldots, x_n, \alpha),$$

where $f_i(t, x_1, \ldots, x_n, \alpha)$ are homogeneous functions of order μ_i, $i = 1, 2, \ldots, n$.

Let $\mu = \max_i \mu_i$ and $\gamma > \max(\mu - 1, v)$. In the system (2.1) perform the change of variables by the formula

$$y = \frac{x}{\|x\|^{\gamma+1}}, \quad \|x\| \neq 0. \tag{2.54}$$

As a result of the change (2.54) we obtain the system

$$\frac{dy}{dt} = f^*(t, y, \alpha), \quad y(t_0) = y_0, \quad t_0 \geq 0. \tag{2.55}$$

Here at $\|y\| \neq 0$ $f^*(t, y, \alpha)$ is determined by the formula

$$f^*(t, y, \alpha) = f\left(t, \frac{y}{\|y\|^{1+\frac{1}{\gamma}}}, \alpha\right) \|y\|^{1+\frac{1}{\gamma}} - \frac{(1+\gamma)y^T f\left(t, \frac{y}{\|y\|^{1+\frac{1}{\gamma}}}, \alpha\right)}{\|y\|^{1-\frac{1}{\gamma}}} y$$

and $f^*(t, y, \alpha) = 0$ at $\|y\| = 0$.
We now prove the following statement.

Proposition 2.8 If in the system (2.1) the vector-function $f \in C(R_+ \times R^n \times R^d, R^n)$ has homogeneous functions of order μ_i, as its components $f_i(t, x_1, \ldots, x_n, \alpha)$, $i = 1, 2, \ldots, n$. Then through the nonlinear transformation (2.54) it is reducible to the form (2.55) and the vector-function $f^*(t, y, \alpha)$ is continuous on $R_+ \times R^n \times R^d$.

Proof The system (2.55) is obtained directly from the system (2.1) by using the nonlinear transformation (2.54). To verify the continuity of the function $f^*(t, y, \alpha)$ on $R_+ \times R^n \times R^d$ it is sufficient to show that the relation $\|f^*(t, y, \alpha)\| \to 0$ holds at $\|y\| \to 0$ uniformly with respect to $t \in R_+$.

Theorem 2.10 *Assume that the pair of moving sets (A^+, B^+) is conditionally invariant and uniformly asymptotically stable with respect to the motions of the system (2.1). Then the pair of moving sets (A_1^-, B_1^-), where*

$$A_1^- = \{y \in R^n \colon \|y\| \geq r^{-\gamma}(\alpha)\},$$
$$B_1^- = \{y \in R^n \colon \|y\| \geq \rho^{-\gamma}(\alpha)\},$$

is conditionally invariant and uniformly asymptotically stable with respect to the motions of the system (2.55).

Proof From (2.54) it follows that

$$\|y(t; t_0, y_0)\| = \|x(t; t_0, x_0)\|^{-\gamma} \quad \text{at all} \quad t \in I, \tag{2.56}$$

where $I = I_1 \cap I_2$ and I_k is the maximal interval of existence of solutions of the systems (2.1) and (2.55) respectively. Therefore if $y_0 \in A_1^{-1}$, then $\|x_0\| = \|y_0\|^{-\frac{1}{\gamma}} \leq r(\alpha)$ and $x(t, t_0, x_0) \in B^+$ at all $t \geq t_0$. From the relation (2.56) if follows that $\|y(t; t_0, y_0)\| = \|x(t; t_0, x_0)\|^{-\gamma} \geq \rho^{-\gamma}(\alpha)$ at all $t \geq t_0$, i. e., $y(t; t_0, y_0) \in B_1^-$. From the conditions of the Theorem 2.10 it follows that for an arbitrary $\tilde{\varepsilon} > 0$ and $t_0 \in R_+$ there exists $\tilde{\delta} = \tilde{\delta}(\tilde{\varepsilon}) > 0$ such that

$$\|x(t; t_0, x_0)\| < \rho + \tilde{\varepsilon} \quad \text{at all} \quad t \geq t_0,$$

if $\|x_0\| < r + \tilde{\delta}$. Then for an arbitrary small $\varepsilon > 0$ we choose $\tilde{\varepsilon} = (\rho^{-\gamma} - \varepsilon)^{-\frac{1}{\gamma}} - \rho > 0$ so that $\tilde{\delta}(\tilde{\varepsilon}) > 0$. According to the given ε we choose $\delta(\varepsilon)$ in the form $\delta(\varepsilon) = r^{-\gamma} - (r + \tilde{\delta})^{-\gamma} > 0$ and show that from the condition $\|y_0\| > r^{-\gamma} - \delta$ it follows that $\|y(t; t_0.y_0)\| > \rho^{-\gamma} - \varepsilon$ at all $t \geq t_0$. Indeed, from the condition $\|y_0\| > r^{-\gamma} - \delta$ it follows that $\|x_0\| = \|y_0\|^{-\frac{1}{\gamma}} < (r^{-\gamma} - \delta)^{-\frac{1}{\gamma}} = r + \tilde{\delta}$. But then $\|x(t; t_0, x_0)\| < \rho + \tilde{\varepsilon}$ at all $t \geq t_0$. According to the relation (2.56) we obtain $\|y(t; t_0, x_0)\| > (\rho + \tilde{\varepsilon})^{-\gamma} = \rho^{-\gamma} - \varepsilon$ at all $t \geq t_0$.

Now prove the asymptotic properties of solutions of the system (2.55). From the conditions of the Theorem 2.10 it follows that there exist $\tilde{\delta}_0 > 0$ and $\tilde{T}(\tilde{\varepsilon}) > 0$ such that $\|x(t; t_0, x_0)\| < \rho + \tilde{\varepsilon}$ at all $t \geq t_0 + \tilde{T}(\tilde{\varepsilon})$, if $\|x_0\| < r_0 + \tilde{\delta}_0$. For an arbitrarily small $\varepsilon > 0$ we choose $\tilde{\varepsilon} = (\rho^{-\gamma} - \varepsilon)^{-\frac{1}{\gamma}} - \rho > 0$, $\delta_0 = r^{-\gamma} - (r + \tilde{\delta}_0)^{-\gamma} > 0$, $T(\varepsilon) = \tilde{T}(\tilde{\varepsilon})$. It is not difficult to see that from the condition $\|y_0\| > r^{-\gamma} - \delta_0$ it follows that $\|x_0\| = \|y_0\|^{-\frac{1}{\gamma}} < (r^{-\gamma} - \delta)^{-\frac{1}{\gamma}} = r + \tilde{\delta}$, and therefore $\|x(t; t_0, x_0)\| < \rho + \tilde{\varepsilon}$ at $t \geq t_0 + \tilde{T}(\tilde{\varepsilon}) = t_0 + T(\tilde{\varepsilon})$. From the relation (2.56) it follows that $\|x(t; t_0, x_0)\|^{-\gamma} = \|y(t; t_0, y_0)\| > (\rho + \tilde{\varepsilon})^{-\gamma} = \rho^{-\gamma} - \varepsilon$ at all $t \geq t_0 + T(\varepsilon)$.

The Theorem 2.10 is proved.

We now continue the study of the behavior of the solutions of the uncertain system (2.1) on the basis of the principle of comparison with the scalar Lyapunov function.

Let $0 < p_0(\alpha) \leq r_0(\alpha) \leq r(\alpha) \leq p(\alpha)$ be some known functions of parameter $\alpha \in S \subseteq R^d$.

Definition 2.18 The moving set $B = \{x \in R^n: p_0(\alpha) \leq \|x\| \leq p(\alpha)\}$ is *conditionally invariant* with respect to the set $A = \{x \in R^n: r_0(\alpha) \leq \|x\| \leq r(\alpha)\}$ and *uniformly asymptotically stable* with respect to solutions of the system (2.1) if the following conditions are satisfied:

(1) From $r_0(\alpha) \leq \|x_0\| \leq r(\alpha)$ follows the estimate $p_0(\alpha) \leq \|x(t, \alpha)\| \leq p(\alpha)$ at all $t \geq t_0$;

(2) for the given $\varepsilon > 0$ and $t_0 \in R_+$:

    (a) there exists $\delta = \delta(\varepsilon) > 0$ such that the condition $r_0(\alpha) - \delta < \|x_0\| < r(\alpha) + \delta$ implies the estimate $p_0(\alpha) - \varepsilon < \|x(t, \alpha)\| < p(\alpha) + \varepsilon$ at all $t \geq t_0$;

    (b) there exist $\delta_0 > 0$ and $\tau = \tau(\varepsilon)$ such that the condition $r_0(\alpha) - \delta_0 < \|x_0\| < r(\alpha) + \delta_0$ implies the estimate $p_0(\alpha) - \varepsilon < \|x(t, \alpha)\| < p(\alpha) + \varepsilon$ at all $t \leq t_0 + \tau$, where $x(t, \alpha) = x(t; t_0, x_0, \alpha)$ is a solution of the system (2.1).

Definition 2.19 The set $K = \{u \in R: \varkappa_0 \leq u \leq \varkappa\}$ is *is invariant and uniformly asymptotically stable* for solutions of the system (2.51), if:

(1) the condition $\varkappa_0 \leq u_0 \leq \varkappa$ implies the estimate $\varkappa_0 \leq u(t) \leq \varkappa$ at all $t_0 \in R_+$;

(2) for the given $\varepsilon^* > 0$ and $t_0 \in R_+$

    (a) there exists $\delta^* = \delta^*(\varepsilon) > 0$ such that the condition $\varkappa_0 - \delta^* < u_0 < \varkappa + \delta^*$ implies the estimate $\varkappa_0 - \varepsilon^* < u(t) < \varkappa + \varepsilon^*$ at all $t \geq t_0$,

    (b) there exist $\delta_0^* > 0$ and $\tau = \tau(\varepsilon^*) > 0$ such that the condition $\varkappa_0 - \delta^* < u_0 < \varkappa + \delta^*$ implies the estimate $\varkappa_0 - \varepsilon^* < u(t) < \varkappa + \varepsilon^*$ at all $t \geq t_0 + \tau$, where $u(t) = u(t; t_0, u_0)$ is a solution of the equation (2.51).

Now we prove the following statement.

Theorem 2.11 *Assume that the equations (2.1) are such that:*

(1) *for any $\alpha \in S \subseteq R^d$ there exist functions $p_0(\alpha) \leq r_0(\alpha) \leq r(\alpha) \leq p(\alpha)$ at all $\alpha \in S$ such that $r(\alpha) \to q - \text{const} > 0$ at $\|\alpha\| \to 0$ and $r_0(\alpha) \to +\infty$ at $\|\alpha\| \to +\infty$;*

(2) *there exists a function* $V \in C(R_+ \times R^n, R_+)$, $V(t,x)$ *locally Lipshitz with respect to* x *at each* $t \in R_+$, *and comparison functions* a_i, b_i *belonging to the* K-*class, such that:*

(a) $b_1(\|x\|) \le V(t,x) \le a_1(\|x\|)$ *at all* $\|x\| > r(\alpha)$,

(b) $b_2(\|x\|) \le V(t,x) \le a_2(\|x\|)$ *at all* $\|x\| < r_0(\alpha)$;

(3) *there exists a function* $G(t, V(t,x), \mu(\alpha))$, $G \in C(R_+^3, R)$, *such that:*

(a) $D^+V(t,x)\big|_{(2.1)} \le G(t, V(t,x), \mu(\alpha))$ *at all* $\|x\| > r(\alpha)$,

(b) $D^+V(t,x)\big|_{(2.1)} = 0$ *at* $\|x\| = r_0(\alpha) = r(\alpha)$,

(c) $D^+V(t,x)\big|_{(2.1)} \ge G(t, V(t,x), \mu(\alpha))$ *at all* $\|x\| < r_0(\alpha)$;

(4) *for each* $r_0(\alpha) \le r(\alpha)$ *there exists* $\varkappa_0 \le \varkappa$ *such that* $\varkappa = a_1(r) = b_1(\rho)$, $\varkappa_0 = b_2(r_0) = a_2(\rho_0)$, *where* $\rho_0 \le r_0 \le r \le \rho$ *and* $\varkappa \to 0$ *at* $r \to 0$ *and* $\varkappa_0 \to +\infty$ *at* $r_0 \to +\infty$;

(5) *the set* K *is invariant and uniformly asymptotically stable.*

Then the set B *is conditionally invariant with respect to the set* A *and uniformly asymptotically stable for the solutions of the system* (2.1).

Proof Let the set B be not conditionally invariant with respect to the set A for the system (2.1). Then for the solution $x(t, \alpha)$ with the initial conditions $r_0(\alpha) \le \|x_0\| \le r(\alpha)$ one can find a value of time $t_2 > t_0$ such that the following conditions are satisfied:

(a) $\|x(t_2, \alpha)\| > \rho(\alpha)$ and $r_0(\alpha) \le \|x(t, \alpha)\|$ at all $t \in [t_0, t_2]$

or

(b) $\|x(t_2, \alpha)\| < \rho_0(\alpha)$ and $\|x(t, \alpha)\| \le r(\alpha)$ at all $t \in [t_0, t_2]$.

According to the conditions (3) (a)–(c) of the Theorem 2.11, from the comparison principle it follows that at all $t \in [t_1, t_2]$

$$V(t, x(t, \alpha)) \le u_M(t; t_0, V(t_0, x(t_0, \alpha)))$$

or

$$V(t, x(t, \alpha)) \ge u_m(t; t_0, V(t_0, x(t_0, \alpha))),$$

where $u_M(t, t_0, u_0)$ and $u_m(t, t_0, u_0)$ are the maximal and the minimal solutions of the comparison equation (2.51) with the initial condition $(t_1, V(t_1, x(t_1, \alpha)))$.

Then in the case (a) we have the inequalities

$$\begin{aligned}
b_1(\rho(\alpha)) &< b_1(\|x(t_2, \alpha)\|) \le V(t_2, x(t_2, \alpha)) \\
&\le u_M(t_2, t_0, a_1(\|x(t_0, \alpha)\|)) = u_M(t_2; t_0, a_1(r(\alpha))) \\
&\le a_1(r(\alpha)) = b_1(\rho(\alpha)),
\end{aligned}$$

or in the case (c), the opposite inequalities

$$a_2(p_0(\alpha)) > a_2(\|x(t_2,\alpha)\|) \geq V(t_2,x(t_2,\alpha))$$
$$\geq u_m(t_2,t_1,b_2(\|x(t_0,\alpha)\|)) = u(t_2;t_0,b_2(r_0(\alpha)))$$
$$\geq b_2(r_0(\alpha)) = a_2(p_0(\alpha)).$$

In both cases we obtained the contradictions which prove the first part of the statement of the Theorem 2.11, i. e., the set B is conditionally invariant with respect to the set A.

Let $\varepsilon > 0$, $t_0 \in R_+$ be given, and the set K be uniformly stable. Then, since $a_1(r(\alpha)) = b_1(p(\alpha)) = \varkappa$ and $\varkappa_0 = b_2(r_0(\alpha)) = a_2(p_0(\alpha))$, for the given $a_2(p_0(\alpha) - \varepsilon)$, $b_1(p(\alpha) + \varepsilon)$ there exist values $\varepsilon_1, \delta_1, \delta > 0$ such as that under the conditions

$$\varkappa_0 + \delta_1 = a_1(r(\alpha) + \delta) < b_1(p(\alpha) + \varepsilon) = \varkappa + \varepsilon_1$$

and

$$\varkappa_0 + \varepsilon_1 = a_2(p_0(\alpha) - \varepsilon) < b_2(r_0(\alpha) - \delta) = \varkappa_0 - \delta_1$$

the condition $\varkappa_0 - \delta_1 < u_0 < \varkappa + \delta_1$ implies the estimate $\varkappa_0 - \varepsilon_1 < u(t) < \varkappa + \varepsilon_1$ at all $t \geq t_0$, where $u(t)$ is any solution of the comparison equation (2.51).

Show that with the chosen value $\delta > 0$ the set B is uniformly stable with respect to the set A, i. e., the condition $r_0(\alpha) - \delta < \|x_0\| < r(\alpha) + \delta$ implies the estimate $p_0(\alpha) - \varepsilon < \|x(t,\alpha)\| < p(\alpha) + \varepsilon$ at all $t \geq t_0$ and at all $\alpha \in \mathcal{S}$.

If this is not so, then there should exist a solution $x(t,\alpha)$ of the system (2.1) with the initial conditions x_0, $r_0(\alpha) - \delta < \|x_0\| < r(\alpha) + \delta$, and points of time $t_2 > t_1 > t_0$ such that at all $t \in [t_0,t_2]$

(a) $\|x(t_2,\alpha)\| \geq p(\alpha) + \varepsilon$ and $\|x(t,\alpha)\| \geq r_0(\alpha)$

or

(b) $\|x(t_2,\alpha)\| \leq p(\alpha) - \varepsilon$ and $\|x(t,\alpha)\| \leq r(\alpha)$.

Consider the case (a). From the condition (3)(a) of the Theorem 2.11 we obtain

$$V(t,x(t,\alpha)) \leq u_M(t;t_0,V(t_0,x(t_0,\alpha))) \quad \text{at all} \quad t \in [t_0,t_2].$$

Hence, according to the condition (2)(a),

$$b_1(p(\alpha) + \varepsilon) \leq b_1(\|x(t_2,\alpha)\|) \leq V(t_2,x(t_2,\alpha)) \leq u_M(t_2,t_0,V(t_0,x_0))$$
$$\leq u_M(t_2,t_0,a_1(r(\alpha) + \delta)) < b_1(p(\alpha) + \varepsilon),$$

which is a contradiction.

In the case (b) we obtain $V(t, x(t, \alpha)) \geq u_m(t; t_0, V(t_0, x(t_0, \alpha)))$ at all $t \in [t_0, t_2]$ and according to the condition (2)(c) of the Theorem 2.11 obtain

$$
\begin{aligned}
a_2(p_0(\alpha) - \varepsilon) > a_2(\|x(t_2, \alpha)\|) &\geq V(t_2, x(t_2, \alpha)) \\
&\geq u_m(t_2, t_0, V(t_0, x_0)) \geq u_m(t_2; t_1, b_2(r(\alpha) - \delta)) \\
&\geq b_2(r_0(\alpha)) = a_2(p_0(\alpha) - \varepsilon),
\end{aligned}
$$

which is a contradiction, too. This proves that the set B is uniformly stable.

Now show that under the conditions of the Theorem 2.11 the set B is uniformly asymptotically stable with respect to the set A.

Let $\varepsilon = p_0(\alpha)$, $\delta_0 = \delta(p_0(\alpha)) > 0$ and the set K be uniformly asymptotically stable. Then for the given $a_2(p_0(\alpha) - \varepsilon)$, $b_1(p(\alpha) + \varepsilon)$ there exists $\tau = \tau(\varepsilon) > 0$ such that the condition $b_2(r_0(\alpha) - \delta_0) < u_0 < a_1(r(\alpha) + \delta_0)$ implies the estimate $a_2(p_0(\alpha) - \varepsilon) < u(t) < b_1(p(\alpha) + \varepsilon)$ at all $t \geq t_0 + \tau$.

Show that with the given choice of $\delta_0 = \delta(p_0(\alpha)) > 0$ under the initial conditions $r_0(\alpha) - \delta_0 < \|x_0\| < r(\alpha) + \delta_0$ for the solution $x(t, \alpha)$ the estimate $p_0(\alpha) - \varepsilon < \|x(t, \alpha)\| < p(\alpha) + \varepsilon$ follows at all $t \geq t_0 + \tau$.

If this is not so, then there should exist a solution $x(t, \alpha)$ of the system (2.1) and a value $t_2 > t_0$ such that:

(a) $\|x(t_2, \alpha)\| \geq p(\alpha) + \varepsilon$ at $t_2 \geq t_0 + \tau$,

(b) $\|x(t_2, \alpha)\| \leq p(\alpha) - \varepsilon$ at $t_2 \geq t_0 + \tau$.

Similarly to the analysis of the uniform stability of the set B with respect to A, we obtain

(a) $b_1(p(\alpha) + \varepsilon) \leq V(t_2, x(t_2, \alpha)) \leq u_M(t_2, t_0, a_1(r(\alpha) + \delta_0)) < b_1(p(\alpha) + \varepsilon)$

and

(b) $a_2(p_0(\alpha) - \varepsilon) \geq V(t_2, x(t_2, \alpha)) \geq u_m(t_2, t_0, b_2(r_0(\alpha) - \delta_0)) > a_2(p_0(\alpha) - \varepsilon)$.

The contradictions (a) and (b) prove the statement of the Theorem 2.11.

Thus, the Theorems 2.1–2.11 form the basis for a new generalization of the direct Lyapunov method for finite-dimensional dynamic systems with uncertain values of their parameters. Those theorems have wide potential for the further development and applications at the analysis of specific problems of mechanics.

Chapter 3

Stability of Uncertain
Controlled Systems

In this chapter the construction of controls in the system (3.1) will be made, which ensure the conditional invariance and the uniform asymptotic stability of solutions of the system (3.1) with respect to the pair of moving sets (A, B). As follows from the previous chapter, this problem can be reduced to the study of the conditional invariance and the uniform asymptotic stability of the solutions of the system (3.1) with respect to the pairs (A^+, B^+) and (A^-, B^-). In its turn, the study of the conditional invariance and the uniform asymptotic stability of the pair (A^-, B^-) is reduced to the analysis of the pair (A_1^+, B_1^+) for an inverse system corresponding to the system (3.1).

3.1   Problem Setting

Consider the class of controlled uncertain systems in the form

$$\frac{dx}{dt} = f(t, x) + \Delta f(t, x, \alpha) + [B(t, x) + \Delta B(t, x, \alpha)]u(t), \qquad (3.1)$$

where $t \in R_+$, $x(t) \in R^n$, $u(t) \in R^m$ is the control, $\alpha \in R^d$ is the uncertainties parameter, $f \in C(R \times R^n, R^n)$, $f = \sum_{i=1}^{s} f_i(t, x)$, $f_i(t, x) \in C(R \times R^n, R)$ are homogeneous functions of order μ_i, $B \in C(R_+ \times R^{m \times n}, R^{m \times n})$, $B_i(t, x) = \sum_{j=1}^{s_2} B_{ij}(t, x)$, and B_{ij} are homogeneous functions of degree $\varkappa_{ij}$,

$$\Delta f \in C(R \times R^n \times R^d, R^n), \qquad \Delta f(t, 0, \alpha) = 0,$$
$$\Delta B \in C(R \times R^n \times R^d, R^{m \times n}), \qquad \Delta B(t, 0, \alpha) = 0.$$

To calculate the value $\sigma = \max\{\mu_i, \varkappa_{ij}\}$, we will call it the *homogeneity degree of the system* (3.1).

For the given functions $r_0(\alpha)$, $r(\alpha)$, $\rho(\alpha)$, $\rho_0(\alpha)$ and the parameter $\alpha \in$

$S \subseteq R^d$ we will consider the moving sets

$$B(\rho_0, \rho) = \{x \in R^n : \ \rho_0 \le \|x\| \le \rho\},$$
$$A(r_0, r) = \{x \in R^n : \ r_0 \le \|x\| \le r\}$$

and the problem of the synthesis of controls in the system (3.1).

3.2 Synthesis of Controls

From the system (3.1) separate the nominal controlled system

$$\frac{dx}{dt} = f(t, x) + B(t, x)u$$

and assume that there exists a control $u_1 \in C(R \times R^n, R^m)$ which stabilizes the equilibrium state $x = 0$ of the uncontrollable nominal system

$$\frac{dx}{dt} = f(t, x), \quad x(t_0) = x_0, \quad t_0 \ge 0,$$

to the uniform asymptotically stable condition.

We denote

$$f_0(t, x) = f(t, x) + B(t, x)u_1(t, x),$$
$$\Delta f_0(t, x, \alpha) = \Delta f(t, x, \alpha) + \Delta B(t, x, \alpha)u_1(t, x).$$

The system (3.1) with the control $u(t, x) = u_1(t, x) + u_2(t, x)$ will take the form

$$\frac{dx}{dt} = f_0(t, x) + \Delta f_0(t, x, \alpha) + [B(t, x) + \Delta B(t, x, \alpha)]u_2, \qquad (3.2)$$

where $f_0(t, 0) = 0$ at all $t \in R_+$. The equilibrium state $x = 0$ of the system

$$\frac{dx}{dt} = f_0(t, x), \quad x(t_0) = x_0, \qquad (3.3)$$

is uniformly asymptotically stable according to the assumption.

Then we will need some assumptions on the systems (3.2), (3.3).

Assumption 3.1 There exist functions $V \in C^{1,1}(R_+ \times R^n, R_+)$, $a, b \in KR$-class and a comparison function $g \in C(R_+ \times R_+, R)$ such that:

(a) $b(\|x\|) \le V(t, x) \le a(\|x\|)$ at $\|x\| \ge r(\alpha)$ and $t \in R_+$;

(b) $DV(t, x)\big|_{(3.3)} \le -g(t, V(t, x))$ at $\|x\| \ge r(\alpha)$ and all $t \in R_+$,
     $g(t, 0) = 0.$

Assumption 3.2 There exist a matrix-valued function $E\colon R \times R^n \times R^d \to R^{m \times m}$ and a constant $\rho_E > -1$ such that

$$\Delta B(t, x, \alpha) = B(t, x) E(t, x, \alpha),$$

$$\min_{\alpha \in S} \lambda_{\min} \left(E + E^{\mathrm{T}} \right) \geq 2\rho_E \quad \text{at} \quad \|x\| \geq r(\alpha).$$

Assumption 3.3 Let $\beta(t, x) = B^{\mathrm{T}}(t, x) \dfrac{\partial V(t, x)}{\partial x}$ and let there exist:

(a) continuous functions $\psi_i\colon R \times R^n \times R_+ \to R_+$, $i = 1, 2, \ldots, s$, such that at $\|x\| \geq r(\alpha)$

$$\max_{\alpha \in S} \left(\frac{\partial V(t, x)}{\partial x} \right)^{\mathrm{T}} \Delta f_0(t, x, \alpha) \leq \sum_{i=1}^{s} \psi_i(t, x, \|\beta\|);$$

(b) for any i $\psi_i(t, x, 0) = 0$ at all $t \in R_+$ and $\|x\| \geq r_0(\alpha)$;

(c) for any $\varepsilon_i > 0$ there exist continuous functions $k_i(t, x, \varepsilon_i) = \varepsilon_i$ such that

$$\psi_i(t, x, k_i) = \varepsilon_i \quad \text{at} \quad \|x\| \geq r_0(\alpha)$$

and

$$\pi_i(t, x, \beta_i, \varepsilon_i) = \beta(t, x) \left[k_i(t, x, \varepsilon_i) \right]^{-2}$$

are continuous on $R_+ \times \operatorname{ext} A(r) \times R_+$.

Following the article by Chen [1996], assume

$$\widehat{\psi}_l(t, x, \beta) = \frac{\psi_l(t, x, \beta)}{1 + \rho_E}, \quad \rho_E > -1$$

at $\|x\| \geq r(\alpha)$.

For the given $\varepsilon_i > 0$, $i = 1, 2, \ldots, s$, and the function $r(\alpha)$ choose the control $u_2(t, x)$ in the form

$$u_2(t, x) = \sum_{i=1}^{s} \psi_i(t, x, \beta), \tag{3.4}$$

where

$$\psi_i(t, x, \beta) = \begin{cases} -\dfrac{\beta(t, x) \widehat{\psi}_i(t, x, \beta)}{\|\beta(t, x)\|^2}, & \widehat{\psi}_i > \varepsilon_i, \\[2ex] -\dfrac{\beta(t, x) \widehat{\psi}_i(t, x, \beta)}{k_i^2(t, x, \varepsilon_i)}, & \widehat{\psi}_i \leq \varepsilon_i, \end{cases}$$

and $(1 + \rho_E) \sum_{i=1}^{s} \varepsilon_i \leq r(\alpha)$ at all $\alpha \in S$.

Now show that the controls (3.4) ensure the conditional invariance and the uniform asymptotic stability of solutions of the system (3.1) with respect to the pair of sets (A^+, B^+). To construct the controls $u_2^*(t, x)$ which ensure the conditional invariance and the uniform asymptotic stability of the pair of sets (A^-, B^-) with respect to solutions of the system (3.1), apply the nonlinear transformation of the system (3.1)

$$y = \frac{x}{\|x\|^{1+\gamma}} \quad \text{at} \quad \|x\| \neq 0.$$

The system (3.1) takes the form

$$\frac{dy}{dt} = f^*(t, y) + \Delta f^*(t, y, \alpha) + [B^*(t, y) + \Delta B^*(t, y, \alpha)]u^*, \qquad (3.5)$$

where $G(\gamma, y, \|y\|) = \left(I\|y\|^{1+\frac{1}{\gamma}} - \frac{(1+\gamma)yy^T}{\|y\|^{1-\frac{1}{\gamma}}} \right)$, I is a unit matrix $n \times n$ and at $\|y\| \neq 0$

$$f^*(t, y) = G(\gamma, y, \|y\|) f\left(t, \frac{y}{\|y\|^{1+\frac{1}{\gamma}}} \right),$$

$$\Delta f^*(t, y, \alpha) = G(\gamma, y, \|y\|) \Delta f\left(t, \frac{y}{\|y\|^{1+\frac{1}{\gamma}}}, \alpha \right),$$

$$B^*(t, y) = G(\gamma, y, \|y\|) B\left(t, \frac{y}{\|y\|^{1+\frac{1}{\gamma}}} \right),$$

$$\Delta B^*(t, y, \alpha) = G(\gamma, y, \|y\|) \Delta B\left(t, \frac{y}{\|y\|^{1+\frac{1}{\gamma}}}, \alpha \right).$$

Obviously, at $y = 0$ we obtain $f^*(t, 0) = \Delta f^*(t, 0, \alpha) = 0$ and $B^*(t, 0) = \Delta B^*(t, 0, \alpha) = 0$ at all $t \in R_+$.

Assume that for the nominal system

$$\frac{dy}{dt} = f^*(t, y) + B^*(t, y)u^*$$

there exists a control $u_1^*(t, y)$ which ensures the uniform asymptotic stability of the equilibrium state $y = 0$ of the system (3.1). Now represent the system (3.5) in the form

$$\frac{dy}{dt} = f_0^*(t, y) + \Delta f_0^*(t, y, \alpha) + [B^*(t, y) + \Delta B^*(t, y, \alpha)]u_2^*, \qquad (3.6)$$

where
$$f_0^*(t, y) = f^*(t, y) + B^*(t, y)u_1^*(t, y),$$
$$\Delta f_0^*(t, y) = \Delta f^*(t, y, \alpha) + \Delta B^*(t, y, \alpha)u_1^*(t, y).$$

Assume that the equilibrium state $y = 0$ of the system

$$\frac{dy}{dt} = f_0^*(t, y), \quad y(t_0) = y_0 \qquad (3.7)$$

is uniformly asymptotically stable.

Now we will need the following assumptions.

Assumption 3.4 There exist functions $V^* \in C^{(1,1)}(R_+ \times R^n, R_+)$, a^*, $b^* \in$ KR-class and a majorizing function $g^*(t, V^*(t, y))$ such that:

(1) $b^*(\|y\|) \leq V^*(t, y) \leq a^*(\|y\|)$ at $\|y\| \geq r_0^{-\gamma}(\alpha)$ and at all $\alpha \in \mathcal{S}$ and $t \in R_+$;

(2) $DV(t, y)|_{(3.7)} \leq -g^*(t, V^*(t, y))$ at $\|y\| \geq r_0^{-\gamma}(\alpha)$ and at all $\alpha \in \mathcal{S}$ and $t \in R_+$.

Assumption 3.5 There exist a function $E: R \times R^n \times R^d \to R^{m \times m}$ and a constant $\rho_{E^*} > -1$ such that

$$\Delta B^*(t, y, \alpha) = B^*(t, y)E^*(t, y, \alpha),$$

$$\min_{\alpha \in \mathcal{S}} \lambda_{\min}\left(E^* + E^{*\mathrm{T}}\right) \geq 2\rho_{E^*}.$$

Assumption 3.6 Let $\beta^*(t, y) = B^{*\mathrm{T}}(t, y)\dfrac{\partial V(t, y)}{\partial y}$ and there exist functions $\psi_l: R_+ \times R^n \times R_+$ such that:

(1) $\max\limits_{\alpha \in \mathcal{S}} \left(\dfrac{\partial V(t, y)}{\partial y}\right)^{\mathrm{T}} \Delta f_0^*(t, y, \alpha) \leq \sum\limits_{l=1}^{s} \psi_l^*(t, y, \beta^*);$

(2) in the domain $\|y\| \geq r_0^{-\gamma}(\alpha)$, $\alpha \in \mathcal{S}$ the following condition is satisfied:

$$\psi_l^*(t, y, 0) = 0, \quad l = 1, 2, \ldots, s;$$

(3) for any $\varepsilon_i^* > 0$ in the domain $\|y\| \geq r_0^{-\gamma}(\alpha)$, $\alpha \in \mathcal{S}$, there exist functions $k_i^* \in C(R_+ \times \mathrm{ext}\, A(r_0^{-\gamma}) \times R_+, R_+)$ such that

$$\psi_i^*(t, y, k_i^*(t, y, \varepsilon_i^*)) = \varepsilon_i^*$$

and the functions

$$\pi_i^*(t, y, \beta^*, \varepsilon_i^*) = \frac{\beta^*(t, y)}{\left(k_i^*(t, y, \varepsilon_i^*)\right)^2}$$

are continuous on $R_+ \times \mathrm{ext}\, A(r_0^{-\gamma}) \times R_+$.

Now consider the functions

$$\widehat{\psi}_i^*(t, y, \|\beta^*\|) = \frac{\psi_i^*(t, y, \beta^*)}{1 + \rho_{E^*}}, \quad \rho_{E^*} > -1,$$

in the domain $\|y\| \geq r_0^{-\gamma}(\alpha)$ at all $\alpha \in \mathcal{S}$.

For the given $\varepsilon_l^* > 0$, $l = 1, 2, \ldots, s$, and the function $r_0^{-\gamma}(\alpha)$ choose the controls $u_2^*(t, x)$ in the form

$$u_2^*(t, y) = \sum_{l=1}^{s} \psi_i^*(t, y, \beta^*(t, y)),$$

where

$$\psi_i^*(t, y, \beta^*) = \begin{cases} -\dfrac{\beta^*(t, y)\widehat{\psi}_i^*(t, y, \beta^*)}{\|\beta^*(t, y)\|} & \text{at} \quad \widehat{\psi}_i > \varepsilon_i, \\[3mm] -\dfrac{\beta^*(t, y)\widehat{\psi}_i^*(t, y, \beta^*)}{(k_i^*(t, y, \varepsilon_i^*))^2} & \text{at} \quad \widehat{\psi}_i \leq \varepsilon_i, \end{cases}$$

and $(1 + \rho_E^*) \sum_{i=1}^{s} \varepsilon_i^* \leq r_0^{-\gamma}(\alpha)$.

The stabilization of motions of the uncertain system (3.1) with respect to the given manifold is effected by using controls synthesized on the basis of the initial and the inverse systems.

The Theorems 2.9 and 2.10 along with the Assumptions 3.1 – 3.6 allow to prove that the controls $u_2(t, x)$ and $u_2^*(t, y) = u_2^*\left(t, \dfrac{x}{\|x\|^{1+\gamma}}\right)$ ensure the conditional invariance and the uniform asymptotic stability of solutions of the system (3.1) with respect to the pair of moving sets (A, B).

Theorem 3.1 *Assume that:*

(1) *for the system (3.1) all the conditions of the Assumptions 3.1 – 3.3 are satisfied;*

(2) *for the system (3.6) all the conditions of the Assumptions 3.4 – 3.6 are satisfied;*

(3) *under the given functions $V(t, x)$, $V^*(t, x)$ and $r_0(\alpha)$, $r(\alpha)$ the following inequalities hold true:*

    (a) $DV(t, x)\big|_{(3.1)} \leq -g(t, V(t, x)) + r(\alpha)$ *at* $\|x\| \geq r(\alpha)$ $t \in R_+$,

    (b) $DV^*(t, y)\big|_{(5.10)} \geq -g^*(t, V^*(t, y)) + r_0^{-\gamma}(\alpha)$ *at* $\|y\| \geq r_0^{-\gamma}(\alpha)$, $t \in R_+$;

(4) *solutions of the comparison equations*

$$\frac{du}{dt} = -g(t, u) + r(\alpha), \qquad\qquad u(t_0) = u_0 \geq 0,$$

$$\frac{dw}{dt} = -g^*(t, w) + r_0^{-\gamma}(\alpha), \qquad\qquad w(t_0) = w_0 \geq 0,$$

exist at all $t \in R_+$, and they are invariant and uniform asymptotically stable with respect to the sets

$$\Omega_1^+ = \{u \in R_+ : \ u \leq H = a(r) = b(\rho)\},$$
$$\Omega_2^+ = \{u \in R_+ : \ u \leq H^* = a(r^{-\gamma}) = b(\rho^{-\gamma})\}$$

respectively.

Then the controls

$$u = u_1(t, x) + u_2(t, x), \quad \|x\| \geq r(\alpha),$$

$$u^* = u_1^*\left(t, \frac{x}{\|x\|^{1+\gamma}}\right) + u_2^*\left(t, \frac{x}{\|x\|^{1+\gamma}}\right), \quad \|x\| \leq r_0(\alpha), \quad \alpha \in \mathcal{S},$$

stabilize the motions of the uncertain system (3.1) *with respect to the conditionally invariant set* $B(\rho_0, \rho)$.

Proof At first consider the dynamics of the uncertain system (3.1) in the domain of the values $x\colon \|x\| \geq r(\alpha)$. With this purpose, apply the Lyapunov function $V(t, x)$ indicated in the Assumption 3.1. For any admissible $\alpha \in \mathcal{S} \subseteq R^d$ we obtain

$$DV(t, x)\big|_{(3.2)} \leq \frac{\partial V}{\partial t} + \left(\frac{\partial V(t, x)}{\partial x}\right)^{\mathrm{T}} [f_0(t, x) + \Delta f_0(t, x)]$$
$$+ \left(\frac{\partial V(t, x)}{\partial x}\right)^{\mathrm{T}} B(t, x)[I + E(t, x, \alpha)]u_2. \tag{3.8}$$

Taking into account the conditions of the Assumptions 3.2, 3.3, from (3.8) we obtain

$$DV(t, x)\big|_{(3.2)} \leq -g(t, V(t, x)) + \sum_{l=1}^{s}\Big(\psi_l - \beta^{\mathrm{T}}(t, x)\beta(t, x)h_l(t, x, \beta)$$
$$- \beta^{\mathrm{T}}(t, x)\Big[\frac{1}{2}(E(t, x, \alpha) + E^{\mathrm{T}}(t, x, \alpha))\Big]\beta(t, x)h_l(t, x, \|\beta\|)\Big), \tag{3.9}$$

where the functions $h_l(t, x, \|\beta\|)$ are determined by the relation

$$\psi_l(t, x, \beta) = -\beta(t, x)h_l(t, x, \|\beta\|), \quad l = 1, 2, \ldots, s.$$

From the inequality (3.9) it follows that

$$DV(t, x)\big|_{(3.2)} \leq -g(t, V(t, x))$$
$$+ \sum_{l=1}^{s}[\psi_l - \|\beta(t, x)\|^2 h_l(t, x, \|\beta\|) - \|\beta(t, x)\|^2 \rho_E h_l(t, x, \|\beta\|)]. \tag{3.10}$$

If $\widehat{\psi}_l(t, x, \|\beta\|) > \varepsilon_l$, $l = 1, 2, \ldots, s$, then

$$\psi_l - \|\beta(t, x)\|^2 h_l - \|\beta(t, x)\|^2 \rho_E h_l$$
$$= \psi_l - \widehat{\psi}_l - \rho_E \widehat{\psi}_l = \psi_l - \widehat{\psi}_l(1 + \rho_E) = 0. \tag{3.11}$$

If $\widehat{\psi}_l(t, x, \|\beta\|) \leq \varepsilon_l$, $l = 1, 2, \ldots, s$, then

$$
\begin{aligned}
\psi_l &- \|\beta(t,x)\|^2 h_l - \|\beta(t,x)\|^2 \rho_E h_l \\
&= \psi_l - \|\beta(t,x)\|^2 k_l^{-2} \widehat{\psi}_l (1 + \rho_E) \\
&\leq \widehat{\psi}_l \left(1 - \frac{\|\beta(t,x)\|^2}{k_l^2} \right)(1 + \rho_E) \leq \widehat{\psi}_l(1 + \rho_E) \leq \varepsilon_l(1 + \rho_E).
\end{aligned}
\tag{3.12}
$$

Taking into account (3.4), and also (3.11), (3.12), from (3.10) we find

$$
DV(t,x)\big|_{(3.2)} \leq -g(t, V(t,x)) + r(\alpha), \quad \alpha \in \mathcal{S}, \tag{3.13}
$$

in the range x: $\|x\| \geq r(\alpha)$.

When the condition (1) of the Theorem 3.1 and the inequality (3.13) are satisfied, all the conditions of the Theorem 2.9 are satisfied as well, therefore the motions of the controlled system (3.1) are conditionally invariant and uniformly asymptotically stable with respect to the pair of moving sets (A^+, B^+).

The conditions for the conditional invariance and the uniform asymptotic stability of the pair of moving sets (A^-, B^-) with respect to the solutions of the system (3.6) follow from the conditional invariance and the uniform asymptotic stability of the pair (A^-, B^-), where

$$
B_1^- = \{ y \in R^n : \|y\| \geq \rho_0^{-\gamma} \},
$$
$$
A_1^- = \{ y \in R^n : \|y\| \geq r_0^{-\gamma} \}
$$

with respect to solutions of the system (3.6).

As above, we obtain the inequality

$$
DV(t,y)\big|_{(3.6)} \leq -g^*(t, V^*(t,y)) + r_0^{-\gamma}, \quad \alpha \in \mathcal{S}.
$$

This inequality, together with the Assumptions 3.4–3.6, is sufficient for the application of the Theorems 2.9 and 2.10.

The Theorem 3.1 is proved.

3.3   Convergence of Controlled Motions to a Moving Set

As an example of application of the Theorem 2.11, consider a controlled uncertain system of the form

$$
\frac{dx}{dt} = f_0(t, x, \alpha) + B(t, x) F(t, x, u), \quad x(t_0) = x_0, \quad t_0 \geq 0. \tag{3.14}
$$

Make the following assumptions about the components of the right-hand part of the system (3.14):

H$_1$. The vector-function f_0 belongs to $C(R_+ \times R^n \times S, R^n)$, $(m \times n)$-matrix B to $C(R_+ \times R^n, R^{n \times m})$ and the vector-function F to $C(R_+ \times R^n \times R^m \times S, R^n)$, where $u \in R^m$.

H$_2$. There exists a function $V \in C^1(R_+ \times R^n, R_+)$ and comparison functions $a, b \in KR$-class, such that:

(a) $b(\|x\|) \leq V(t, x)$ at $\|x\| \geq r(\alpha)$,

(b) $V(t, x) \leq a(\|x\|)$ at $\|x\| \leq r_0(\alpha)$.

H$_3$. There exist comparison functions $C_1, C_2 \in KR$-class, such that:

(a) $DV(t, x)\big|_{f_0} \leq -C_1(V(t, x))$ at $\|x\| > r(\alpha)$,

(b) $DV(t, x)\big|_{f_0} = 0$ at $\|x\| = r(\alpha)$,

(c) $DV(t, x)\big|_{f_0} \geq -C_2(V(t, x))$ at $\|x\| < r_0(\alpha)$,

where $r_0(\alpha) \leq r(\alpha)$ at all $\alpha \in S$.

H$_4$. There exist functions $\beta_1, \beta_2 \in C(R_+ \times R^n, R_+)$ and $\overline{\beta}_1, \overline{\beta}_2 \in C(R_+ \times R^n, R_+)$ such that $u^T F(t, x, u, \alpha) \geq -\beta_1(t, x)\|u\| + \beta_2(t, x)\|u\|^2$ at $\|x\| > r(\alpha)$ and $\beta_1 \leq \beta_2 \overline{\beta}_1$, $\beta_1 \leq \overline{\beta}_2$.

H$_5$. There exist functions $\gamma_1, \gamma_2 \in C(R_+ \times R^n, R_+)$ such that

$$u^T F(t, x, u, \alpha) \leq -\gamma_1(t, x)\|u\| + \gamma_2(t, x)\|u\|^2 \quad \text{at} \quad \|x\| < r_0(\alpha)$$

and $\gamma_1 \geq \gamma_2 \overline{\beta}_1$, $\gamma_1 \geq \overline{\beta}_2$, $\gamma_2 \geq \overline{\beta}_1$.

H$_6$. There exists a set $P = \{p_\mu \in C(R_+ \times R^n, R^n), \mu > 0\}$ which satisfies the conditions:

(a) $\|\nabla(t, x)\| p_\mu = -\|p_\mu\| \nabla(t, x)$, where

$$\nabla(t, x) = B^T(t, x)\left(\frac{\partial V}{\partial x_1}, \ldots, \frac{\partial V}{\partial x_n}\right)^T,$$

$\eta(t, x) = \overline{\beta}_2 \nabla(t, x)$, and if $\|x\| > r(\alpha)$, then $\|p_\mu\| \geq \overline{\beta}_1(1 - \frac{r(\alpha)}{\|\eta\|})$;

(b) if $\|x\| < r_0(\alpha)$ and $\|\eta(t, x)\| > 0$, then $\|p_\mu\| \leq \overline{\beta}_1\left(1 - \frac{r_0(\alpha)}{\|\eta(t, x)\|}\right)$ at all $\alpha \in S$.

Prove the following statement.

Theorem 3.2 *Assume that for the uncertain controlled system (3.14) all the conditions of the Assumptions H_1–H_6 are satisfied, and in addition, $C_2^{-1}(u) \leq C_1^{-1}(u)$.*

Then the set B is conditionally invariant with respect to the set A, and the motions of the system (3.14) are uniformly asymptotically stable with respect to the set B.

Proof At first consider the dynamics of the system (3.14) in the domain $\|x\| > r(\alpha)$ at all $\alpha \in \mathcal{S}$. According to the conditions $H_3(a)$ and H_4, obtain the estimates

$$DV(t,x)\big|_{f_0} \leq -C_1(V(t,x))$$

and

$$
\begin{aligned}
\nabla(t,x)F(t,x,p_\mu,\alpha) &= -\frac{\|\nabla(t,x)\|}{\|p_\mu(t,x)\|}F(t,x,p_\mu,\alpha)p_\mu \\
&\leq \|\nabla(t,x)\|\big[\beta_1(t,x) - \beta_2(t,x)\|p_\mu(t,x)\|\big] \\
&\leq \|\nabla(t,x)\|\left[\beta_1(t,x) - \beta_2(t,x)\overline{\beta}_1(t,x)\left(1 - \frac{r(\alpha)}{\|\eta(t,x)\|}\right)\right] \\
&\leq \|\nabla(t,x)\|\beta_1(t,x)\frac{r(\alpha)}{\|\eta(t,x)\|} \leq r(\alpha)
\end{aligned}
$$

at all $\alpha \in \mathcal{S}$. Hence in the domain $\|x\| > r(\alpha)$ the estimate

$$DV(t,x)\big|_{(3.14)} \leq -C_1(V(t,x)) + r(\alpha) \tag{3.15}$$

is true at all $\alpha \in \mathcal{S}$.

Similarly to the above, in the domain $\|x\| < r_0(\alpha)$ the following inequalities hold true:

$$DV(t,x)\big|_{f_0} \geq -C_2(V(t,x))$$

and

$$
\begin{aligned}
\nabla(t,x)F(t,x,p_\mu,\alpha) &= -\frac{\|\nabla(t,x)\|}{\|p_\mu(t,x)\|}F(t,x,p_\mu,\alpha)p_\mu \\
&\geq \|\nabla(t,x)\|\big[\gamma_1(t,x) - \gamma_2(t,x)\|p_\mu(t,x)\|\big] \\
&\geq \|\nabla(t,x)\|\left[\gamma_1(t,x) - \gamma_2(t,x)\overline{\beta}_1(t,x)\left(1 - \frac{r_0(\alpha)}{\|\eta(t,x)\|}\right)\right] \\
&\geq \|\nabla(t,x)\|\gamma_1(t,x)\frac{r_0(\alpha)}{\|\eta(t,x)\|} \geq r_0(\alpha)
\end{aligned}
$$

at all $\alpha \in \mathcal{S}$. Therefore in the domain $\|x\| < r_0(\alpha)$ we obtain the estimate

$$DV(t,x)\big|_{(3.14)} \geq -C_2(V(t,x)) + r_0(\alpha) \tag{3.16}$$

at all $\alpha \in \mathcal{S}$.

Along with the inequalities (3.15), (3.16), consider the comparison equations

$$\frac{du}{dt} = -C_1(u) + r(\alpha), \qquad u(t_0) = u_0 \geq 0,$$

$$\frac{d\omega}{dt} = -C_2(\omega) + r_0(\alpha), \qquad \omega(t_0) = \omega_0 \geq 0.$$

According to the Theorem 2.11, obtain $u = C_1^{-1}(r) = \varkappa$ and $\omega = C_2^{-1}(r_0) = \varkappa_0$. Taking into account the conditions H_2, the conditions of the Theorem 3.2 and the properties of the functions a, b, C_1, C_2, and also the fact that $\rho_0(\alpha) \leq r_0(\alpha) \leq \rho(\alpha)$ at all $\alpha \in \mathcal{S}$, we find $\varkappa = a(r) = b(\rho)$, $\varkappa_0 = b(r_0) = a(\rho_0)$ and $\varkappa_0 \leq \varkappa$. Then, repeating the reasoning of the proof of the Theorem 2.11, we conclude that the set $K = \{u \in R_+ : \varkappa_0 \leq u \leq \varkappa\}$ is invariant, and the motions of the system (3.14) are uniformly asymptotically stable with respect to that set.

3.4 Stabilization of Rotary Motions of a Rigid Body in an Environment with Indefinite Resistance

Consider the problem of stabilization of rotary motions of a rigid body in an environment with incomplete information about the resistance forces. It is assumed that the rigid body is controlled by three bounded forces at three unknown components (but with known boundaries of change) of the resistance forces. It is necessary to construct controls under which the motions will be stabilized to the surface of a moving ellipsoid whose surface is an invariant set for the motions of the considered system.

Let $\hat{b}_1$, $\hat{b}_2$, $\hat{b}_3$ be orthogonal unit vectors fixed in the rigid body in the centre of gravity and parallel to the main axes of inertia. The control forces and the resistances have the form $u_1\hat{b}_1$, $u_2\hat{b}_2$, $u_3\hat{b}_3$ and $e_1\hat{b}_1$, $e_2\hat{b}_2$, $e_3\hat{b}_3$, respectively. Let the angular velocity of the rotary motion of the rigid body be determined by the expression

$$\omega = x_1\hat{b}_1 + x_2\hat{b}_2 + x_3\hat{b}_3.$$

Then the rotary motion of the rigid body is described by the dynamic Euler equations in the following form (see Corless and Leitmann [1996]):

$$\frac{dx_1}{dt} = \left(\frac{I_2 - I_3}{I_1}\right)x_2x_3 + \frac{u_1}{I_1} + \frac{e_1(t)}{I_1},$$

$$\frac{dx_2}{dt} = \left(\frac{I_3 - I_1}{I_2}\right)x_3x_1 + \frac{u_2}{I_2} + \frac{e_2(t)}{I_2}, \qquad (3.17)$$

$$\frac{dx_3}{dt} = \left(\frac{I_2 - I_3}{I_3}\right)x_1x_2 + \frac{u_3}{I_3} + \frac{e_3(t)}{I_3},$$

where I_1, I_2, $I_3 > 0$ are the principal moments of inertia of the body with respect to its centre of gravity. If $x_1 = x_2 = x_3 = 0$, then the body does not rotate.

Assume that at all $t \in R_+$ the function which describes the unknown resistance of the environment satisfies the condition

$$|e_1(t)| \le \rho, \quad |e_2(t)| \le \rho, \quad |e_3(t)| \le \rho, \quad 0 < \rho < 1. \tag{3.18}$$

The controls $u_1(t)$, $u_2(t)$, $u_3(t)$ will be sought in the class of bounded functions for which the following inequalities hold true:

$$|u_1(t)| \le 1, \quad |u_2(t)| \le 1, \quad |u_3(t)| \le 1 \quad \text{at all} \quad t \in R_+.$$

Using the change of variables

$$y_i = I_i^{1/2} x_i, \quad i = 1, 2, 3,$$

reduce the system (3.17) to the form

$$\begin{aligned}
\frac{dy_1}{dt} &= \left(\frac{I_2 - I_3}{I_1 \sqrt{I_2 I_3}} \right) y_2 y_3 + \frac{u_1}{\sqrt{I_1}} + \frac{e_1(t)}{\sqrt{I_1}}, \\
\frac{dy_2}{dt} &= \left(\frac{I_3 - I_1}{I_2 \sqrt{I_1 I_3}} \right) y_1 y_3 + \frac{u_2}{\sqrt{I_2}} + \frac{e_2(t)}{\sqrt{I_2}}, \\
\frac{dy_3}{dt} &= \left(\frac{I_1 - I_2}{I_3 \sqrt{I_1 I_2}} \right) y_1 y_2 + \frac{u_3}{\sqrt{I_3}} + \frac{e_3(t)}{\sqrt{I_3}}.
\end{aligned} \tag{3.19}$$

Use the canonical Lyapunov function

$$V(y) = \frac{1}{2} (y_1^2 + y_2^2 + y_3^2), \quad y \in R^3,$$

and show that the controls

$$u_i = -\text{sign } y_i (\|y\| - r(\rho)), \quad i = 1, 2, 3, \tag{3.20}$$

stabilize the motion of the system (3.19) with respect to the moving surface of the ellipsoid

$$A(r) = \{y \in R^3 : I_1 y_1^2 + I_2 y_2^2 + I_3 y_3^2 = r^2(\rho)\}. \tag{3.21}$$

Estimate the change of the function $V(y)$:

$$\begin{aligned}
\left. \frac{dV}{dt} \right|_{(3.19)} &= \frac{1}{\sqrt{I_1}} (u_1 y_1 + e_1(t) y_1) + \frac{1}{\sqrt{I_2}} (u_2 y_2 + e_2(t) y_2) + \\
&\quad + \frac{1}{\sqrt{I_3}} (u_3 y_3 + e_3(t) y_3).
\end{aligned} \tag{3.22}$$

Taking into account the condition (3.18) and the control (3.20), for the expression (3.22) obtain

$$\left.\frac{dV}{dt}\right|_{(3.19)} \leq (\rho - 1)\left(\frac{1}{\sqrt{I_1}} + \frac{1}{\sqrt{I_2}} + \frac{1}{\sqrt{I_3}}\right) < 0$$

in the domain $\operatorname{ext} A(r)$: $\|y\| > r(\rho)$,

$$\left.\frac{dV}{dt}\right|_{(3.19)} = 0$$

On the surface of the ellipsoid $A(r)$: $\|y\| = r(\rho)$,

$$\left.\frac{dV}{dt}\right|_{(3.19)} \geq (1 - \rho)\left(\frac{1}{\sqrt{I_1}} + \frac{1}{\sqrt{I_2}} + \frac{1}{\sqrt{I_3}}\right) > 0$$

in the domain $\operatorname{int} A(r)$: $\|y\| < r(\rho)$.

Therefore, for the system (3.19) and the function $V(y)$ all the conditions of the Theorem 2.4 are satisfied, and the controls

$$u_i(x) = -\operatorname{sign} x_i(I_1 x_1^2 + I_2 x_2^2 + I_3 x_3^2 - r^2(\rho)) \qquad (3.23)$$

stabilize the motion of the rigid body with respect to the surface of the ellipsoid (3.21), i.e., at $y_0 \in \operatorname{ext} A(r)$ solutions of the system (3.19) are attracted to the surface $A(r)$, and at $y_0 \in \operatorname{int} A(r)$ they are repelled from the equilibrium state $y_1 = y_2 = y_3 = 0$.

A special case of the problem (3.17) is the problem of the exponential convergence of the motions of the system (3.17) to the surface of a stationary sphere (see Corless and Leitmann [1996]).

Construct the function $V(x) = x^{\mathrm{T}} P x$, where

$$P = \begin{pmatrix} I_1 & 0 & 0 \\ 0 & I_2 & 0 \\ 0 & 0 & I_3 \end{pmatrix}.$$

It is known that

$$I_m \|x\|^2 \leq x^{\mathrm{T}} P x \leq I_M \|x\|^2$$

at all $x \in R^3$, where $I_M = \max\{I_1, I_2, I_3\}$, $I_m = \min\{I_1, I_2, I_3\}$.

Let q be the index of the exponential convergence of motions of the system (3.17). Select $\gamma \geq q I_M$ and calculate the constants $\widetilde{\gamma}$ and ε according to the results obtained by Corless and Leitmann [1996]:

$$\widetilde{\gamma} = \gamma(1 - \rho)^{-1},$$
$$0 < \varepsilon < \varepsilon^* = (1 - \rho)^2 I_m (3\rho I_M^2 \alpha)^{-1}.$$

Recall the notation of the function

$$\operatorname{sat}(y) = \begin{cases} y & \text{at} \quad \|y\| \leq 1, \\ \|y\|^{-1} y & \text{at} \quad \|y\| > 1. \end{cases}$$

According to the results of Corless and Leitmann [1996], the controls

$$u_i(x) = -\rho\,\text{sat}\left(\varepsilon^{-1}x_i\right) - (1-\rho)\text{sat}\left(\tilde{\gamma}x_i\right), \quad i = 1, 2, 3, \tag{3.24}$$

guarantee the exponential convergence of the motions of the system (3.17) to the surface of the sphere of radius

$$r_\varepsilon = \left(\frac{3\rho\varepsilon}{\alpha I_M}\right)^{\frac{1}{2}} \tag{3.25}$$

for all initial values $x_i(0)$, $i = 1, 2, 3$ located in the sphere of radius

$$r^* = (1-\rho)I_m^{1/2}I_M^{3/2}\alpha. \tag{3.26}$$

It is not difficult to show that the sphere of radius (3.25) is inscribed into the moving ellipsoid (3.21), while the sphere of radius (3.26) is circumscribed around it.

Thus, the controls (3.23) and (3.24) solve the problem of the stabilization of the motion of a rigid body under the condition of an uncertain value of resistance of the environment in the two cases: the convergence of the motions to the moving ellipsoid (3.21) and their convergence to the stationary surface of a sphere of radius (3.25).

3.5 Stability of an Uncertain Linear System with Neuron Control

Consider the linear approximation of the controlled uncertain system

$$\frac{dx}{dt} = f(x, u, \alpha),$$

where $x \in R^n$, $u \in R^m$, $f \in C(R^n \times R^m \times S, R^n)$ and $f(0, 0, \alpha) = 0$ at any $\alpha \in S$, in the form

$$\frac{dx}{dt} = (A_0 + \alpha A)x + Bu. \tag{3.27}$$

Here $A_0 \in R^{n \times n}$, $B \in R^{n \times m}$ are constant nominal matrices, $A \in R^{n \times n}$ is a matrix of uncertainties, and $\alpha \in S$: $|\alpha| \leq \mu \in R_+$, $\mu > 0$ is a constant.

It is assumed that the pair (A_0, B) is controlled.

Let W_1 and W_2 be some weight matrices characterizing respectively the closed and the output layers of the continual neuron net.

The control vector-function $u(t)$ is chosen in the form

$$u(t) = W_2 F(W_1 x(t)) = W_2 F(h), \tag{3.28}$$

where $F(h)$ is a vector-function describing the continuous activation of neurons, with the components $f_i(h_i)$, $i = 1, 2, \ldots, n$. In the theory of neuron nets, functions of the form

(1) $f_i(h_i) = 1/(1 + \exp(-h_i))$;

(2) $f_i(h_i) = \tanh(h_i)$;

(3) $f_i(h_i) = \text{arctg}(h_i)$ and other functions are applied .

Further in this section, components of the vector-function $F(h)$ are functions of the form (2).

We now determine the conditions for the stability of the zero solution of the uncertain system (3.27) with the neuron control (3.28). At first, we prove the following statement.

Lemma 3.1 *Let the constant $(n \times n)$-matrix A and the positive-definite symmetrical $(n \times n)$-matrix P be given. Then in R^n the matrix inequality*

$$A^T P + PA \leq A^T PA + P \qquad (3.29)$$

holds true.

Proof Denote $M = A^T PA + P - A^T P - PA$ and consider the quadratic form $x^T M x$ at any $(x \neq 0) \in R^n$. Since the relation

$$x^T [A^T PA + P - A^T P - PA] x = \| P^{1/2} Ax - P^{1/2} x \|^2 \geq 0,$$

holds true, then $x^T M x \geq 0$ at all $(x \neq 0) \in R^n$. Hence follows the inequality (3.29).

Theorem 3.3 *Assume that for the uncertain system (3.27) there exist positive-definite symmetrical matrices $P \in R^{n \times n}$ and $Q \in R^{n \times n}$, a matrix $S \in R^{n \times p}$ and a constant $0 < \beta < 1$ such that*

$$(A_0 + \beta BW_2 W_1)^T P + P(A_0 + \beta BW_2 W_1)$$
$$= -Q - SS^T - \mu(A^T PA + P), \qquad (3.30)$$
$$PBW_2 = -W_1^T - \sqrt{2} SI,$$

where I is a unit matrix of dimension $p \times p$.

Then the neuron control (3.28) ensures the stability of the equilibrium state $x = 0$ of the system (3.27).

Proof Let for the nominal system corresponding to (3.27) a quadratic form $V(x) = x^T Px$ be constructed with the matrix P satisfying the conditions indicated in the Theorem 3.3. Having calculated the full derivative of this function along the solutions of the system (3.27) with the control (3.28), we obtain

$$
\begin{aligned}
DV(x)\big|_{(3.27)} &= [(A_0 + \alpha A)x + BW_2 F(W_1 x)]^\mathrm{T} Px \\
&\quad + x^\mathrm{T} P[(A_0 + \alpha A)x + BW_2 F(W_1 x)] \\
&= [(A_0 + \beta BW_2 W_1 + \alpha A)x + BW_2 \tilde{F}(h)]^\mathrm{T} Px \\
&\quad + x^\mathrm{T} P[(A_0 + \beta BW_2 W_1 + \alpha A)x + BW_2 \tilde{F}(h)],
\end{aligned}
\tag{3.31}
$$

where $h = W_1 x$ and $\tilde{F}(h) = F(h) - \beta h$.

Denote $\tilde{A} = A_0 + \beta BW_2 W_1$ and, taking into account the matrix inequality from the lemma 3.1, transform the expression (3.31) to the form

$$
\begin{aligned}
DV(x)\big|_{(3.27)} &= [(\tilde{A} + \alpha A)x + BW_2 \tilde{F}(h)]^\mathrm{T} Px \\
&\quad + x^\mathrm{T} P[(\tilde{A} + \alpha A)x + BW_2 \tilde{F}(h)] \\
&= x^\mathrm{T}(\tilde{A} + \alpha A)^\mathrm{T} Px + \tilde{F}^\mathrm{T}(h)W_2^\mathrm{T} B^\mathrm{T} Px \\
&\quad + x^\mathrm{T} P(\tilde{A} + \alpha A)x + x^\mathrm{T} PB^\mathrm{T} W_2 \tilde{F}(h) \\
&= x^\mathrm{T}(\tilde{A}^\mathrm{T} P + P\tilde{A})x + \alpha x^\mathrm{T}(A^\mathrm{T} P + PA)x + 2x^\mathrm{T} PBW_2 \tilde{F}(h) \\
&\leq x^\mathrm{T}(\tilde{A}^\mathrm{T} P + P\tilde{A})x + \alpha x^\mathrm{T}(A^\mathrm{T} P + P)x + 2x^\mathrm{T} PBW_2 \tilde{F}(h) \\
&\leq x^\mathrm{T}(\tilde{A}^\mathrm{T} P + P\tilde{A})x + \mu(A^\mathrm{T} P + P)x + 2x^\mathrm{T} PBW_2 \tilde{F}(h).
\end{aligned}
$$

Hence, taking into account the conditions (3.30), we obtain the estimate

$$
\begin{aligned}
DV(x)\big|_{(3.27)} &\leq -x^\mathrm{T} Qx - x^\mathrm{T} SS^\mathrm{T} x - 2h^\mathrm{T}\tilde{F}(h) - 2\sqrt{2} x^\mathrm{T} S\tilde{F}(h) \\
&= -x^\mathrm{T} Qx - \|x^\mathrm{T} S + \sqrt{2}\tilde{F}^\mathrm{T}(h)\|^2 - 2[h^\mathrm{T}\tilde{F}(h) - \tilde{F}^\mathrm{T}(h)\tilde{F}(h)] \\
&= -x^\mathrm{T} Qx - \|x^\mathrm{T} S + \sqrt{2}\tilde{F}^\mathrm{T}(h)\|^2 - 2\sum_{i=1}^{p} \tilde{f}_i(h_i)(h_i - \tilde{f}_i(h_i)).
\end{aligned}
\tag{3.32}
$$

Note that the first two summands in the right-hand part of the inequality (3.32) are not positive. Consider the third summand. The values $h_i \in R$ and therefore $h_i \geq 0$ or $h_i \leq 0$. Let $h_i \geq 0$, then $h_i - \tilde{f}_i(h_i) = (h_i - \tanh(h_i)) + \beta h_i \geq 0$. Since $0 < \beta < 1$, then $\tilde{f}_i(h_i) = \tanh(h_i) - \beta h_i \geq 0$ for all $0 < h_i < h^*$, where $h^* = \min h_i^*$, h_i^* is the root of the equations $\tanh(h_i) - \beta h_i = 0$ at any $i = 1, 2, \ldots, p$. Hence it follows that $\tilde{f}_i(h_i)(h_i - \tilde{f}_i(h_i)) \geq 0$ at $0 < h_i < h_i^*$.

The case $h_i \leq 0$ is analyzed in a similar way. As a result we obtain $\sum_{i=1}^{p} \tilde{f}_i(h_i)(h_i - \tilde{f}_i(h_i)) \leq 0$, if $\max_i |h_i| = \|W_1 x\| < h^*$.

Thus, there exists $\varepsilon > 0$ such that

$$
DV(x)\big|_{(3.27)} \leq 0
$$

at all x: $\|x\| < \varepsilon$. This condition, together with the condition for the positive definiteness of the matrix P, is sufficient for the stability by Lyapunov of the equilibrium state $x = 0$ of the system (3.27).

3.6 Conditions for Parametric Quadratic Stabilizability

Consider the nonlinear system of differential equations

$$\dot{x} = [A + \Delta A(\alpha)]x + B\Phi(u), \qquad (3.33)$$

where $x(t) \in R^n$ is the system status, $u(t) \in R^m$ is a control at a time t, $\alpha \in R^l$ is a vector-parameter. The constant matrices $A \in R^{n \times n}$ and $B \in R^{n \times m}$ represent the known part of the system, $\Delta A(\alpha) \in R^{n \times n}$ denotes the indefinite members and is a continuous matrix-valued function α, $\Phi \colon R^m \to R^m$ is a nonlinear continuous function. Assume that at any given initial state $x_0 = x(t_0)$, the fixed value of parameter $\alpha \in R^l$ and the continuous control u the system of equations (3.33) has a unique solution $x(t; x_0, \alpha, u)$.

The control u is considered to be linear with respect to the state vector $x(t)$, i.e. $u = Kx + r$, where $K \in R^{m \times n}$ is a constant matrix and $r \in R^m$ is a correction function.

With regard to the system (3.33) make the following assumption.

Assumption 3.7 The system of equations (3.33) is such that:

(1) the function $\Phi(u) = (\Phi_1(u), \dots, \Phi_m(u))$ is determined and continuous on some open set $\Gamma \subseteq R^m$ together with the partial derivatives $\dfrac{\partial \Phi_i}{\partial u_j}$, $i, j = 1, \dots, m$;

(2) the point $u = 0$ belongs to the set Γ, and

$$\Phi(0) = 0 \quad \text{and} \quad \left.\frac{\partial \Phi(u)}{\partial u}\right|_{u=0} \neq 0;$$

(3) the matrix

$$C = A + B\left.\frac{\partial \Phi(u)}{\partial u}\right|_{u=0} K$$

is stable;

(4) for all p from the domain $P \subseteq R^l$ the following estimate is true:

$$\|\Delta A(p)\| \leq \eta < \frac{1}{2\|C^{-1}\|};$$

(5) there exists a value of the parameter α^*, belonging to the domain P, such that $\Delta A(\alpha^*) = 0$.

Under the conditions of the Assumption 3.7 the system (3.33) for the values of parameters α^*, $r^* = 0$ has the equilibrium state $x^* = 0$, which is stable.

According to Ohta, Šiljak [1] give the definition of the global PQ-stabilizability.

Definition 3.1 The system (3.33) is called *globally PQ-stabilizable by the control* $u = Kx + r$, if there exists a matrix $K \in R^{m \times n}$ a symmetrical positive-definite matrix $P \in R^{n \times n}$ and a number $\beta > 0$ such that for all $(\alpha, r) \in P \times R \subseteq R^l \times R^m$ the following conditions are satisfied:

(1) there exists a unique equilibrium state $x^e(\alpha, r)$ of the system

$$\dot{x} = [A + \Delta A(\alpha)]x + B\Phi(Kx + r); \qquad (3.34)$$

(2) the derivative of the quadratic Lyapunov function

$$V(x, x^e(\alpha, r)) = (x - x^e(\alpha, r))^T P(x - x^e(\alpha, r))$$

along the solutions of the system (3.34) satisfies the inequality

$$DV(x, x^e(\alpha, r))\big|_{(3.34)} \le -\beta \|x - x^e(\alpha, r)\|^2$$

for all $(x, \alpha, r) \in R^n \times P \times R$.

Taking into account the conditions of the Assumption 3.7, find the domain $\Omega_\alpha \times \Omega_r \subset R^l \times R^m$, for all values of parameters, from which the condition (1) of the Definition 3.1 is satisfied, and determine the sufficient conditions for the global PQ-stabilizability of the system (3.33) with respect to the found domain by the control $u = Kx + r$, i.e., the conditions under which the condition (2) of the Definition 3.1 is satisfied.

Let $r = (r_1, \ldots, r_s)$, where r_i, $i = 1, \ldots, s$, are some subvectors of the vector r with dimensions n_i respectively. The domain

$$\Pi = \left\{ (x, \alpha, r) \mid \Omega_x \colon \|x\| \le a, \quad \Omega_\alpha = P, \quad \Omega_r = \prod_{i=1}^{s} \Omega_{r_i}, \right.$$

$$\left. \Omega_{r_i} \colon \|r_i\| \le b_i, \quad i = 1, \ldots, s \right\}$$

such that for all (α, r) from $\Omega_\alpha \times \Omega_r$ there exists $x^e(\alpha, r)$ — the unique equilibrium state of the system (3.34), which belongs to Ω_x, can be determined by using the approach indicated in Appendix.

The equation

$$[A + \Delta A(\alpha)]x + B\Phi(Kx + r) = 0, \qquad (3.35)$$

from which the desired equilibrium state is determined, appears in the form

$$x = C^{-1}(Cx - [A + \Delta A(\alpha)]x - B\Phi(Kx + r))$$

and considers the iterative process

$$x_{n+1} = C^{-1}(Cx_n - [A + \Delta A(\alpha)]x_n - B\Phi(Kx_n + r)). \qquad (3.36)$$

Applying the theorem of convergence of the general iterative process in pseudometric space (see Collatz [1]) to (3.36), we ascertain that the equation (3.35) has a unique solution if the iterative process (3.36) converges. For this it is sufficient that the following conditions be satisfied:

$$\|B\| \max_{\Omega_u} \left\| \left.\frac{\partial \Phi(u)}{\partial u}\right|_u - \left.\frac{\partial \Phi(u)}{\partial u}\right|_{u=0} \right\| \|K\| \le \frac{1}{2\|C^{-1}\|} - \eta,$$

where $\Omega_u = \{u \colon \|U_i\| \le c_i, \; i = 1,\dots,s\}$, U_i, $i = 1,\dots,s$, are subvectors of the vector u, and

$$\|\Phi(r)\| \le \frac{a}{2\|C^{-1}\|\,\|B\|} \,. \tag{3.37}$$

Since

$$U_i = \begin{pmatrix} K_{n_1+\cdots+n_{i-1}+1} \\ \cdots\cdots\cdots\cdots \\ K_{n_1+\cdots+n_i} \end{pmatrix} x + r_i,$$

where K_j is the j-th row of the matrix K, $i = 1,\dots,s$, $j = 1,\dots,m$, using the inequalities

$$\left\| \begin{pmatrix} K_{n_1+\cdots+n_{i-1}+1} \\ \cdots\cdots\cdots\cdots \\ K_{n_1+\cdots+n_i} \end{pmatrix} \right\| a + b_i \le c_i, \quad i = 1,\dots,s,$$

and the estimate (3.37), we can estimate the boundary of the domain Π. Note the spectral norm of the matrix.

Let for the equation (3.35) the domain Π be determined. The following statements should be proved.

Theorem 3.4 *Let the vector-function $\Phi(u)$ in the system (3.33) satisfy the inequality*

$$\left\| \left.\frac{d\Phi(u)}{du}\right|_u - \left.\frac{d\Phi(u)}{du}\right|_{u=0} \right\| < \frac{1}{\|B\|\,\|K\|}\left(\frac{\lambda_{\min}(Q)}{2\|P\|} - \eta\right). \tag{3.38}$$

For all $u \in R^m$, where $\lambda_{\min}(Q)$ is the minimum eigenvalue of the matrix Q, Q is an arbitrary symmetrical positive-definite matrix of dimensions $n \times n$, P is a symmetrical positive-definite matrix which is a solution of the matrix Lyapunov equation

$$C^{\mathrm{T}}P + PC = -Q. \tag{3.39}$$

Then the system (3.33) is globally PQ-stabilizable by the control $u = Kx + r$.

Remark 3.1 The domain of change of the parameters of the system (3.33) is the domain $\Omega_\alpha \times \Omega_r$, where Ω_r is determined by the method indicated in Appendix for the equation (3.35), and

$$\Omega_\alpha = \left\{ \alpha \in R^l \,\big|\, \|\Delta A(\alpha)\| \le \eta < \frac{\lambda_{\min}(Q)}{2\|P\|} \right\}.$$

Since, according to the condition (4) of the Assumption 3.7, $\eta < \dfrac{1}{2\|C^{-1}\|}$, such a domain can be determined.

Proof Let the function $\Phi(u)$ satisfy the inequality (3.38). Determine the domain $\Omega_\alpha \times \Omega_r$, for each value of parameter, from which the equation (3.35) has a unique solution, and the domain Ω_x to which it belongs. With respect to the domain of change of parameters of the system $\Omega_\alpha \times \Omega_r$ the condition (1) of the Definition 3.1 is satisfied. By the change of variables

$$z = x - x^e(\alpha, r)$$

we reduce the system (3.33) to the form

$$\frac{dz}{dt} = [A + \Delta A(\alpha)](z + x^e(\alpha, r)) + B\Phi(Kz + Kx^e(\alpha, r) + r). \qquad (3.40)$$

and show that the derivative of the quadratic function

$$V(z) = z^{\mathrm{T}} P z, \qquad (3.41)$$

where P is determined from the equation (3.39), along the solutions of the system (3.40) is negative for the values $(\alpha, r) \in \Omega_\alpha \times \Omega_r$ at all $x \in R^n$. In other words, (3.41) is a Lyapunov function which allows determining the global asymptotic stability of the zero equilibrium state of the system (3.40), i.e., the equilibrium state $x^e(\alpha, r)$ of the system (3.33). Calculate

$$
\begin{aligned}
\left.\frac{dV(z)}{dt}\right|_{(3.40)} &= z^{\mathrm{T}} P([A + \Delta A(\alpha)](z + x^e(\alpha, r)) \\
&\quad + B\Phi(Kz + Kx^e(\alpha, r) + r)) + ([A + \Delta A(\alpha)](z + x^e(\alpha, r)) \\
&\quad + B\Phi(Kz + Kx^e(\alpha, r) + r))^{\mathrm{T}} P z \\
&= z^{\mathrm{T}} P \left([A + \Delta A(\alpha)] + B \left. \frac{\partial \Phi(u)}{\partial u} \right|_{\tilde{u}} K \right) z \\
&\quad + z^{\mathrm{T}} \left([A + \Delta A(\alpha)] + B \left. \frac{\partial \Phi(u)}{\partial u} \right|_{\tilde{u}} K \right)^{\mathrm{T}} P z \\
&= z^{\mathrm{T}} (C^{\mathrm{T}} P + P C) z \\
&\quad + z^{\mathrm{T}} \left(\Delta A(\alpha) + B \left(\left. \frac{\partial \Phi(u)}{\partial u} \right|_{\tilde{u}} - \left. \frac{\partial \Phi(u)}{\partial u} \right|_{u=0} \right) K \right)^{\mathrm{T}} P z \\
&\quad + z^{\mathrm{T}} P \left(\Delta A(\alpha) + B \left(\left. \frac{\partial \Phi(u)}{\partial u} \right|_{\tilde{u}} - \left. \frac{\partial \Phi(u)}{\partial u} \right|_{u=0} \right) K \right) z,
\end{aligned}
$$

where $\tilde{u}$ is some point of the space R^m. Hence obtain the estimate

$$
\left.\frac{dV(z)}{dt}\right|_{(3.40)} \leq \left(-\lambda_{\min}(Q) + 2\|P\| \left(\eta \right.\right.
$$
$$
\left.\left. + \|B\| \, \|K\| \left\| \left. \frac{\partial \Phi(u)}{\partial u} \right|_{\tilde{u}} - \left. \frac{\partial \Phi(u)}{\partial u} \right|_{u=0} \right\| \right) \right) \|z\|^2 = \alpha(\tilde{u}) \|z\|^2.
$$

Taking into account the inequality (3.38), find

$$\alpha(\tilde{u}) < 0$$

for all $\tilde{u} \in R^m$. Thus, the condition (2) of the Definition 3.1 is satisfied, and the system (3.33) is globally PQ-stabilizable by the control $u = Kx + r$.

The Theorem 3.4 is proved.

Theorem 3.5 *Let the function $\Phi(u)$ in the system (3.33) satisfy the inequality*

$$\left\| \left. \frac{\partial \Phi(u)}{\partial u} \right|_u - \left. \frac{\partial \Phi(u)}{\partial u} \right|_{u=0} \right\| \leq \frac{1}{\|B\| \|K\|} \left(\frac{\lambda_{\min}(Q)}{4\|P\| \|C\| \|C^{-1}\|} - \eta \right) \quad (3.42)$$

for all $u \in R^m$. Then the system (3.33) is globally PQ-stabilizable by the control $u = Kx + r$, the domain of change of parameters of the system (3.33) is the domain $P \times \Omega_r$, $\Omega_r = \{r \in R^m: \|r\| \leq b\}$, where b is an arbitrarily large predetermined positive number,

$$P = \left\{ \alpha \in R^l: \|\Delta A(\alpha)\| \leq \eta < \frac{\lambda_{\min}(Q)}{4\|P\| \|C\| \|C^{-1}\|} \right\}.$$

Proof Let the function $\Phi(u)$ satisfy the inequality (3.42), and b be an arbitrary positive number. Since $\|C\| \|C^{-1}\| \geq 1$, then

$$\frac{1}{\|B\| \|K\|} \left(\frac{\lambda_{\min}(Q)}{4\|P\| \|C\| \|C^{-1}\|} - \eta \right) \leq \frac{1}{\|B\| \|K\|} \left(\frac{\lambda_{\min}(Q)}{4\|P\|} - \eta \right)$$

$$< \frac{1}{\|B\| \|K\|} \left(\frac{\lambda_{\min}(Q)}{2\|P\|} - \eta \right)$$

and the condition (6) of the Theorem 3.5 is satisfied. Then under the condition (3.42) the system (3.33) is globally PQ-stabilizable by the control $u = Kx + r$.

Show that the domain of change of the system parameters is the domain $P \times \Omega_r$. Let

$$\Pi = \{(x, \alpha, r) \mid \Omega_x: \|x\| \leq a, \ \Omega_\alpha = P, \ \Omega_r: \|r\| \leq b\},$$

where

$$a = \frac{2\|C^{-1}\|b}{\|K\|} \left(\frac{\lambda_{\min}(Q)}{4\|P\| \|C\| \|C^{-1}\|} - \eta \right) + 2\|C^{-1}\| \|B\| \left\| \left. \frac{d\Phi(u)}{du} \right|_{u=0} \right\| b.$$

According to the method used for the estimation of the domain Π, at all $(\alpha, r) \in \Omega_\alpha \times \Omega_r$ the condition for the existence of the unique solution of the equation (3.35), belonging to Ω_x, is the correctness of the inequalities

$$\|C^{-1}\| \left(\eta + \|B\| \|K\| \left\| \left. \frac{\partial \Phi(u)}{\partial u} \right|_u - \left. \frac{\partial \Phi(u)}{\partial u} \right|_{u=0} \right\| \right) \leq \frac{1}{2} \quad (3.43)$$

for all $u \in \Omega_u$, $\Omega_u = \{u \in R^m : \|u\| \leq c\}$, and

$$\|\Phi(r)\| \leq \frac{a}{2\|C^{-1}\| \, \|B\|} \tag{3.44}$$

for all $r \in \Omega_r$. Having estimated the value c using the inequality (3.43), from the inequalities (3.44) and

$$\|K\|a + b \leq c \tag{3.45}$$

we can estimate the domain Π. However, under the condition (3.42) the inequality (3.43) is true at all $u \in R^m$, i.e., with an arbitrarily large c. So the chosen values a and b satisfy the inequality (3.45) and can be applied for the estimation of the domain Π. The inequality (3.45) is satisfied for all $r \in \Omega_r$:

$$\|\Phi(r)\| = \left\| \Phi(0) + \left. \frac{\partial \Phi(u)}{\partial u} \right|_{u=\tilde{r}} r \right\| \leq \left\| \left. \frac{\partial \Phi(u)}{\partial u} \right|_{u=\tilde{r}} \right\| b$$

$$= \left\| \left. \frac{\partial \Phi(u)}{\partial u} \right|_{u=\tilde{r}} - \left. \frac{\partial \Phi(u)}{\partial u} \right|_{u=0} + \left. \frac{\partial \Phi(u)}{\partial u} \right|_{u=0} \right\| b$$

$$\leq \left\| \left. \frac{\partial \Phi(u)}{\partial u} \right|_{u=\tilde{r}} - \left. \frac{\partial \Phi(u)}{\partial u} \right|_{u=0} \right\| b + \left\| \left. \frac{\partial \Phi(u)}{\partial u} \right|_{u=0} \right\| b$$

$$\leq \frac{b}{\|B\| \, \|K\|} \left(\frac{\lambda_{\min}(Q)}{4\|P\| \, \|C\| \, \|C^{-1}\|} - \eta \right) + \left\| \left. \frac{\partial \Phi(u)}{\partial u} \right|_{u=0} \right\| b$$

$$= \frac{1}{2\|C^{-1}\| \, \|B\|} \left(\frac{2\|C^{-1}\|b}{\|K\|} \left(\frac{\lambda_{\min}(Q)}{4\|P\| \, \|C\| \, \|C^{-1}\|} - \eta \right) \right.$$

$$\left. + 2\|C^{-1}\| \, \|B\| \left\| \left. \frac{\partial \Phi(u)}{\partial u} \right|_{u=0} \right\| b \right) = \frac{a}{2\|C^{-1}\| \, \|B\|}.$$

Therefore the condition (3.44) is satisfied for all $r \in \Omega_r$, and for all $(\alpha, r) \in \Omega_\alpha \times \Omega_r$ there exists a unique solution of the equation

$$[A + \Delta A(\alpha)]x + B\Phi(Kx + r) = 0,$$

belonging to Ω_x. That is, the domain of change of the system parameters is $P \times \Omega_r$.

The Theorem 3.5 is proved.

Example 3.1 Let an airplane have a bicycle scheme of the landing gear with the controlled forewheel (see Letov [1964], Neimark and Fufaev [1967]).

Let Θ denote the angle of deflection of the longitudinal plane of the airplane from the vertical, Ψ denote the angle of rotation of the forewheel with respect to the line connecting the points M_1 and M_2, where M_1 and M_2 are the points of contact of the wheels with the plane of rolling motion.

Then assume that the plane of rolling motion is absolutely coarse and the equation of the motion of the airplane with respect to the line $M_1 M_2$ has the form

$$\ddot{\Theta} = \frac{mg - C_y^* V^2}{md} \Theta - \left(\frac{bV}{cd} \dot{\Psi} + \frac{V^2}{cd} \Psi \right) - L\dot{\Theta}. \tag{3.46}$$

If the mass of the airplane is $m = 1000$ kg, the velocity is $V = 5$ m/s, the distance from the centre of gravity G of the airplane to the line $M_1 M_2$ $d = 1$ m, the distance of the projection of the point G onto the line $M_1 M_2$ from the point M_1 of contact of the rear wheel $b = 5$ m, the distance between the points M_1 and M_2 $c = 10$ m, the coefficient responsible for the bearing strength, $C_y^* = 100$, and the coefficient meeting the damping moment, $L = 0.1$, then the equation (3.46) takes the form

$$\ddot{\Theta} = 7.3\Theta - 2.5\dot{\Psi} + 2.5\Psi - 0.1\dot{\Theta}. \tag{3.47}$$

Let the forewheel be automatically controlled by the actuating motor, the equations of its motions being as follows:

$$\dot{\Psi} = w\dot{\Theta} + f(\sigma),$$
$$\sigma = a\Theta + E\dot{\Theta} - \frac{1}{l}\dot{\Psi}. \tag{3.48}$$

Here $w = kV > 0$ ($k = $ const), $a > 0$, $E > 0$, $l > 0$ are parameters of the control system. Choose $k = 1$. Performing the change of variables $\Theta = \eta_1$, $\dot{\Theta} = \eta_2$, $\Psi = \eta_3$, reduce the equations (3.47), (3.48) to the form

$$\dot{\eta}_1 = \eta_2,$$
$$\dot{\eta}_2 = 7.3\eta_1 - 12.6\eta_2 - 2.5\eta_3 - 2.5f(\sigma),$$
$$\dot{\eta}_3 = 5\eta_2 + f(\sigma),$$
$$\sigma = a\eta_1 + E\eta_2 - \frac{1}{l}\eta_3,$$

or in the matrix form

$$\dot{\eta} = A\eta + Bf(\sigma),$$
$$\sigma = K\eta, \tag{3.49}$$

where $\eta = (\eta_1, \eta_2, \eta_3)^T$,

$$A = \begin{pmatrix} 0 & 1 & 0 \\ 7.3 & -12.6 & -2.5 \\ 0 & 5 & 0 \end{pmatrix}, \quad B = \begin{pmatrix} 0 \\ -2.5 \\ 1 \end{pmatrix}, \quad K = \begin{pmatrix} a & E & -\frac{1}{l} \end{pmatrix}.$$

Assuming that in the system (3.49) there are some uncertainties due both to the system parameters and to the control, the system (3.49) can be represented in the form

$$\dot{\eta} = [A + \Delta A(\alpha)]\eta + Bf(\sigma),$$
$$\sigma = K\eta + r, \tag{3.50}$$

where $\Delta A(\alpha) \in R^{3\times 3}$, $r \in R^1$. The essence of the problem is to choose the parameters of the control system a, E, l, with the given function $f(\sigma)$, so that the control $\sigma = K\eta + r$, where $K = \begin{pmatrix} a & E & -\frac{1}{l} \end{pmatrix}$, would ensure the global PQ-stabilizability of the system (3.50), and also to obtain the estimates of the domain of values of the parameters α and r, at which the stability of the system (3.50) will remain.

Let $f(\sigma) = \dfrac{399}{400}\sigma + \dfrac{1}{400}\sin\sigma$. According to the conditions for the function $f(\sigma)$ obtain $f(0) = 0$ and $f'(0) = 1 \neq 0$. For the stable matrix

$$C = \begin{pmatrix} 0 & 1 & 0 \\ -12.7 & -12.85 & -2.25 \\ 8 & 5.1 & -0.1 \end{pmatrix}.$$

According to the equation from the condition (3) of the Assumption 3.7, calculate the matrix $K = \begin{pmatrix} 8 & 0.1 & -0.1 \end{pmatrix}$, i.e., the parameters of the control system $a = 8$, $E = 0.1$, $l = 10$. For the matrix

$$Q = \begin{pmatrix} 1 & 0 & 0 \\ 0 & 1 & 0 \\ 0 & 0 & 1 \end{pmatrix}$$

the matrix P has the form

$$P = \begin{pmatrix} 2.776 & 0.446 & 0.646 \\ 0.446 & 0.154 & 0.202 \\ 0.646 & 0.202 & 0.454 \end{pmatrix}.$$

Let the domain Ω_α be such that for all α from Ω_α the following estimate would hold true:

$$\|\Delta A(\alpha)\| \leq \frac{1}{20} < \min\left(\frac{\lambda_{\min}(Q)}{2\|P\|}, \frac{1}{2\|C^{-1}\|}\right) \tag{3.51}$$

$$= \min(0.165, 0.208) = 0.165.$$

Applying the approach from the Appendix, calculate the domain

$$\Pi = \{(x, \alpha, r)|: \ \|x\| \leq 0.362, \ \alpha \in \Omega_\alpha, \ |r| \leq 0.03\}. \tag{3.52}$$

According to the Theorem 3.5, the sufficient condition for the global PQ-stabiliziability of the system (3.50) by the control $\sigma = K\eta + r$ is satisfied. The estimates of uncertainties (3.51), (3.52) secure the stability of the considered system.

Chapter 4

Stability of Quasilinear Uncertain Systems

Quasilinear systems with uncertain values of parameters are typical models of many events and processes in technical applications. In this chapter we will analyze the problem of the stability of solutions of a quasilinear uncertain system consisting of two subsystems. It is assumed that the linear approximation of each of the subsystems is reducible to the diagonal form or to the Jordan form. The conditions for the stability of solutions with respect to a moving invariant set are determined on the basis of the results of Chapter 2 in such a form that for their actual verification, the sign-definiteness of special matrices is analyzed.

Along with the matrix-valued Lyapunov function, the vector Lyapunov function is also used in this chapter as a particular case of the matrix-valued function. Several results are illustrated by specific examples.

4.1 Uncertain Quasilinear System and Its Transformation

Consider a perturbed motion described by the quasilinear equations

$$\frac{dz}{dt} = Pz + Q(z, w, \alpha),$$

$$\frac{dw}{dt} = Hw + G(z, w, \alpha). \tag{4.1}$$

Here $z \in R^n$, $w \in R^m$, $\alpha \in S \subseteq R^d$ is the uncertainness parameter of the system (4.1), P, H are constant matrices of the respective dimensions, $Q: R^n \times R^m \times R^d \to R^n$ and $G: R^n \times R^m \times R^d \to R^m$ have as their components power series of integer positive powers of z and w, beginning from terms of not lower than the second order and converging absolutely in the product of the arbitrarily large open connected neighbourhoods $\mathcal{N}_z$ and $\mathcal{N}_w$ of the states $z = 0$ and $w = 0$ at any values of the uncertainness parameter $\alpha \in S \subseteq R^d$.

Using the linear nonsingular transformations $z = Tx$ and $w = Ry$

(det $T \neq 0$, det $R \neq 0$) reduce the linear part of the system (4.1) to the diagonal form

$$\frac{dx}{dt} = Ax + f_1(x, y, \alpha),$$

$$\frac{dy}{dt} = By + f_2(x, y, \alpha),$$

(4.2)

where $x \in R^n$, $y \in R^m$, $A = \text{diag}\{\lambda_1, \ldots, \lambda_n\}$, $B = \text{diag}\{\beta_1, \ldots, \beta_m\}$, $f_1(x, y, \alpha) = Q(Tz, Rw, \alpha)$, $f_2(x, y, \alpha) = G(Tz, Rw, \alpha)$.

For the components x_s and y_k of the vectors x and y consider the variables

$$\begin{aligned}
x_s &= r_s \exp(i\theta_s), & \overline{x}_s &= r_s \exp(-i\theta_s), \\
y_k &= \rho_k \exp(i\varphi_k), & \overline{y}_k &= \rho_k \exp(-i\varphi_k), \\
s &= 1, 2, \ldots, n, & k &= 1, 2, \ldots, m, \\
\theta_s &\in (0, \pi), & \varphi_k &\in (0, \pi).
\end{aligned}$$

Hence obtain

$$\begin{aligned}
3r_s &= x_s \exp(-i\theta_s), & r_s &= \overline{x}_s \exp(i\theta_s), & s &= 1, 2, \ldots, n, \\
\rho_k &= y_k \exp(-i\varphi_k), & \rho_k &= \overline{y}_k \exp(i\varphi_k), & k &= 1, 2, \ldots, m,
\end{aligned}$$

and therefore

$$\frac{dr_s}{dt} = \frac{1}{2} \left(\frac{dx_s}{dt} e^{-i\theta_s} + \frac{d\overline{x}_s}{dt} e^{i\theta_s} \right), \qquad s = 1, 2, \ldots, n,$$

$$\frac{d\rho_k}{dt} = \frac{1}{2} \left(\frac{dy_k}{dt} e^{-i\varphi_k} + \frac{d\overline{y}_k}{dt} e^{i\varphi_k} \right), \qquad k = 1, 2, \ldots, m.$$

Taking into account the systems of equations (4.2), obtain

$$\frac{dr_s}{dt} = \text{Re } \lambda_s r_s + \frac{1}{2} \left(f_{1s} e^{-i\theta_s} + \overline{f}_{1s} e^{i\theta_s} \right), \qquad s = 1, 2, \ldots, n,$$

$$\frac{d\rho_k}{dt} = \text{Re } \beta_k \rho_k + \frac{1}{2} \left(f_{2k} e^{-i\varphi_k} + \overline{f}_{2k} e^{i\varphi_k} \right), \qquad k = 1, 2, \ldots, m.$$

(4.3)

In the systems (4.3) the equations corresponding to the complex conjugate roots will repeat because the respective variables have the same absolute values. Therefore the number of different equations in the systems (4.3) will be $n_1 < n$ and $m_1 < m$, respectively.

Remark 4.1 If in the system (4.2) the functions f_1 and f_2 are represented in the form

$$f_1(x, y, \alpha) = f_1^*(x, y, 0) + \Delta f_1(\cdot, \alpha),$$

$$f_2(x, y, \alpha) = f_2^*(x, y, 0) + \Delta f_2(\cdot, \alpha),$$

where $\Delta f_1 = f_1(x, y, \alpha) - f_1^*(x, y, 0)$, $\Delta f_2 = f_2(x, y, \alpha) - f_2^*(x, y, 0)$, $\alpha \in \mathcal{S} \subseteq$

R^d, then the system (4.3) is transformed to the following one:

$$\frac{dr_s}{dt} = \text{Re } \lambda_s\, r_s + \frac{1}{2}\left(f_{1s}^* e^{-i\theta_s} + \overline{f}_{1s}^* e^{i\theta_s}\right) + \frac{1}{2}\left(\Delta f_{1s} e^{-i\theta_s} + \Delta \overline{f}_{1s} e^{i\theta_s}\right),$$

$$\frac{d\rho_k}{dt} = \text{Re } \beta_k\, \rho_k + \frac{1}{2}\left(f_{2k}^* e^{-i\varphi_k} + \overline{f}_{2k}^* e^{i\varphi_k}\right) + \frac{1}{2}\left(\Delta f_{2k} e^{-i\varphi_k} + \Delta \overline{f}_{2k} e^{i\varphi_k}\right).$$

In this system the impact of the uncertainties $\alpha \in S \subseteq R^d$ is "concentrated" in the last summands of the right-hand parts of the equations. Sometimes such a transformation of the system can facilitate the estimation of the impact of the "uncertainness" upon the dynamics of the system (4.1).

For the system (4.3) we will consider the moving set $A^*(\varkappa)$ determined by the formula

$$A^*(\varkappa) = \{r \in R^n, \ \rho \in R^m: \ \|r\| + \|\rho\| = \varkappa(\alpha)\}.$$

Here it is assumed that $\varkappa(\alpha) > 0$ and, in addition, $\varkappa(\alpha) \to \varkappa_0$ ($\varkappa_0 = \text{const} > 0$) at $\|\alpha\| \to 0$ and $\varkappa(\alpha) \to \infty$ at $\|\alpha\| \to \infty$.

4.2 Application of the Canonical Matrix-Valued Function

We will analyze the system (4.3), using the matrix-valued function

$$U(r, \rho) = \begin{pmatrix} u_{11}(r) & u_{12}(r, \rho) \\ u_{21}(r, \rho) & u_{22}(\rho) \end{pmatrix}, \qquad (4.4)$$

whose elements $u_{ij}(\cdot)$, $i, j = 1, 2$, are determined by the variables r and ρ of the system (4.3):

$$u_{11}(r) = \sum_{s=1}^{n_1} \nu_s r_s^2, \quad s = 1, 2, \ldots, n_1,$$

$$u_{22}(\rho) = \sum_{k=1}^{m_1} \psi_k \rho_k^2, \quad k = 1, 2, \ldots, m_1, \qquad (4.5)$$

$$u_{12}(r, \rho) = u_{21}(r, \rho) = \gamma \sum_{s=k}^{\min(n_1, m_1)} r_s \rho_k,$$

where ν_s, ψ_k are some positive constants and γ is an arbitrary constant.

The function (4.4) with the elements (4.5) will be called canonical, taking into account that its elements are constructed on the basis of the transformation of the linear approximation of the system (4.2) to the canonical form.

Denote $\nu_m = \min_s(\nu_s)$, $\nu_M = \max_s(\nu_s)$, $\psi_m = \min_k(\psi_k)$, $\psi_M = \max_k(\psi_k)$.

It is easy to verify that for the functions (4.5) the following estimates are true:

$$\nu_m\|r\|^2 \le u_{11}(r) \le \nu_M\|r\|^2 \qquad \text{at all} \quad r \in R^{n_1},$$
$$\psi_m\|\rho\|^2 \le u_{22}(\rho) \le \psi_M\|\rho\|^2 \qquad \text{at all} \quad \rho \in R^{m_1},$$
$$-\gamma\|r\|\,\|\rho\| \le u_{12}(r,\rho) \le \gamma\|r\|\,\|\rho\| \qquad \text{at all} \quad (r,\rho) \in R^{n_1} \times R^{m_1}.$$

Now apply the function

$$V(r,\rho) = \eta^T U(r,\rho)\eta, \quad \eta \in R_+^2, \quad \eta > 0, \tag{4.6}$$

for which the following estimates hold true:

$$\begin{aligned}
\text{(a)} \quad & v^T H^T A_1 H v \le V(r,\rho) \quad \text{at} \quad \|r\| + \|\rho\| \ge \varkappa(\alpha), \\
\text{(b)} \quad & V(r,\rho) \le v^T H^T A_2 H v \quad \text{at} \quad \|r\| + \|\rho\| \le \varkappa(\alpha),
\end{aligned} \tag{4.7}$$

where $v^T = (\|r\|, \|\rho\|)$, $H = \text{diag}(\eta_1, \eta_2)$,

$$A_1 = \begin{pmatrix} \nu_m & -\gamma \\ -\gamma & \psi_m \end{pmatrix}, \qquad A_2 = \begin{pmatrix} \nu_M & \gamma \\ \gamma & \psi_M \end{pmatrix}.$$

To apply the direct Lyapunov's method generalized in Chapter 2 to the analysis of solutions of the system (4.3) on the basis of the function (4.6), at first determine the structure of its full derivative in view of the system (4.3). For that we introduce some assumptions.

Assumption 4.1 There exist constants μ_{jl}, $j = 1, 2$, $l = 1, 2, 3, 4$, such that at $\|r\| + \|\rho\| > \varkappa(\alpha)$ the following inequalities are true:

$$\text{(a)} \quad \sum_{s=1}^{n_1} \nu_s \left(f_{1s} e^{-i\theta_s} + \overline{f}_{1s} e^{i\theta_s} \right) r_s \le \mu_{11}\|r\|^2 + \mu_{12}\|\rho\|\,\|r\|,$$

$$\text{(b)} \quad \sum_{k=1}^{m_1} \psi_k \left(f_{2k} e^{-i\varphi_k} + \overline{f}_{2k} e^{i\varphi_k} \right) \rho_k \le \mu_{21}\|\rho\|^2 + \mu_{22}\|\rho\|\,\|r\|,$$

$$\text{(c)} \quad \sum_{s=k} \gamma \left(f_{1s} e^{-i\theta_s} + \overline{f}_{1s} e^{i\theta_s} \right) \rho_k \le \mu_{13}\|\rho\|\,\|r\| + \mu_{14}\|\rho\|^2,$$

$$\text{(d)} \quad \sum_{s=k} \gamma \left(f_{2k} e^{-i\varphi_k} + \overline{f}_{2k} e^{i\varphi_k} \right) r_s \le \mu_{23}\|\rho\|\,\|r\| + \mu_{24}\|r\|^2.$$

Assumption 4.2 There exist constants $\overline{\mu}_{jl}$, $j = 1, 2$, $l = 1, 2, 3, 4$, such that at $\|r\| + \|\rho\| < \varkappa(\alpha)$ the inequalities (a)–(d) from the Assumption 4.1 with the sign "$\ge$" instead of "$\le$" hold true.

Denote $a = \max_s (\text{Re}\,\lambda_s)$, $b = \max_k (\text{Re}\,\beta_k)$, $\bar{a} = \min_s (\text{Re}\,\lambda_s)$, $\bar{b} = \min_k (\text{Re}\,\beta_k)$. If all the conditions of the Assumptions 4.1 and 4.2 are satisfied, then for the fully derived elements of the matrix-valued function (4.4) along the solutions of the system (4.3) the following estimates hold true:

(a) at all $\|r\| + \|\rho\| > \varkappa(\alpha)$, $\alpha \in \mathcal{S} \subseteq R^d$

$$\dot{u}_{11}(r) \leq (2\nu_M a + \mu_{11})\|r\|^2 + \mu_{12}\|r\|\,\|\rho\|,$$

$$\dot{u}_{22}(\rho) \leq (2\psi_M b + \mu_{21})\|\rho\|^2 + \mu_{22}\|r\|\,\|\rho\|,$$

$$\dot{u}_{12}(r,\rho) \leq \frac{1}{2}\mu_{24}\|r\|^2 + \left[\gamma(a+b) + \frac{1}{2}(\mu_{13}+\mu_{23})\right]\|r\|\|\rho\| + \frac{1}{2}\mu_{14}\|\rho\|^2;$$

(b) at all $\|r\| + \|\rho\| < \varkappa(\alpha)$, $\alpha \in \mathcal{S} \subseteq R^d$

$$\dot{u}_{11}(r) \geq (2\nu_m \overline{a} + \overline{\mu}_{11})\|r\|^2 + \overline{\mu}_{12}\|r\|\,\|\rho\|,$$

$$\dot{u}_{22}(\rho) \geq (2\psi_m \overline{b} + \overline{\mu}_{21})\|\rho\|^2 + \overline{\mu}_{22}\|r\|\,\|\rho\|,$$

$$\dot{u}_{12}(r,\rho) \geq \frac{1}{2}\overline{\mu}_{24}\|r\|^2 + \left[\gamma(\overline{a}+\overline{b}) + \frac{1}{2}(\overline{\mu}_{13}+\overline{\mu}_{23})\right]\|r\|\|\rho\| + \frac{1}{2}\overline{\mu}_{14}\|\rho\|^2.$$

Lemma 4.1 *Let all the conditions of the Assumptions 4.1 and 4.2 be satisfied. Then for the full derivative of the function (4.6) in view of the system (4.3) the following estimates hold true:*

(a) *at* $\|r\| + \|\rho\| > \varkappa(\alpha)$ *and at all* $\alpha \in \mathcal{S} \subseteq R^d$

$$\left.\frac{dV(r,\rho)}{dt}\right|_{(4.3)} \leq v^{\mathrm{T}} C v, \tag{4.8}$$

where $v^{\mathrm{T}} = (\|r\|, \|\rho\|)$, $C = \begin{pmatrix} c_{11} & c_{12} \\ c_{21} & c_{22} \end{pmatrix}$,

$$c_{12} = c_{21},$$

$$c_{11} = \eta_1^2(2\nu_M a + \mu_{11}) + \eta_1\eta_2\mu_{24},$$

$$c_{22} = \eta_2^2(2\psi_M b + \mu_{21}) + \eta_1\eta_2\mu_{14},$$

$$c_{12} = \frac{1}{2}\eta_1^2\mu_{12} + \frac{1}{2}\eta_2^2\mu_{22} + \eta_1\eta_2\left[\gamma(a+b) + \frac{1}{2}(\mu_{13}+\mu_{23})\right];$$

(b) *at* $\|r\| + \|\rho\| < \varkappa(\alpha)$ *and at all* $\alpha \in \mathcal{S} \subseteq R^d$

$$\left.\frac{dV(r,\rho)}{dt}\right|_{(4.3)} \geq v^{\mathrm{T}} D v, \tag{4.9}$$

where $D = \begin{pmatrix} d_{11} & d_{12} \\ d_{21} & d_{22} \end{pmatrix}$,

$$d_{12} = d_{21},$$

$$d_{11} = \eta_1^2(2\nu_m \overline{a} + \overline{\mu}_{11}) + \eta_1\eta_2\overline{\mu}_{24},$$

$$d_{22} = \eta_2^2(2\psi_m \overline{b} + \overline{\mu}_{21}) + \eta_1\eta_2\overline{\mu}_{14},$$

$$d_{12} = \frac{1}{2}\eta_1^2\overline{\mu}_{12} + \frac{1}{2}\eta_2^2\overline{\mu}_{22} + \eta_1\eta_2\left[\gamma(\overline{a}+\overline{b}) + \frac{1}{2}(\overline{\mu}_{13}+\overline{\mu}_{23})\right].$$

The proof is made by the substitution of estimates for the components of the matrix-valued function into the expression (4.6).

The estimates (4.7) and the inequalities (4.8), (4.9) allow to determine the sufficient conditions for the uniform asymptotic stability of solutions of the system (4.1) with respect to the moving invariant set $A^*(\varkappa)$.

Theorem 4.1 *Assume that in the system* (4.2) *the functions* $f_1(z, w, \alpha)$ *and* $f_2(z, w, \alpha)$ *are continuous on* $R^n \times R^m \times R^d$ *and*

(1) *for any* $\alpha \in S \subseteq R^d$ *there exists a function* $\varkappa(\alpha) > 0$ *such that the set* $A^*(\varkappa)$ *is not empty at all* $\alpha \in S \subseteq R^d$;

(2) *for the transformed system* (4.3) *the matrix-valued function* (4.4) *with the elements* (4.5) *is constructed;*

(3) *all the conditions of the Assumptions 4.1 and 4.2 are satisfied;*

(4) *in the inequalities* (4.7)(a), (b) *the matrices* A_1 *and* A_2 *are positive-definite;*

(5) *the following conditions are satisfied:*

(a) $\quad \dfrac{dV(r,\rho)}{dt}\bigg|_{(4.3)} < 0 \quad at \quad \|r\| + \|\rho\| > \varkappa(\alpha),$

(b) $\quad \dfrac{dV(r,\rho)}{dt}\bigg|_{(4.3)} = 0 \quad at \quad \|r\| + \|\rho\| = \varkappa(\alpha),$

(c) $\quad \dfrac{dV(r,\rho)}{dt}\bigg|_{(4.3)} > 0 \quad at \quad \|r\| + \|\rho\| < \varkappa(\alpha);$

(6) *for the functions* $a(\|r\| + \|\rho\|) = \lambda_m(H^{\mathrm{T}} A_1 H)(\|r\|, \|\rho\|)^{\mathrm{T}}(\|r\|, \|\rho\|)$ *and* $b(\|r\| + \|\rho\|) = \lambda_M(H^{\mathrm{T}} A_2 H)(\|r\|, \|\rho\|)^{\mathrm{T}}(\|r\|, \|\rho\|)$ *the relation*

$$a(\varkappa(\alpha)) = b(\varkappa(\alpha))$$

is true at any $\varkappa(\alpha) > 0$.

Then the solutions $(x(t, \alpha), y(t, \alpha))$ of the system (4.2) are uniformly asymptotically stable with respect to the invariant moving set $A^*(\varkappa)$.

The proof is similar to that of the Theorem 2.2 and is not given here.

4.3   Isolated Quasilinear Systems

Let the perturbed motion be described by the system

$$\frac{dz}{dt} = Pz + Q(z, \alpha), \tag{4.10}$$

where $z \in R^n$, P is a constant real $(n \times n)$-matrix and $Q \colon R^n \times S \to R^n$, $S \subseteq R^d$.

As in Section 4.2, reduce the system (4.10) to the form

$$\frac{dx}{dt} = P^* x + Q^*(x, \alpha). \tag{4.11}$$

Here $x \in R^n$, $P^* = A^{-1}PA = [\delta_{sr}\lambda_r + \delta_{s-1,r}\mu_r]$, $+$ $s, r = 1, 2, \ldots, n$, A is an $(n \times n)$-matrix of nonsingular transformation $z = Ax$ (i.e. $\det A \neq 0$), $Q^* = A^{-1}Q(Ax, \alpha)$, λ_r — are eigenvalues of the matrix P, μ_r are sufficiently small positive numbers.

Determine the moving set

$$A^*(\varkappa) = \Big\{ x \in R^n \colon \ \sum_{s=1}^{n} |x_s| = \varkappa(\alpha) \Big\}, \tag{4.12}$$

where $\varkappa(\alpha) \to \varkappa_0$ at $\|\alpha\| \to 0$.

The analysis of the behavior of solutions of the system (4.11) with respect to the set (4.12) will be conducted, using the positive-definite function

$$V = \|x\| = \sum_{s=1}^{n} |x_s| = \sum_{s=1}^{n} r_s. \tag{4.13}$$

Since

$$\frac{dr_s}{dt} = \mathrm{Re}\,\lambda_s r_s + \frac{1}{2}\,\mu_{s-1}(x_{s-1}e^{-i\theta_s} + \bar{x}_{s-1}e^{i\theta_s}) + \frac{1}{2}(Q_s^* e^{-i\theta_s} + \bar{Q}_s^* e^{i\theta_s}), \tag{4.14}$$

the estimates for the time derivative of the function V in the domains $\mathrm{ext}\,A^*(\varkappa)$ and $\mathrm{int}\,A^*(\varkappa)$ will be as follows:

(a) at $\|x\| > \varkappa(\alpha)$, $\alpha \in S \subseteq R^d$,

$$\frac{dV}{dt} \leq (\Delta + \mu)V + f(V, \alpha), \tag{4.15}$$

where $\Delta = \max_s \{\mathrm{Re}\,\lambda_s\}$, $\mu \geq \mu_r$, $r = 1, 2, \ldots, n$; $f(V, \alpha)$ is a polynomial with respect to V, estimating the last summand in the right-hand part of the expression (4.14), $f(0, \alpha) = 0$ at any $\alpha \in S$;

(b) at $\|x\| < \varkappa(\alpha)$, $\alpha \in \mathcal{S} \subseteq R^d$, obtain

$$\frac{dV}{dt} \geq (\Delta^* - \mu^*)V - f(V, \alpha), \tag{4.16}$$

where $\Delta^* = \min_s \{\text{Re } \lambda_s\}$, $\mu^* \leq \mu_r$, $r = 1, 2, \ldots, n$.

On the basis of the estimates (4.15) and (4.16) one can formulate the following statement.

Theorem 4.2 *Assume that in the system* (4.10) *the function* $Q(z, \alpha)$ *is continuous on* $R^n \times \mathcal{S}$ *and the following conditions are satisfied:*

(1) *for any* $\alpha \in \mathcal{S} \subseteq R^d$ *there exists a function* $\varkappa(\alpha) > 0$ *at any* $\alpha \in \mathcal{S}$ *such that the set* (4.12) *is not empty at all* $\alpha \in \mathcal{S}$;

(2) *there exist functions* $W_1(x)$ *and* $W_2(x)$ *such that at all* $\alpha \in \mathcal{S} \subseteq R^d$

   (a) $(\Delta + \mu)V + f(V, \alpha) \leq W_1(x)$ *at* $\|x\| > \varkappa(\alpha)$,
   (b) $(\Delta^* - \mu^*)V - f(V, \alpha) \geq W_2(x)$, *at* $\|x\| < \varkappa(\alpha)$,
   (c) $\text{Re } \lambda_s r_s + \frac{1}{2} \mu_{s-1}(x_{s-1}e^{-i\theta_s} + \bar{x}_{s-1}e^{i\theta_s}) + \frac{1}{2}(Q_s^* e^{-i\theta_s} + \bar{Q}_s^* e^{i\theta_s}) = 0$ *at* $\|x\| = \varkappa(\alpha)$;

(3) *the following inequalities are true*

   (a) $W_1(x) < 0$ *at* $\|x\| > \varkappa(\alpha)$,
   (b) $W_2(x) > 0$ *at* $\|x\| < \varkappa(\alpha)$.

Then the motions of the system (4.10) *are uniformly asymptotically stable with respect to the moving set* $A^*(\varkappa)$.

The proof of this theorem will not be given here as it is the corollary of the proof of the Theorem 4.1.

Example 4.1 Consider the system of equations of perturbed motion

$$\begin{aligned}\frac{dx}{dt} &= \mu x + y - g(x, y, \alpha)x(x^2 + y^2), \\ \frac{dy}{dt} &= \mu y - x - g(x, y, \alpha)y(x^2 + y^2), \quad \alpha \in \mathcal{S},\end{aligned} \tag{4.17}$$

where $\mu = \text{const} > 0$ and $g(x, y, \alpha) > 0$ is a function characterizing the "uncertainties" of the system (4.17).

In the system (4.17) perform the change of variables

$$x = -r\cos\theta, \quad y = r\sin\theta$$

and reduce it to the form

$$\frac{dr}{dt} = \mu r - g(r, \theta, \alpha)r^3, \quad \frac{d\theta}{dt} = 1, \tag{4.18}$$

where
$$g^m \leq g(r, \theta, \alpha) \leq g^M \tag{4.19}$$

at all $(r, \theta, \alpha) \in R_+ \times [0, 2\pi] \times S$ and $g^m < g^M$ are given constants.

Note that the solution $r = 0$ of the equations of the first approximation (4.18) is unstable in the Lyapunov sense, since the linear approximation $\dfrac{dr}{dt} = \mu r$ has an eigenvalue $\lambda = \mu > 0$.

Along with the system (4.18) consider the function $V = r^2$.

For the derivative dV/dt, in view of the system (4.18),

$$\left. \frac{dV}{dt} \right|_{(4.18)} = 2r \frac{dr}{dt} = 2r^2 [\mu - g(r, \theta, \alpha)r^2], \quad \alpha \in S.$$

Hence at any function $g(r, \theta, \alpha)$ satisfying the condition (4.19) at all $t \geq 0$, the following inequalities hold true

$$\left. \frac{dV}{dt} \right|_{(4.18)} < 0 \quad \text{at} \quad r^2 > \frac{\mu}{g(r, \theta, \alpha)},$$

$$\left. \frac{dV}{dt} \right|_{(4.18)} = 0 \quad \text{at} \quad r^2 = \frac{\mu}{g(r, \theta, \alpha)},$$

$$\left. \frac{dV}{dt} \right|_{(4.18)} > 0 \quad \text{at} \quad r^2 < \frac{\mu}{g(r, \theta, \alpha)},$$

then the moving set

$$\mathcal{A}^*(\varkappa) = \left\{ r \colon r^2 = \frac{\mu}{g(r, \theta, \alpha)} \right\}, \quad \alpha \in S,$$

is uniformly asymptotically stable.

Now consider the motion of the system (4.18) with respect to the domains:

$$S_1 = \{ r \colon r^2 < H \}, \quad 0 < H < \infty,$$

$$S_2 = \{ r \colon r^2 \leq \delta \}, \quad \delta = \left(\frac{\mu}{g^M} \right)^{1/2},$$

$$S_3 = \{ r \colon r^2 \geq \eta \}, \quad \eta = \left(\frac{\mu}{g^m} \right)^{1/2},$$

under the limitations (4.19).

Let the motion of the system (4.18) begin outside the ring of radius $r_0 + \sigma$, where $r_0 = \left(\dfrac{\mu}{g^m} \right)^{1/2}$, σ is an arbitrarily small constant. Since

$$\left. \frac{dV}{dt} \right|_{(4.18)} = 2\mu V - 4g(r, \theta, \alpha)V^2,$$

then the estimate of the time interval within which the solutions of the system (4.18) will reach the moving surface

$$r^2 = \frac{\mu}{g(r, \theta, \alpha)}$$

is obtained from the inequality

$$\tau \leq \int_{\varkappa_1}^{\varkappa} \frac{dc}{2\mu c - 4g^m c^2}, \qquad (4.20)$$

where $\varkappa_1 < \varkappa$, $\varkappa_1 = \frac{1}{2} r^2$ and $\varkappa = \frac{1}{2}(r_0 + \sigma)^2$. From the estimate (4.20) it follows that

$$\tau \leq \frac{1}{2\mu} \ln \left| \frac{(r_0 + \sigma)^2}{r^2} \frac{(r^2 - r_0^2)}{2 r_0 \sigma + \sigma^2} \right|.$$

In a similar way we estimate the time interval sufficient for the solutions beginning in the domain $r^* - \sigma \geq 0$, where $r^* = \left(\frac{\mu}{g^M}\right)^{1/2}$, to reach the moving surface $r^2 = \dfrac{\mu}{g(r, \theta, \alpha)}$.

Note that the function $g(r, \theta, \alpha)$, $\alpha \in \mathcal{S} \subseteq R^d$, is not assumed to be continuously differentiable, therefore the equation (4.18) can effectively be analyzed by the qualitative method, while its direct integration is difficult.

4.4   Quasilinear Systems with Nonautonomous Uncertainties

Consider the system (4.10) whose uncertainties $\alpha = \alpha(t)$ are time-variable, and the vector-function $\Phi(z, \alpha(t))$ at all $\alpha(t) \in \mathcal{S}$ can be represented in the form

$$\Phi(z, \alpha(t)) = \Phi(z) + R(z, \alpha(t)),$$

where the components of $\Phi(z)$ are power-series of z, beginning from the terms not lower than the second order and converging absolutely in the neighborhood $D(a) \subset R^n$ of the equilibrium state $z = 0$; the vector-function $R(z, \alpha(t))$ has the sense of uncertain nonautonomous continuously acting perturbations.

Assume that $R(z, \alpha(t)) \neq 0$, $t \geq 0$, and for the components R_i the estimates $|R_i(z, \alpha(t))| \leq \tilde{\sigma}_i(t)$ hold at all $(z, \alpha) \in D(a) \times \mathcal{S}$, where $\tilde{\sigma}_i(t) \in C(R, R_+)$, $i = 1, 2, \ldots, n$, are positive bounded functions on any finite interval.

In this case the system (4.10) takes the form

$$\frac{dz}{dt} = Pz + \Phi(z) + R(z, \alpha(t)). \qquad (4.21)$$

Analyze the behavior of the solutions of this system in the neighbourhood of the state $z = 0$.

As above, using the transformation $x = Tz$ ($\det T \neq 0$), reduce the system (4.21) to the diagonal form

$$\frac{dx_s}{dt} = \lambda_s x_s + X_s(x_1, \ldots, x_n) + \Pi_s(x_1, \ldots, x_n, \alpha(t)), \qquad (4.22)$$

where X_s and Π_s are components of the vector-functions $T^{-1}\Phi(T^{-1}x)$ and $T^{-1}R(t, T^{-1}x, \alpha(t))$ respectively.

Represent the variables x_i in the form

$$x_s = p_s \exp(i\theta_s), \qquad \overline{x}_s = p_s \exp(-i\theta_s),$$

and obtain

$$\frac{dp_s}{dt} = \frac{1}{2}(\dot{x}_s \exp(-i\theta_s) + \dot{\overline{x}}_s \exp(i\theta_s)), \qquad s = 1, 2, \ldots, n. \qquad (4.23)$$

Taking into account the equations (4.22), write the relations (4.23) in the form

$$\frac{dp_s}{dt} = \alpha_s p_s + \frac{1}{2}(X_s e^{-i\theta_s} + \overline{X}_s e^{i\theta_s}) + \frac{1}{2}(\Pi_s e^{-i\theta_s} + \overline{\Pi}_s e^{i\theta_s}), \qquad (4.24)$$

where $\alpha_s = \operatorname{Re} \lambda_s$, $s = 1, 2, \ldots, n$.

For the derivative of the function

$$V = \sum_{s=1}^{n'} |x_s| = \sum_{s=1}^{n'} p_s, \qquad n' \leq n,$$

along the solutions of the system (4.24) one can easily obtain the expression

$$\frac{dV}{dt} = \sum_{s=1}^{n'} \alpha_s p_s + \frac{1}{2}\sum_{s=1}^{n'}(X_s e^{-i\theta_s} + \overline{X}_s e^{i\theta_s}) + \frac{1}{2}\sum_{s=1}^{n'}(\Pi_s e^{-i\theta_s} + \overline{\Pi}_s e^{i\theta_s})$$

and the estimate

$$\frac{dV}{dt} \leq \alpha V + \beta V^2 + \Pi(t), \qquad t \in R_+, \qquad (4.25)$$

in the domain $D(a) \subset R^n$. Here $\alpha = \max_s \alpha_s$ and $\beta = \sum_{s=1}^{n'} \beta_s$, β_s are some positive constants, $\Pi(t) = \sum_{s=1}^{n'}\sum_{i=1}^{n} |t_{si}|\,\tilde{\sigma}_i(t)$.

To the inequality (4.25) the following comparison equation corresponds

$$\frac{du}{dt} = \alpha u + \beta u^2 + \Pi(t), \qquad u(t_0) = u_0 \geq 0, \qquad (4.26)$$

which is not integrable in quadratures.

By an appropriate choice of the constants $b > 0$, $c > 0$ we can secure

$$\alpha u + \beta u^2 - b(e^{-cu} - 1) \leq 0 \tag{4.27}$$

in the domain $D(a^*) \subseteq D(a)$.

Then, taking into account (4.27), determine the solution of the comparison equation (4.26) by the solution of the equation

$$\frac{dw}{dt} = b(e^{-cw} - 1) + \Pi(t), \quad w(t_0) = w_0 \geq 0. \tag{4.28}$$

This equation has a solution of the form

$$w(t) = \psi(t) + \frac{1}{c} \ln\left\{ \exp(cw_0) + bc \int_0^t \exp[-c\psi(s)] \, ds \right\},$$

where $\psi(t) = \int_0^t \Pi(s)ds - bt$ at all $t \in R_+$.

From the inequalities (4.25) and (4.27) it follows that at the initial values $t_0 = 0$ and $w_0 \leq V_0$ the estimate $V(x(t)) \leq w(t)$ holds true in the domain $D(a^*)$ at all $t \geq 0$.

Hence the unperturbed motion of the uncertain system (4.21) has a certain type of stability in the domain $D(a^*)$, if at $|V_0| < \delta$, $\delta = \delta(a^*)$ the inequality holds true:

$$\psi(t) + \frac{1}{c} \ln\left\{ \exp(cw_0) + bc \int_0^t \exp[-c\psi(s)] \, ds \right\} < a^*, \tag{4.29}$$

where $a^* = \min\{a, -\frac{3\alpha}{8\beta}\}$ and a is a predefined quantity determining the open domain $D(a) \subseteq R^n$.

Now analyze the dynamic behavior of solutions of the system (4.21) under some concrete assumptions.

Case A. Assume that in the system (4.21) the vector-function $R(z, \alpha(t)) \equiv 0$ at all $t \geq 0$. In this case the function $\Pi(t) = 0$ and the condition (4.29) take the form

$$-bt + \frac{1}{c} \ln\{\exp(cw_0) + \exp(cbt) - 1\} < a^*.$$

Hence any solution of the system (4.21) with initial conditions from the domain $w_0 \leq V_0$ asymptotically approximates the equilibrium state $z = 0$ with the estimate $V(x(t)) \leq w(t, w_0)$, where

$$w(t, w_0) = \frac{1}{c} \ln\{\exp[c(w_0 - bt)] - \exp(-cbt) + 1\} \to 0 \quad \text{at} \quad t \to +\infty.$$

In fact this case corresponds to the absence of uncertainties in the system (4.21).

Case B. Let the uncertainties be "small" in the system (4.21), and the estimating function $\Pi(t)$ be bound by some constant $g > 0$. If $\Pi(t) \leq g$ at all $t \geq 0$, then from the estimate (4.29) obtain

$$\exp[-c(b-g)t]\,[\exp(cw_0) - b/(b-g)] < \exp(ca^*) - b/(b-g). \qquad (4.30)$$

Let the uncertainties $\alpha(t) \in \mathcal{S}$ be such that

$$\exp(ca^*) - b/(b-g) \geq 0.$$

Then the inequality

$$g \leq b[1 - \exp(-ca^*)]$$

is the estimate of uncertainties in the domain $v = \sum\limits_{s=1}^{n'} |z_s^o| \leq w_0$, under which the inequality (4.30) holds true at all $t \geq 0$.

Thus, if the uncertainties $\alpha(t) \in \mathcal{S}$ in the system (4.21) are such that

$$\Pi(t) = \sum_{s=1}^{n'}\sum_{i=1}^{n} |t_{si}|\,|R_i(z, \alpha(t))| \leq g$$

at all $t \geq 0$, then any solution $z(t, z_0)$, remaining in the domain $v \leq a^*$, at all $t \geq 0$ will asymptotically approximate the equilibrium state $z = 0$ according to the estimate $v(t, v_0) \leq w(t, w_0)$, where

$$w(t, w_0) = \frac{1}{c}\ln\{\exp[-c(a-g)t](\exp(cw_0) - b/(b-g)) + b/(b-g)\}$$

and $\lim\limits_{t \to \infty} w(t, w_0) = \frac{1}{c}\ln(b/(b-g))$.

Case C. Let the motion of the system (4.21) begin from the zero state, i. e., at $t_0 = 0$ the initial values $z(t_0) = 0$ and $w_0 = 0$ for the comparison equation (4.28). The solution of the system (4.21) is estimated by the inequality $v(z(t)) \leq w(t)$ at all $t \geq 0$, where

$$w(t) = \psi(t) + \frac{1}{c}\ln\left\{1 + bc\int_0^t \exp[-c\psi(s)]\,ds\right\}.$$

In the system (4.21) such uncertainties $\alpha(t) \in \mathcal{S}$, are admissible that $w(t) < \tilde{a}^*$ at all $t \geq 0$, where $\tilde{a}^* \leq a^*$.

4.5 Synchronizing of Motions in Uncertain Quasilinear Systems

The problem of synchronizing of motions of uncertain systems is connected with the requirements for engineering applications in such areas as safe information transfer, the modeling of neuron nets, the description of processes in laser dynamics, etc.

In the general context of the qualitative analysis of solutions of uncertain systems, in this paragraph the direct Lyapunov's method based on the simplest auxiliary function is applied to the above mentioned problem.

Consider two uncertain systems of the form

$$\frac{dx}{dt} = (A + \Delta A_1(t))x + f(t, x) + r_1(t), \tag{4.31}$$

where $x \in R^n$, A is an $(n \times n)$-constant matrix, $\Delta A_1(t)$ is an $(n \times n)$-matrix of uncertainties, satisfying the condition $\|\Delta A_1(t)\| \le \delta_1$, $\delta_1 - \text{const} > 0$, $r_1(t)$ is an external perturbation with the estimate $\|r_1(t)\| \le d_1$, $d_1 = \text{const} > 0$ at all $t \ge t_0$ and

$$\frac{d\overline{x}}{dt} = (A + \Delta A_2(t))\overline{x} + f(t, \overline{x}) + r_2(t) + B(\overline{y} - y) + \sigma,$$
$$y = Kx, \quad \overline{y} = K\overline{x}. \tag{4.32}$$

Here $\overline{x} \in R^n$, $\Delta A_2(t)$ is an $(n \times n)$-matrix of uncertainties with the estimate $\|\Delta A_2(t)\| \le \delta_2$, $\delta_1 = \text{const} > 0$, $r_2(t)$ is an external perturbation with the estimate $\|r_2(t)\| \le d_2$, $d_2 = \text{const} > 0$ at all $t \ge t_0$, $K \in R^{1 \times n}$ is the vector coefficient of the strengthening of feedback, $B \in R^{n \times 1}$ the control vector such that the pair (A, B) is controlled, $\sigma \in R^{n \times 1}$ is a nonlinear input vector.

Assumption 4.3 The reference (master) (4.31) and the operative (slave) (4.32) systems are such that:

(1) the vector-function $f(t, x)$ is continuous in the domain $R_+ \times D$, $D \subset R^n$, and satisfies the Lipshitz condition

$$\|f(t, x) - f(t, y)\| \le L\|x - y\|,$$

at all $(x, y) \in D$ with the constant $L > 0$;

(2) there exist positive constants δ_3 and d such that

    (a) $\|\Delta A_1(t) - \Delta A_2(t)\| \le \delta_3$ at any $\Delta A_i(t) \in \mathcal{S}$, $i = 1, 2$,

    (b) $\|r_1(t) - r_2(t)\| \le d$ at all $t \ge t_0$.

Now let $e = \bar{x} - x$, $e \in R^n$ denote the mismatch between the state vectors of the systems (4.31), and upon simple transformations obtain

$$\frac{de}{dt} = (A + BK)e + f(t, \bar{x}) - f(t, x) + $$
$$+ r_2(t) - r_1(t) + \Delta A_2(t)\bar{x} - \Delta A_1(t)x + \sigma(e). \tag{4.33}$$

Set limitations for the vectors K, B and the nonlinear input σ, under which the solutions of the operative system (4.32) asymptotically approximate the solutions of the reference system (4.31), i.e., there occurs synchronization of motions in the sense of the limit relation

$$\lim_{t \to \infty} \|e(t, t_0, e_0)\| = 0 \quad \text{at} \quad (t_0, e_0) \in \text{int} \, (R_+ \times D).$$

Now we will need the following Cauchy inequality with ε.

Lemma 4.2 (Ladyzhenskaya [1973]) *Let x and y be vectors whose dimensions are matched with those of the matrices P and $Q(t)$, $\|Q(t)\| \leq r$, $r = \text{const} > 0$, at all $t \geq t_0$. Then for the bilinear form $2x^T P^T Q(t)y$ the following estimate holds true*

$$2x^T P^T Q(t)y \leq \varepsilon x^T P^T Px + \frac{r^2}{\varepsilon} y^T y,$$

where $\varepsilon > 0$ is an arbitrary constant.

Proof From

$$\left(\sqrt{\varepsilon}Px - \frac{1}{\sqrt{\varepsilon}}Q(t)y\right)^T \left(\sqrt{\varepsilon}Px - \frac{1}{\sqrt{\varepsilon}}Q(t)y\right) \geq 0,$$

the required estimate follows:

$$x^T P^T Q(t)y + y^T Q^T(t)Px \leq \varepsilon x^T P^T Px + \frac{1}{\varepsilon} y^T Q^T(t)Q(t)y$$
$$\leq \varepsilon x^T P^T Px + \frac{r^2}{\varepsilon} y^T y.$$

We now show that the following statement is true.

Theorem 4.3 *Let the system (4.32) be such that:*

(1) *the pair (A, B) is controllable;*

(2) *all the conditions of the Assumption 4.3 are satisfied;*

(3) *there exist positive constants* ε_i, $i = 1, 2, 3, 4$, *and* $\eta > 0$ *such that*

$$P(A + BK) + (A + BK)^{\mathrm{T}}P + \sum_{i=1}^{4} \varepsilon_i P^2 + (\varepsilon_1^{-1}\delta_2^2 + \varepsilon_3^{-1}L^2)I + 2\eta P < 0,$$

(4.34)

$$\sigma(e) = \frac{-(\varepsilon_2^{-1}\delta_3^2 x^{\mathrm{T}} x + \varepsilon_4^{-1}d^2)}{2\|e\|^2} P^{-1}e,$$

(4.35)

where P is a positive-definite symmetrical matrix, and I is a unit matrix.

Then the solution $e(t, t_0, x_0)$ of the system (4.33) exponentially decreases, and the motions of the pair of systems (4.31) and (4.32) are exponentially synchronized.

Proof For the matrix P satisfying the inequality (4.34) consider the quadratic form $V(e) = e^{\mathrm{T}}Pe$ and calculate its full derivative in view of the system (4.33)

$$\begin{aligned}
\frac{dV}{dt}\bigg|_{(4.33)} &= \big((A + BK)e + f(t, \bar{x}) - f(t, x) + r_2(t) - r_1(t) + \Delta A_2(t)\bar{x} \\
&\quad - \Delta A_1(t)x + \sigma\big)^{\mathrm{T}}Pe + e^{\mathrm{T}}P\big((A + BK)e + f(t, \bar{x}) - f(t, x) \\
&\quad + r_2(t) - r_1(t) + \Delta A_2(t)\bar{x} - \Delta A_1(t)x + \sigma\big) \\
&= e^{\mathrm{T}}\big(P(A + BK) + (A + BK)^{\mathrm{T}}P\big) + \Delta A_2^{\mathrm{T}}P + P\Delta A_2(t)e \\
&\quad + 2(f(t, \bar{x}) - f(t, x))^{\mathrm{T}}Pe + 2e^{\mathrm{T}}P\sigma \\
&\quad + 2e^{\mathrm{T}}P(\Delta A_2(t) - \Delta A_1(t))x + 2e^{\mathrm{T}}P(r_2(t) - r_1(t)).
\end{aligned}$$

(4.36)

Then, applying the estimate from the Lemma 4.2, for the summands of the right-hand part of the relation (4.36) obtain

$$\begin{aligned}
2e^{\mathrm{T}}P\Delta A_2(t)e &\le \varepsilon_1 e^{\mathrm{T}}P^2 e + \varepsilon_1^{-1}\delta_2^2 e^{\mathrm{T}}e, \\
2e^{\mathrm{T}}P(\Delta A_2(t) - A_1(t))x &\le \varepsilon_2 e^{\mathrm{T}}P^2 e + \varepsilon_2^{-1}\delta_3^2 x^{\mathrm{T}}x, \\
2(f(t, \bar{x}) - f(t, x))^{\mathrm{T}}Pe &\le \varepsilon_3 e^{\mathrm{T}}P^2 e + \varepsilon_3^{-1}L^2 e^{\mathrm{T}}e, \\
2e^{\mathrm{T}}P(r_2(t) - r_1(t)) &\le \varepsilon_4 e^{\mathrm{T}}P^2 e + \varepsilon_4^{-1}d^2.
\end{aligned}$$

(4.37)

Taking into account the inequalities (4.37), obtain the estimate of the full derivative of the function $V(e)$:

$$\begin{aligned}
\frac{dV}{dt}\bigg|_{(4.33)} &\le e^{\mathrm{T}}\Big(P(A + BK) + (A + BK)^{\mathrm{T}}P + \sum_{i=1}^{4}\varepsilon_i P^2 \\
&\quad + (\varepsilon_1^{-1}\delta_2^2 + \varepsilon_3^{-1}L^2)I\Big)e \le -2\eta e^{\mathrm{T}}Pe = -2\eta V(e) < 0.
\end{aligned}$$

Hence

$$V(e(t)) \le V(e_0)\exp(-2\eta(t - t_0)), \quad t \ge t_0.$$

Thus, for any $\varepsilon > 0$ and $\delta(\varepsilon) = \dfrac{\lambda_m^{1/2}(P)}{\lambda_M^{-1/2}(P)} \varepsilon$ the estimate

$$\|e(t, t_0, e_0)\| \leq \varepsilon \exp(-\eta(t - t_0)), \quad t \geq t_0, \tag{4.38}$$

is true if $\|e_0\| < \delta(\varepsilon)$.

The estimate (4.38) proves the statement of the Theorem 4.3.

Remark 4.2   From the expression (4.35) it follows that $\sigma(e) \to \infty$ at $e(t, t_0, e_0) \to \infty$. However, if for the error $e(t)$ some lower boundary β is taken, then the expression $\sigma(e)$ can be substituted by $\sigma^*(t)$:

$$\sigma^*(t) = \begin{cases} \sigma(e) & \text{at } \|e\| \geq \beta, \\[2mm] \dfrac{-(\varepsilon_2^{-1}\delta_3^2 x^{\mathrm{T}} x + \varepsilon_4^{-1} d^2)}{2r^2} P^{-1} e & \text{at } \|e\| < \beta. \end{cases}$$

The Theorem 4.3 implies the following statement.

Corollary 4.1 *Let for the system* (4.33) *the conditions of the 4.3 be satisfied, and*

(1) *the pair* (A, B) *be controllable;*

(2) *there exist positive constants* ε_i, $i = 1, 2, 3, 4$, *and matrices* $X > 0$ *and* W *of the respective dimensions, such that*

$$\begin{pmatrix} AX + XA^{\mathrm{T}} + BW^{\mathrm{T}} + WB^{\mathrm{T}} + 2\eta X + \sum_{i=1}^{4} \varepsilon_i I & \delta_2 X & LX \\ \delta_2 X & -\varepsilon_1 I & 0 \\ LX & 0 & -\varepsilon_3 I \end{pmatrix} < 0.$$

Then the solution $e(t, t_0, e_0)$ *of the system* (4.33) *with the feedback* $K = W^{\mathrm{T}} X^{-1}$ *and the nonlinear input* $\sigma^*(e)$ *exponentially decrease and the motions of the pair of the systems* (4.31) *and* (4.32) *are exponentially synchronized.*

Example 4.2   Consider the uncertain Lorenz system (see Lorenz [1963])

$$\begin{aligned} \frac{dx_1}{dt} &= \sigma_1(y_1 - x_1), \\ \frac{dy_1}{dt} &= -r_1 x_1 - x_1 z_1 - y_1, \\ \frac{dz_1}{dt} &= x_1 y_1 - b_1 z_1, \end{aligned} \tag{4.39}$$

where the parameters σ_1, r_1, b_1 take on values from some compact sets. Along with the system (4.39), consider the system

$$\frac{dx_2}{dt} = \sigma_2(y_2 - x_2),$$

$$\frac{dy_2}{dt} = r_2 x_2 - x_1 z_1 - y_2 + u(x_2 - x_1), \qquad (4.40)$$

$$\frac{dz_2}{dt} = x_1 y_2 - b_2 z_2,$$

where σ_2, r_2 and b_2 are the uncertainties parameters of the system.

Let $\tilde{\sigma} = \sigma_2 - \sigma_1$, $\tilde{r} = r_2 - r_1$ and $\tilde{b} = b_2 - b_1$ be the uncertainties in the determination of parameters of the Lorenz system, and $e_1 = x_2 - x_1$ and let $e_2 = y_2 - y_1$, $e_3 = z_2 - z_1$ be the mismatch between the state vectors of the systems (4.39) and (4.40). It is not difficult to obtain the system

$$\frac{de_1}{dt} = \sigma_2(e_2 - e_1) + \tilde{\sigma}(y_1 - x_1),$$

$$\frac{de_2}{dt} = r_2 e_1 + \tilde{r} x_1 - x_1 e_3 - e_2 + u e_1, \qquad (4.41)$$

$$\frac{de_3}{dt} = x_1 e_1 - b_2 e_3 - \tilde{b} z_1.$$

Let the system

$$\frac{d\tilde{\sigma}}{dt} = -(y_1 - x_1)e_1,$$

$$\frac{d\tilde{r}}{dt} = -x_1 e_2, \qquad (4.42)$$

$$\frac{d\tilde{b}}{dt} = z_1 e_3$$

describe the variation of fluctuations in the values of parameters of the systems (4.39) and (4.40). Along with the system (4.41) consider the function $V(e) = \frac{1}{2}(e_1^2 + e_2^2 + e_3^2 + \tilde{\sigma}^2 + \tilde{r}^2 + \tilde{b}^2)$ and its full derivative, taking into account the equations (4.42), in the form

$$\left.\frac{dV}{dt}\right|_{(4.41)} = -\sigma_2 e_1^2 + \sigma_2 e_2 e_1 + r_2 e_1 e_2 - e_2^2 + u e_1 e_2 - b_2 e_3^2.$$

If the control parameter u is chosen in the form $u = -(\sigma_2 + r_2)$, then

$$\left.\frac{dV}{dt}\right|_{(4.41)} = -\sigma_2 e_1^2 - e_2^2 - b_2 e_3^2. \qquad (4.43)$$

The function $V(e)$ is positive-definite, while the function (4.43) at $\sigma_2 > 0$ and $b_2 > 0$ is negative-definite. Therefore $\lim \|e(t, t_0, e_0)\| = 0$ at $t \to \infty$ and motions of the systems (4.39) and (4.40) are asymptotically synchronized under the above mentioned conditions.

Example 4.3 Consider an uncertain system of the form (see Chen [1996], Ming-Chung, Ho et al. [2007])

$$\frac{dx}{dt} = a(y - x),$$

$$\frac{dy}{dt} = (c - a)x - xz + cy, \qquad (4.44)$$

$$\frac{dz}{dt} = xy - bz,$$

where $a > 0$, $b > 0$, $c > 0$ and $c < a < 2c$.

Then for the system (4.44) consider two systems: the reference system

$$\frac{dx_1}{dt} = a_1(y_1 - x_1),$$

$$\frac{dy_1}{dt} = (c_1 - a_1)x_1 - x_1 z_1 + c_1 y_1, \qquad (4.45)$$

$$\frac{dz_1}{dt} = x_1 y_1 - b_1 z_1,$$

where the parameters a_1, b_1, c_1 are uncertain, and the operative system

$$\frac{dx_2}{dt} = a_2(y_2 - x_2),$$

$$\frac{dy_2}{dt} = (c_2 - a_2)x_2 - x_1 z_2 + c_2 y_2 + u(y_2 - y_1), \qquad (4.46)$$

$$\frac{dz_2}{dt} = x_1 y_2 - b_2 z_2,$$

where the parameters a_2, b_2, c_2 also are uncertain, and the control $u \in R$.

Denote $\tilde{a} = a_2 - a_1$, $\tilde{b} = b_2 - b_1$, $\tilde{c} = c_2 - c_1$; $e_1 = x_2 - x_1$, $e_2 = y_2 - y_1$, $e_3 = z_2 - z_1$ and, using simple calculation, obtain the system

$$\frac{de_1}{dt} = a_2(e_2 - e_1) + \tilde{a}(y_1 - x_1),$$

$$\frac{de_2}{dt} = (c_2 - a_2)e_1 + (\tilde{c} - \tilde{a})x_1 - x_1 e_3 + c_2 e_2 + \tilde{c}y_1 + ue_2, \qquad (4.47)$$

$$\frac{de_3}{dt} = x_1 e_2 - b_2 e_3 - \tilde{b}z_1.$$

The system of equations describing the parameter variation in the systems (4.45) and (4.46) has the form

$$\frac{d\tilde{a}}{dt} = x_1 e_1 - (y_1 - x_1)e_1,$$

$$\frac{d\tilde{b}}{dt} = z_1 e_3, \qquad (4.48)$$

$$\frac{d\tilde{c}}{dt} = -e_2(x_1 + y_1).$$

For the function $V(e) = \frac{1}{2}(e_1^2 + e_2^2 + e_3^2 + \tilde{a}^2 + \tilde{b}^2 + \tilde{c}^2)$, taking into account the equations (4.48),

$$\left.\frac{dV}{dt}\right|_{(4.47)} = -a_2 e_1^2 + c_2 e_1 e_2 + (c_2 + u)e_2^2 - b_2 e_3^2.$$

If we choose $u = -\dfrac{c_2^2 + 4a_2 c_2}{4a^2} - k^2 (k > 0)$, then $-a_2 e_1^2 + c_2 e_1 e_2 + (c_2 + u)e_2^2 < 0$ and therefore $dV/dt|_{(4.47)} < 0$. Since $V(e) > 0$, $\lim \|e(t, t_0, e_0)\| = 0$ at $t \to \infty$ and the motions of the systems (4.45) and (4.46) are asymptotically synchronized.

Chapter 5

Stability of Large-Scale Uncertain Systems

Different engineering systems with branched and flexible structures, such as aerospace systems, power networks, economic and ecological systems, are described by systems of equations of large dimensions, whose parameters are defined uncertainly. The presence of "uncertainties" in the mathematical model of a system of such kind essentially complicates the analysis of the stability of its operation.

In this chapter we will discuss some approaches to the solution of the above mentioned problem. They are based on the results of Chapter 2 and use the ideas of hierarchical and vector Lyapunov functions.

5.1 Description of a Large-Scale System

Assume that the system (2.1) has all criteria of a large-scale system, among which there is the high dimension of the system of equations of perturbed motion and the presence of subsystems interacting with each other. Let the system (2.1) be decomposed in some way into s interacting subsystems

$$\frac{dx_i}{dt} = f_i(t, x_i) + g_i(t, x_1, \ldots, x_s, \alpha), \quad i = 1, 2, \ldots, s, \tag{5.1}$$

where $x_i \in R^{n_i}$, $t \in R_+$, $f_i \in C(R_+ \times R^{n_i}, R^{n_i})$, $g_i \in C(R_+ \times R^{n_1} \times \ldots \times R^{n_s} \times \mathcal{S}, R^{n_i})$, and $n_1 + n_2 + \cdots + n_s = n$. As in the case of the system (2.1), assume that $f_i(t, 0) = 0$ and $g_i(t, 0, \ldots, 0, \alpha) = 0$ at $x_i = 0$, $i = 1, 2, \ldots, s$, and at any $\alpha \in \mathcal{S}$.

From the general form of the system (5.1) it follows that its nonlinear approximation (at $g_i(t, x_1, \ldots, x_s, \alpha) \equiv 0$) does not contain the uncertainties parameter $\alpha \in \mathcal{S} \subseteq R^d$, whereas the functions $g_i(t, x_1, \ldots, x_s, \alpha)$ are defined uncertainly. For example,

$$g_i(t, x_1, \ldots, x_s, \alpha) = \sum_{j=1, i \neq j}^{s} C_{ij}(\alpha) x_j, \quad i = 1, 2, \ldots, s, \tag{5.2}$$

where $C_{ij}(\alpha)$ are $n_i \times n_j$-matrices whose nonzero elements depend on the parameter $\alpha \in \mathcal{S} \subseteq R^d$ or

$$g_i(t, x_1, \ldots, x_s, \alpha) = \sum_{j=1, i \neq j}^{s} g_{ij}(t, x_j, \alpha), \quad i = 1, 2, \ldots, s, \qquad (5.3)$$

where $g_{ij} \in C(R_+ \times R^{n_j} \times \mathcal{S}, R^{n_i})$. The system (5.1), taking into account (5.2) and (5.3), takes the form

$$\frac{dx_i}{dt} = f_i(t, x_i) + \sum_{j=1, i \neq j}^{s} C_{ij}(\alpha) x_j, \quad \alpha \in \mathcal{S} \subseteq R^d,$$

or

$$\frac{dx_i}{dt} = f_i(t, x_i) + \sum_{j=1, i \neq j}^{s} g_{ij}(t, x_j, \alpha), \quad \alpha \in \mathcal{S} \subseteq R^d,$$

respectively.

We now determine the sufficient conditions of the uniform asymptotic stability of solutions of the system (5.1) with respect to the moving invariant set $A(r)$.

These conditions are connected with the existence of the auxiliary functions $v_i(t, x_i)$ for the subsystems

$$\frac{dx_i}{dt} = f_i(t, x_i), \quad i = 1, 2, \ldots, s, \qquad (5.4)$$

with special properties, and qualitative properties of the functions $g_i(t, x_1, \ldots, x_s, \alpha)$, $i = 1, 2, \ldots, s$, defined uncertainly.

Remark 5.1 The procedure of the mathematical decomposition of the system (2.1) into s interrelated subsystems (5.1) is a nontrivial problem. For its solution (in the case when the uncertainties parameter α in the system (2.1) is missing) special approaches have been developed (see Gruijć, Martynyuk, Ribbens-Pavella [1984]). Those approaches, though slightly modified, can be applied in the analysis of the uncertain systems (5.1).

Assume that for each of the subsystems (5.4) an auxiliary function $v_i(t, x_i)$, $i = 1, 2, \ldots, s$ is constructed. The function

$$V(t, x, c) = c^{\mathrm{T}} L(t, x), \qquad (5.5)$$

where $c_i \in R_+^s$ and $L = [v_1(t, x_1), \ldots, v_s(t, x_s)]^{\mathrm{T}}$, $v_i \in C(R_+ \times R^{n_i}, R_+)$. Under the known assumptions on the functions v_i (see Michel and Miller [1977, p. 26]) the function (5.5) is positive-definite and decreasing. Note that as a function $L(t, x)$ we can take the matrix-valued function, where $v_{ij}(t, x_i, x_j) \equiv 0$ at all $(i \neq j) \in [1, s]$, i.e.,

$$L(t, x) = \mathrm{diag}[v_{11}(t, x_1), \ldots, v_{ss}(t, x_s)].$$

The set $A(r)$ for the system (5.1) will be determined as follows:

$$A(r) = A_1(r) \cup A_2(r) \cup \cdots \cup A_s(r).$$

The properties of the motion of the subsystems (5.4) and the limitations on the change of the functions $g_i(t, x_1, \ldots, x_s, \alpha)$, $\alpha \in S \subseteq R^d$, are the main factors influencing the dynamics of the uncertain system (5.1).

5.2 Stability of Solutions with Respect to a Moving Set

To solve the problem we will utilize the following statements.

Theorem 5.1 *Let the following conditions be satisfied:*

(1) *for each $\alpha \in S \subseteq R^d$ there exists a function $r = r(\alpha) > 0$ such that the set $A(r)$ is not empty at all $\alpha \in S \subseteq R^d$;*

(2) *for each isolated subsystem (5.4) there exist continuously differentiable functions $v_i \colon R_+ \times R^{n_i} \to R_+$, functions $\psi_{i1}, \psi_{i2}, \psi_{i3} \in KR$-class, and constants $(\underline{\sigma}_i, \overline{\sigma}_i) \in R$ such that:*

(a) $\psi_{i1}(\|x_i\|) \le v_i(t, x_i)$ *at* $\|x_i\| > \frac{1}{s}r(\alpha)$,

(b) $v_i(t, x_i) \le \psi_{i2}(\|x_i\|)$ *at* $\|x_i\| \le \frac{1}{s}r(\alpha)$,

(c) $Dv_i(t, x_i)\big|_{(5.4)} < \overline{\sigma}_i \psi_{i3}(\|x_i\|)$ *at* $\|x_i\| > \frac{1}{s}r(\alpha)$, $\alpha \in S \subseteq R^d$,

(d) $Dv_i(t, x_i)\big|_{(5.4)} > \underline{\sigma}_i \psi_{i3}(\|x_i\|)$ *at* $\|x_i\| \le \frac{1}{s}r(\alpha)$, $\alpha \in S \subseteq R^d$;

(3) *at the given functions v_i and ψ_{i3} there exist constants $(\underline{a}_{ij}, \overline{a}_{ij}) \in R$ such that:*

(a) $\left(\dfrac{\partial v_i(t, x_i)}{\partial x_i} \right)^{\mathrm{T}} g_i(t, x_1, \ldots, x_s, \alpha) < [\psi_{i3}(\|x_i\|)]^{1/2} \sum_{j=1}^{s} \overline{a}_{ij}$

$\times [\psi_{j3}(\|x_j\|)]^{1/2}$ *at* $\|x_i\| > \frac{1}{s}r(\alpha)$,

(b) $Dv_i(t, x_i)|_{(5.4)} + \left(\dfrac{\partial v_i(t, x_i)}{\partial x_i} \right)^{\mathrm{T}} g_i(t, x_1, \ldots, x_s, \alpha) = 0$

at $\|x_i\| = \frac{1}{s}r(\alpha)$,

(c) $\left(\dfrac{\partial v_i(t, x_i)}{\partial x_i} \right)^{\mathrm{T}} g_i(t, x_1, \ldots, x_s, \alpha) > [\psi_{i3}(\|x_i\|)]^{1/2} \sum_{j=1}^{s} \underline{a}_{ij}$

$\times [\psi_{j3}(\|x_j\|)]^{1/2}$ *at* $\|x_i\| < \frac{1}{s}r(\alpha)$;

(4) *for the given constants* $(\underline{\sigma}_i, \overline{\sigma}_i)$ *there exists a vector* $c^{\mathrm{T}} = (c_1, \ldots, c_s) > 0$
such that:

(a) *the matrix* $\overline{Q} = [\overline{q}_{ij}]$ *with the elements*

$$\overline{q}_{ij} = \begin{cases} c_i(\overline{\sigma}_i + \overline{a}_{ij}) & at \ i = j, \\ (c_i\overline{a}_{ij} + c_j\overline{a}_{ij})/2 & at \ i \neq j \end{cases}$$

is negative-definite,

(b) *the matrix* $\underline{Q} = [\underline{q}_{ij}]$ *with the elements*

$$\underline{q}_{ij} = \begin{cases} c_i(\underline{\sigma}_i + \underline{a}_{ij}) & at \ i = j, \\ (c_i\underline{a}_{ij} + c_j\underline{a}_{ij})/2 & at \ i \neq j \end{cases}$$

is positive-definite;

(5) *for the given function* $r(\alpha) > 0$ *and the functions*

$$\psi_1(\|x\|) \leq \sum_{i=1}^{s} c_i\psi_{i1}(\|x_i\|) \quad and \quad \psi_2(\|x\|) \geq \sum_{i=1}^{s} c_i\psi_{i2}(\|x_i\|)$$

the following relation is true:

$$\psi_1(r) = \psi_2(r).$$

Then the set $A(r)$ *is invariant for the solutions of the system* (5.1), *and the motions* $x(t, \alpha)$ *of the system* (5.1) *are uniformly asymptotically stable with respect to* $A(r)$.

Proof Consider the motion of the system (5.1), beginning at $t_0 \in R_+$ and $\|x_0\| < r(\alpha) + \delta$.

For the system (5.1) apply the Lyapunov function (5.5):

$$V(t, x, c) = \sum_{i=1}^{s} c_i v_i(t, x_i), \qquad (5.6)$$

where $v_i(t, x_i)$ satisfies the conditions (2) (a), (b) of the Theorem 5.1, and $c_i > 0$ at all $i = 1, 2, \ldots, s$. Hence the function $V(t, x, c)$ satisfies the estimates

(a) $\displaystyle\sum_{i=1}^{s} c_i\psi_{i1}(\|x_i\|) \leq V(t, x, c) \quad at \quad \|x\| > r(\alpha),$

(b) $\displaystyle V(t, x, c) \leq \sum_{i=1}^{s} c_i\psi_{i2}(\|x_i\|) \quad at \quad \|x\| \leq r(\alpha).$

Since according to the Theorem 5.1 the functions ψ_{i1}, ψ_{i2} belong to the *KR*-class, it is possible to select functions $\psi_1(\|x\|)$ and $\psi_2(\|x\|)$ of the *KR*-class, such that

$$\psi_1(\|x\|) \leq \sum_{i=1}^s c_i \psi_{i1}(\|x_i\|) \quad \text{at} \quad \|x_i\| > \frac{1}{s} r(\alpha),$$

$$\psi_2(\|x\|) \geq \sum_{i=1}^s c_i \psi_{i2}(\|x_i\|) \quad \text{at} \quad \|x_i\| \leq \frac{1}{s} r(\alpha)$$

and therefore

(a) $\psi_1(\|x\|) \leq V(t, x, c) \quad \text{at} \quad \|x\| > r(\alpha),$

(b) $V(t, x, c) \leq \psi_2(\|x\|) \quad \text{at} \quad \|x\| \leq r(\alpha).$

Now show that under the conditions (2)(c), (3)(a), and (4)(a) of the Theorem 5.1 the function (5.6) along the solutions of the system (5.1) is monotone decreasing at all $t \in R_+$ and at $\|x\| > r(\alpha)$.

Indeed,

$$\left. \frac{dV}{dt} \right|_{(5.1)} = \sum_{i=1}^s \left\{ c_i \left[\frac{\partial v_i(t, x_i)}{\partial t} + \left(\frac{\partial v_i(t, x_i)}{\partial x_i} \right)^{\mathrm{T}} f_i(t, x_i) \right] \right.$$

$$\left. + c_i \left[\left(\frac{\partial v_i(t, x_i)}{\partial x_i} \right)^{\mathrm{T}} g_i(t, x_1, \ldots, x_s, \alpha) \right] \right\}$$

$$= \sum_{i=1}^s \left\{ c_i D v_i(t, x_i) |_{(5.4)} + c_i \left[\left(\frac{\partial v_i(t, x_i)}{\partial x_i} \right)^{\mathrm{T}} g_i(t, x_1, \ldots, x_s, \alpha) \right] \right\}$$

$$< \sum_{i=1}^s c_i \bar{\sigma}_i \psi_{i3}(\|x_i\|) + c_i [\psi_{i3}(\|x_i\|)]^{1/2} \sum_{j=1}^s \bar{a}_{ij} [\psi_{j3}(\|x_j\|)]^{1/2}$$

(5.7)

at $\|x_i\| > \frac{1}{s} r(\alpha)$ and $\alpha \in \mathcal{S}$.

Denote $w^{\mathrm{T}} = [\psi_{13}(\|x_1\|)^{1/2}, \ldots, \psi_{s3}(\|x_s\|)^{1/2}]$ and let $\overline{K} = [\bar{k}_{ij}]$ be an $(s \times s)$-matrix with the elements

$$\bar{k}_{ij} = \begin{cases} c_i [\bar{\sigma}_i + \bar{a}_{ii}] & \text{at } i = j, \\ c_i \bar{a}_{ij} & \text{at } i \neq j. \end{cases}$$

From (5.7) obtain

$$\left. \frac{dV}{dt} \right|_{(5.1)} < w^{\mathrm{T}} \left(\frac{1}{2} [\overline{K} + \overline{K}^{\mathrm{T}}] \right) w = w^{\mathrm{T}} \overline{Q} w$$

(5.8)

at $\|x\| > r(\alpha) \quad \text{and} \quad \alpha \in \mathcal{S}.$

Since the matrix $\overline{Q}$ is symmetrical, all its eigenvalues are real. According to the condition (4)(a) the matrix $\overline{Q}$ is negative-definite, and therefore all its eigenvalues are negative, so $\lambda_M(\overline{Q}) < 0$. Taking this into account, from the inequality (5.8) obtain the estimate

$$\left.\frac{dV}{dt}\right|_{(5.1)} < \lambda_M(\overline{Q})w^{\mathrm{T}}w = \lambda_M(\overline{Q}) \sum_{i=1}^{s} \psi_{i3}(\|x_i\|) \tag{5.9}$$

$$\text{at} \quad \|x_i\| > \frac{1}{s}r(\alpha) \quad \text{and} \quad \alpha \in \mathcal{S}.$$

Since ψ_{i3} belongs to the KR-class, one can find a function $\Psi_3(\|x\|)$ of the class KR such that

$$\Psi_3(\|x\|) \le \sum_{i=1}^{s} \psi_{i3}(\|x_i\|) \quad \text{at} \quad \|x\| > r(\alpha) \quad \text{and} \quad \|x_i\| > \frac{1}{s}r(\alpha),$$

which gives the possibility to transform the estimate (5.9) to the following one:

$$\left.\frac{dV}{dt}\right|_{(5.1)} < -\lambda_M(\overline{Q})\Psi_3(\|x\|) \quad \text{at all} \quad \|x\| > r(\alpha) \quad \text{and} \quad \alpha \in \mathcal{S}. \tag{5.10}$$

Now show that under the conditions (2)(d), (3)(b) and (4)(b) of the Theorem 5.1 along any solution of the system (5.1) the function $V(t,x,c)$ increases, and therefore the motion of the system (5.1) verges to the boundary of the set $A(r)$ from within.

Indeed, obtain

$$\left.\frac{dV}{dt}\right|_{(5.1)} > \sum_{i=1}^{s} c_i\underline{\sigma}_i\psi_{i3}(\|x_i\|) + c_i[\psi_{i3}(\|x_i\|)]^{1/2} \sum_{j=1}^{s} \underline{a}_{ij}[\psi_{j3}(\|x_j\|)]^{1/2}$$

at $\|x_i\| < \frac{1}{s}r(\alpha)$ and $\alpha \in \mathcal{S}$.

Similarly to (5.8) we obtain

$$\left.\frac{dV}{dt}\right|_{(5.1)} > w^{\mathrm{T}}\left(\frac{1}{2}[\underline{K} + \underline{K}^{\mathrm{T}}]\right)w = w^{\mathrm{T}}\underline{Q}w \quad \text{at} \quad \|x\| < r(\alpha) \quad \text{and} \quad \alpha \in \mathcal{S}. \tag{5.11}$$

From the fact that the matrix $\underline{Q}$, according to the condition (4)(b) of the Theorem 5.1, is positive-definite it follows that all its eigenvalues are positive, and the minimum eigenvalue $\lambda_m(\underline{Q}) > 0$. Therefore the inequality (5.11) is true subsequent to the inequality

$$\left.\frac{dV}{dt}\right|_{(5.1)} > \lambda_m(\underline{Q})w^{\mathrm{T}}w = \lambda_m(\underline{Q}) \sum_{i=1}^{s} \psi_{i3}(\|x_i\|) \tag{5.12}$$

$$\text{at} \quad \|x_i\| < \frac{1}{s}r(\alpha) \quad \text{and} \quad \alpha \in \mathcal{S}.$$

Owing to the fact that the functions $\psi_{i3}(\|x_i\|)$ at all $i \in [1, s]$ belong to the KR-class, one can find a function $\Phi_3(\|x\|) \geq \sum_{i=1}^{s} \psi_{i3}(\|x_i\|)$ at $\|x_i\| \leq \frac{1}{s} r(\alpha)$ such that the inequality (5.12) will take the form

$$\frac{dV}{dt}\bigg|_{(5.1)} > \lambda_m(\underline{Q})\Phi_3(\|x\|) \tag{5.13}$$

at all $\|x\| < r(\alpha)$ and $\alpha \in \mathcal{S}$.

The inequality (5.13) is sufficient for the condition (3)(b) of the Theorem 2.3 to be satisfied. Along with the other conditions of the Theorem 5.1, the inequalities (5.10), (5.13) allow to apply the reasoning of the proof of the Theorem 2.3 (Case (b)).

Thus, the motions $x(t, \alpha)$ of the system (5.1) uniformly asymptotically approximate the set $A(r)$ from within, which completes the proof of the Theorem 5.1.

To estimate the impact of the uncertain value of the functions $g_i(t, x_1, \ldots, x_s, \alpha)$, $\alpha \in \mathcal{S} \subseteq R^d$, upon the dynamics of the system (5.1), in the inequalities (3)(a), (b) some constants $(\overline{a}_{ij}, \underline{a}_{ij}) \in R$ were used. A small generalization of the Theorem 5.1 is the statement in which instead of the constants $(\overline{a}_{ij}, \underline{a}_{ij}) \in R$ some functions are used: $\overline{a}_{ij}(t, x, \alpha)\colon R_+ \times \text{ext}\, A(r) \times \mathcal{S} \to R$, $\underline{a}_{ij}(t, x, \alpha)\colon R_+ \times \text{int}\, A(r) \times \mathcal{S} \to R$.

Theorem 5.2 *Let the following conditions be satisfied:*

(1) *the conditions (1), (2), (5) of the Theorem 5.1;*

(2) *at the given functions v_i and ψ_{i3} there exist functions $\overline{a}_{ij}(t, x, \alpha)\colon R_+ \times \text{ext}\, A(r) \times R^d \to R$ and $\underline{a}_{ij}(t, x, \alpha)\colon R_+ \times \text{int}\, A(r) \times R^d \to R$ such that:*

(a) $\left(\dfrac{\partial v_i(t, x_i)}{\partial x_i}\right)^{\mathrm{T}} g_i(t, x_1, \ldots, x_s, \alpha) < [\psi_{i3}(\|x_i\|)]^{1/2} \displaystyle\sum_{j=1}^{s} \overline{a}_{ij}(t, x, \alpha)$

$\times [\psi_{j3}(\|x_j\|)]^{1/2}$ *at* $\|x\| > r(\alpha)$, $\alpha \in \mathcal{S} \subseteq R^d$,

(b) $Dv_i(t, x_i)|_{(5.4)} + \left(\dfrac{\partial v_i(t, x_i)}{\partial x_i}\right)^{\mathrm{T}} g_i(t, x_1, \ldots, x_s, \alpha) = 0$

at $\|x_i\| = \dfrac{1}{s} r(\alpha)$,

(c) $\left(\dfrac{\partial v_i(t, x_i)}{\partial x_i}\right)^{\mathrm{T}} g_i(t, x_1, \ldots, x_s, \alpha) > [\psi_{i3}(\|x_i\|)]^{1/2} \displaystyle\sum_{j=1}^{s} \underline{a}_{ij}(t, x, \alpha)$

$\times [\psi_{j3}(\|x_j\|)]^{1/2}$ *at* $\|x\| < r(\alpha)$ $\alpha \in \mathcal{S} \subseteq R^d$;

(3) *for the given constants $(\underline{\sigma}_i, \overline{\sigma}_i)$ there exists a vector $c^{\mathrm{T}} = (c_1, \ldots, c_s) > 0$ and constants $(\overline{\Delta}, \underline{\Delta}) > 0$ such that:*

(a) *the matrix* $\overline{Q}(t,x,\alpha) + \overline{\Delta}E$ *negative-definite at all* $\|x\| > r(\alpha)$, $\alpha \in \mathcal{S} \subseteq R^d$, *where* E *is a unit* $(s \times s)$-*matrix and the matrix* $\overline{Q}(t,x,\alpha)$ *has the elements*

$$\overline{q}_{ij} = \begin{cases} c_i(\overline{\sigma}_i + \overline{a}_{ij}(t,x,\alpha)) & at \ i = j, \\ [c_i\overline{a}_{ij}(t,x,\alpha) + c_j\overline{a}_{ij}(t,x,\alpha)]/2 & at \ i \neq j, \end{cases}$$

(b) *the matrix* $\underline{Q}(t,x,\alpha) - \underline{\Delta}E$ *positively-definite at all* $\|x\| < r(\alpha)$, $\alpha \in \mathcal{S} \subseteq R^d$, *where the matrix* $\underline{Q}(t,x,\alpha)$ *has the elements*

$$\underline{q}_{ij} = \begin{cases} c_i(\underline{\sigma}_i + \underline{a}_{ij}(t,x,\alpha)) & at \ i = j, \\ [c_i\underline{a}_{ij}(t,x,\alpha) + c_j\underline{a}_{ij}(t,x,\alpha)]/2 & at \ i \neq j; \end{cases}$$

(4) *the condition* (5) *of the Theorem* 5.1.

Then the set $A(r)$ *is invariant for the solutions of the system* (5.1), *and the motions* $x(t,\alpha)$ *of the system* (5.1) *are uniformly asymptotically stable with respect to the set* $A(r)$.

Proof At first prove the uniform asymptotic stability of the motions of the system (5.1) beginning from the outside of the set $A(r)$.

As in the proof of the Theorem 5.1, we will apply the function (5.6) which under the condition (1) of the Theorem 5.2 is the Lyapunov function. Denote an $(s \times s)$-matrix with the elements

$$\overline{l}_{ij}(t,x,\alpha) = \begin{cases} c_i[\overline{\sigma}_i + \overline{a}_{ij}(t,x,\alpha)] & at \ i = j, \\ c_i\overline{a}_{ij}(t,x,\alpha) & at \ i \neq j \end{cases}$$

by $\overline{L}(t,x,\alpha) = [\overline{l}_{ij}(t,x,\alpha)]$ and keep the value of the vector w^{T} from the previous proof.

Similarly to the inequality (5.7) find the estimate

$$\left.\frac{dV}{dt}\right|_{(5.1)} \leq w^{\mathrm{T}}\overline{L}(t,x,\alpha)w = w^{\mathrm{T}}\left[\frac{1}{2}\left(\overline{L}(t,x,\alpha) + \overline{L}^{\mathrm{T}}(t,x,\alpha)\right)\right]w$$

$$= w^{\mathrm{T}}\overline{Q}(t,x,\alpha)w \leq -\overline{\Delta}w^{\mathrm{T}}w = -\overline{\Delta}\sum_{j=1}^{s}\psi_{i3}(\|x_j\|) \qquad (5.14)$$

at $\|x_j\| > \frac{1}{s}r(\alpha)$, $\alpha \in \mathcal{S} \subseteq R^d$.

Since the functions $\psi_{i3}(\|x_i\|)$ belong to the KR-class at all $I \in [1,s]$, there exists a function $\Psi_3(\|x\|) \leq \sum_{i=1}^{s}\psi_{i3}(\|x_i\|)$ at $\|x_i\| > \frac{1}{s}r(\alpha)$, $\alpha \in \mathcal{S}$, and the inequality (5.14) takes the form

$$\left.\frac{dV}{dt}\right|_{(5.1)} < -\overline{\Delta}\Psi_3(\|x\|) \qquad (5.15)$$

at all $t \in R_+$ and $\|x\| > r(\alpha)$, $\alpha \in \mathcal{S}$.

Along with the other conditions of the Theorem 5.2, the inequality (5.15) secures the asymptotic stability of motions with respect to the moving set $A(r)$, beginning from the outside of this set.

Now consider the case of stability of motions of the system (5.1), beginning from within the set $A(r)$.

Under the conditions (2)(d) of the Theorem 5.1 and (2)(b) and (3)(b) of the Theorem 5.2 obtain

$$\left.\frac{dV}{dt}\right|_{(5.1)} \geq w^{\mathrm{T}} \underline{L}(t, x, \alpha) w = w^{\mathrm{T}} \left[\frac{1}{2}(\underline{L}(t, x, \alpha) + \underline{L}^{\mathrm{T}}(t, x, \alpha))\right] w$$

$$= w^{\mathrm{T}} \underline{Q}(t, x, \alpha) w \geq -\Delta w^{\mathrm{T}} w = \underline{\Delta} \sum_{j=1}^{s} \psi_{i3}(\|x_j\|) \tag{5.16}$$

at $\|x_j\| < \frac{1}{s} r(\alpha)$, $\alpha \in \mathcal{S}$.

The inequality (5.16) will hold true subsequent to the inequality

$$\left.\frac{dV}{dt}\right|_{(5.1)} > \underline{\Delta}\Phi_3(\|x\|) \tag{5.17}$$

at all $\|x\| < r(\alpha)$, $\alpha \in \mathcal{S}$, where Φ belonging to the KR-class was determined above.

The increase of the function $V(t, x, \alpha)$ along the solutions of the system (5.1), beginning within the set $A(r)$, means that at $t \to +\infty$ the motion $x(t, \alpha)$ asymptotically approximates the boundary of the set $A(r)$ from within.

The Theorem 5.2 is proved.

5.3 Application of the Hierarchical Lyapunov Function

In this section we give the conditions for the stability of solutions of the uncertain system (2.1) with respect to a moving invariant set on the basis of the hierarchical Lyapunov function. This function is constructed on the basis of the two-level decomposition of the system (2.1).

Assume that the system (2.1) tolerates mathematical decomposition into m interacting subsystems

$$\widehat{S}_i: \quad \frac{dx_i}{dt} = \hat{f}_i(t, x_i) + g_i(t, x_1, \dots, x_m, \alpha), \tag{5.18}$$

where $x_i \in R^{n_i}$, $\hat{f}_i \in C(R_+ \times R^{n_i}, R^{n_i})$, $g_i \in C(R_+ \times R^{n_1} \times \dots \times R^{n_m} \times \mathcal{S}, R^{n_i})$, $\sum_{i=1}^{m} n_i = n$, $\mathcal{S} \subseteq R^d$.

The transformation of the system (2.1) to the form (5.18) will be called the first level mathematical decomposition of the system (2.1).

Along with the system (5.18) consider (i, j)-pairs $(i \neq j) \in [1, m]$ of the independent subsystems

$$
S_{ij}: \quad
\begin{cases}
\dfrac{dx_i}{dt} = w_i(t, x_i, x_j), \\[2mm]
\dfrac{dx_j}{dt} = w_j(t, x_i, x_j),
\end{cases}
\tag{5.19}
$$

where $w_i \in C(R_+ \times R^{n_i} \times R^{n_j}, R^{n_i})$, $w_j \in C(R_+ \times R^{n_i} \times R^{n_j}, R^{n_j})$.

It is easy to see that if $(x_i, x_j) = (0, \ldots, 0, x_i^\mathrm{T}, 0, \ldots, 0, x_j^\mathrm{T}, 0, \ldots, 0)^\mathrm{T}$ at all $(i \neq j) \in [1, m]$, then

$$
\begin{aligned}
w_i &= f_i(t, 0, \ldots, 0, x_i^\mathrm{T}, 0, \ldots, 0, x_j^\mathrm{T}, 0, \ldots, 0, 0), \\
w_j &= f_j(t, 0, \ldots, 0, x_i^\mathrm{T}, 0, \ldots, 0, x_j^\mathrm{T}, 0, \ldots, 0, 0) \quad \forall\, (i \neq j) \in [1, m].
\end{aligned}
\tag{5.20}
$$

Now denote

$$
\begin{aligned}
H_i(t, x_1, \ldots, x_m, \alpha) &= f_i(t, x_1, \ldots, x_m, \alpha) - w_i(t, x_i, x_j), \\
H_j(t, x_1, \ldots, x_m, \alpha) &= f_j(t, x_1, \ldots, x_m, \alpha) - w_j(t, x_i, x_j).
\end{aligned}
\tag{5.21}
$$

Taking into account (5.19)–(5.21), write the system (2.1) in the form

$$
\widehat{S}_{ij}: \quad
\begin{cases}
\dfrac{dx_i}{dt} = w_i(t, x_i, x_j) + H_i(t, x_1, \ldots, x_m, \alpha), \\[2mm]
\dfrac{dx_j}{dt} = w_j(t, x_i, x_j) + H_j(t, x_1, \ldots, x_m, \alpha), \quad i \neq j.
\end{cases}
\tag{5.22}
$$

The transformation of the system (2.1) to the form (5.22) will be called the second level mathematical decomposition of the system (2.1).

If $g_i \equiv 0$, $H_i(t, \cdot) \equiv 0$, $H_j(t, \cdot) \equiv 0$ At all $i = 1, 2, \ldots, m$ in the system (2.1), then the subsystems

$$
S_i: \quad \frac{dx_i}{dt} = \hat{f}_i(t, x_i), \quad x_i(t_0) = x_{i0}, \quad i = 1, 2, \ldots, m,
\tag{5.23}
$$

together with (i, j)-pairs (5.22) of second-level independent subsystems are some nonlinear approximation of the uncertain system (2.1), obtained as a result of hierarchical decomposition.

Construct the two-index system of functions

$$
U(t, x) = [v_{ij}(t, \cdot)], \quad i, j = 1, 2, \ldots, m,
$$

in which the elements

$$
v_{ii}(t, x_i) \in C(R_+ \times R^{n_i}, R_+), \quad i = 1, 2, \ldots, m,
$$

are constructed on the basis of the S_i-subsystems (5.23), and the elements

$$v_{ij}(t, x_i, x_j) \in C(R_+ \times R^{n_i} \times R^{n_j}, R), \quad i \neq j \in 1, 2, \ldots, m,$$

are constructed on the basis of S_{ij}-pairs of the subsystems (5.19).

The function

$$V(t, x, d) = d^{\mathrm{T}} U(t, x) d, \quad d \in R_+^m, \tag{5.24}$$

is hierarchical and under certain conditions can be the Lyapunov function for the analysis of the behavior of the solutions of the uncertain system (2.1).

Assumption 5.1 There exist:

(1) open time-invariant neighbourhoods $\mathcal{N}_i \subseteq R^{n_i}$ of the equilibrium states $x_i = 0$, $i = 1, 2, \ldots, m$;

(2) the functions $(\varphi_i, \psi_i) \in KR$ at all $i = 1, 2, \ldots, m$ and $(\varphi_{ij}, \psi_{ij}) \in KR$ at all $(i \neq j) \in [1, m]$ such that the following inequalities are true:

   (a) $\varphi_i(\|x_i\|) \leq v_{ii}(t, x_i) \leq \psi_i(\|x_i\|)$, $i = 1, 2, \ldots, m$, at all $(t, x_i) \in R_+ \times R^{n_i}$,

   (b) $\varphi_{ij}(\|x_{ij}\|) \leq v_{ij}(t, x_i, x_j) \leq \psi_{ij}(\|x_{ij}\|)$ at all $(i \neq j) \in [1, m]$ and $(t, x_i, x_j) \in R_+ \times R^{n_i} \times R^{n_j}$, where $x_{ij} = (x_i^{\mathrm{T}}, x_j^{\mathrm{T}})^{\mathrm{T}} \in R^{n_i + n_j}$.

Lemma 5.1 *If all the conditions of the Assumption 5.1 are satisfied, then the function (5.24) is locally Lipshitz with respect to x, positive-definite, decreasing, and radially unbounded.*

This follows from the fact that for the function

$$V(t, x, e) = e^{\mathrm{T}} U(t, x) e, \quad e = (1, \ldots, 1)^{\mathrm{T}} \in R_+^m,$$

it is possible to indicate functions $(\gamma_1, \gamma_2) \in KR$-class, such that the following bilateral inequality is true:

$$\gamma_1(\|x\|) \leq V(t, x, e) \leq \gamma_2(\|x\|) \quad \forall (t, x) \in R_+ \times R^n. \tag{5.25}$$

Now represent the function (5.24) in the form

$$V(t, x, d) = \sum_{i=1}^{m} \left[d_i^2 v_{ii}(t, x_i) + \sum_{\substack{j=1 \\ i \neq j}}^{m} d_i d_j v_{ij}(t, x_i, x_j) \right]$$

and note that $v_{ij}(t, \cdot) = v_{ji}(t, \cdot)$ at all $(i \neq j) \in [1, m]$. Calculate $D^+ V(t, x, d)$ as follows:

$$D^+ V(t, x, d) = \sum_{i=1}^{m} \left[d_i^2 D^+ v_{ii}(t, x_i) \big|_{(5.18)} + \sum_{\substack{j=1 \\ i \neq j}}^{m} d_i d_j D^+ v_{ij}(t, x_i, x_j) \big|_{(5.22)} \right],$$

where $D^+ v_{ii}(t, x_i)\big|_{(5.18)}$ and $D^+ v_{ij}(t, x_i, x_j)\big|_{(5.22)}$ are determined by the known rules of calculation of the upper right Dini derivative.

The behavior of solutions of isolated subsystems will be analyzed with respect to the moving sets

$$A_i(r) = \left\{ x_i \in R^{n_i}: \ \|x_i\| = \frac{1}{m} r(\alpha) \right\}, \quad i = 1, 2, \ldots, m.$$

Denote the internal and the external parts of the spaces R^{n_i} with respect to the moving sets $A_i(r)$ by int $A_i(r)$ and ext $A_i(r)$.

Assumption 5.2 The independent subsystems (5.23) of the first level of decomposition and the functions $g_i(t, x_1, \ldots, x_m)$, $i = 1, 2, \ldots, m$, such that:

(1) there exist functions $v_{ii}(t, x_i)$ satisfying the conditions 2(a) of the Assumption 5.1;

(2) there exist functions $\varkappa_i(\|x_i\|) \in KR$-class and constants ρ_{ii}^0, $\mu_{ik}(\alpha)$, $i, k = 1, 2, \ldots, m$, such that:

   (a) $D^+ v_{ii}(t, x_i)\big|_{(5.23)} \leq \rho_{ii}^0 \varkappa_i(\|x_i\|)$ at all $(t, x_i) \in R_+ \times \text{ext } A_i(r)$, $i = 1, 2, \ldots, m$,

   (b) $D^+ v_{ii}(t, x_i)\big|_{(5.18)} - D^+ v_{ii}(t, x_i)\big|_{(5.23)} \leq \varkappa_i^{1/2}(\|x_i\|) \sum\limits_{k=1}^m \mu_{ik}(\alpha) \times \varkappa_k^{1/2}(\|x_k\|)$ at all $(t, x_i, \alpha) \in R_+ \times \text{ext } A_i(r) \times S$, $i = 1, 2, \ldots, m$;

(3) there exist constants β_{ii}^0, $\gamma_{ik}(\alpha)$, $i, k = 1, 2, \ldots, m$, such that:

   (a) $D^+ v_{ii}(t, x_i)\big|_{(5.23)} \geq \beta_{ii}^0 \varkappa_i(\|x_i\|)$ at all $(t, x_i) \in R_+ \times \text{int } A_i(r)$, $i = 1, 2, \ldots, m$,

   (b) $D^+ v_{ii}(t, x_i)\big|_{(5.18)} - D^+ v_{ii}(t, x_i)\big|_{(5.23)} \geq \varkappa_i^{1/2}(\|x_i\|) \sum\limits_{k=1}^m \gamma_{ik}(\alpha) \times \varkappa_k^{1/2}(\|x_k\|)$ at all $(t, x_i, \alpha) \in R_+ \times \text{int } A_i(r) \times S$, $i = 1, 2, \ldots, m$.

Assumption 5.3 The independent S_{ij}-pairs of the subsystems (5.19) of the second order decomposition and the functions $H_i(t, x_1, \ldots, x_m, \alpha)$, $H_j(t, x_1, \ldots, x_m, \alpha)$ such that:

(1) there exist functions $v_{ij}(t, \cdot)$ $(i \neq j)$ satisfying the conditions (2b) of the Assumption 5.1;

(2) there exist functions $\varkappa_i(\|x_i\|) \in KR$-class and constants ρ_{ij}^1, ρ_{ij}^2, ρ_{ij}^3, $\nu_{kp}^{ij}(\alpha)$, $i, j, k, p = 1, 2, \ldots, m$, such that:

   (a) $D^+ v_{ij}(t, x_i, x_j)\big|_{(2.4)} \leq \rho_{ij}^1 \varkappa_i(\|x_i\|) + 2\rho_{ij}^2 \varkappa_i^{1/2}(\|x_i\|) \varkappa_j^{1/2}(\|x_j\|) + \rho_{ij}^3 \varkappa_i(\|x_j\|)$ at all $(t, x_i, x_j) \in R_+ \times \text{ext } A_i(r) \times \text{ext } A_j(r)$,

(b) $D^+v_{ij}(t,x_i,x_j)|_{(5.22)} - D^+v_{ij}(t,x_i,x_j)|_{(5.19)} \leq \sum_{k,p=1}^{m} \nu_{kp}^{ij}(\alpha) \times$

$\varkappa_k^{1/2}(\|x_k\|)\varkappa_p^{1/2}(\|x_p\|)$ at all $(t,x_i,x_j,\alpha) \in R_+ \times \mathrm{ext}\, A_i(r) \times$ $\mathrm{ext}\, A_j(r) \times \mathcal{S}$;

(3) there exist constants $\theta_{ij}^1, \theta_{ij}^2, \theta_{ij}^3, \xi_{kp}^{ij}(\alpha), i,j,k,p = 1,2,\ldots,m$, such that:

(a) $D^+v_{ij}(t,x_i,x_j)|_{(5.19)} \geq \theta_{ij}^1 \varkappa_i(\|x_i\|) + 2\theta_{ij}^2 \varkappa_i^{1/2}(\|x_i\|)\varkappa_j^{1/2}(\|x_j\|) +$ $\theta_{ij}^3 \varkappa_i(\|x_j\|)$ at all $(t,x_i,x_j) \in R_+ \times \mathrm{int}\, A_i(r) \times \mathrm{int}\, A_j(r)$,

(b) $D^+v_{ij}(t,x_i,x_j)|_{(5.22)} - D^+v_{ij}(t,x_i,x_j)|_{(2.4)} \geq \sum_{k,p=1}^{m} \xi_{kp}^{ij}(\alpha) \times$

$\varkappa_k^{1/2}(\|x_k\|)\varkappa_p^{1/2}(\|x_p\|)$ at all $(t,x_i,x_j,\alpha) \in R_+ \times \mathrm{int}\, A_i(r) \times$ $\mathrm{int}\, A_j(r) \times \mathcal{S}$.

Now we can make the following statement.

Lemma 5.2 *If all the conditions of the Assumptions 5.2 and 5.3 are satisfied, then for the Dini derivative $D^+v(t,x,d)$ of the function $v(t,x,d)$ with the elements $v_{ij}(t,\cdot)$ satisfying the conditions of the Assumption 5.1, the following estimates hold true:*

(a) *at $\|x\| > r(\alpha)$*

$$D^+V(t,x,d) \leq \varkappa^{\mathrm{T}}(\|x\|)Q(\alpha)\varkappa(\|x\|), \quad \alpha \in \mathcal{S} \subseteq R^d, \qquad (5.26)$$

(b) *at $\|x\| < r(\alpha)$*

$$D^+V(t,x,d) \geq \varkappa^{\mathrm{T}}(\|x\|)R(\alpha)\varkappa(\|x\|), \quad \alpha \in \mathcal{S} \subseteq R^d. \qquad (5.27)$$

Here $Q(\alpha) = \frac{1}{2}(B(\alpha) + B^{\mathrm{T}}(\alpha))$ and $B(\alpha) = [b_{ij}(\alpha)]$, where

$$b_{kk}(\alpha) = d_k^2(\rho_{kk}^0 + \mu_{kk}(\alpha)) + d_k\left(\sum_{\substack{j=1\\j\neq k}}^{m} d_j\rho_{kj}^1 + \sum_{\substack{i=1\\i\neq k}}^{m} d_i\rho_{ik}^3\right) + \sum_{i,j=1}^{m} d_id_j\nu_{ik}^{ij}(\alpha),$$

$$b_{kl}(\alpha) = d_k^2\mu_{kl}(\alpha) + 2d_kd_l\rho_{kl}^2 + \sum_{\substack{i,j=1\\i\neq j}}^{m} d_id_j\nu_{kl}^{ij}(\alpha).$$

Similarly $R(\alpha) = \frac{1}{2}(C(\alpha) + C^{\mathrm{T}}(\alpha))$ and $C(\alpha) = [c_{ij}(\alpha)]$, where

$$c_{kk}(\alpha) = d_k^2(\beta_{kk}^0 + \gamma_{kk}(\alpha)) + d_k\left(\sum_{\substack{j=1\\j\neq k}}^{m} d_j\theta_{kj}^1 + \sum_{\substack{i=1\\i\neq k}}^{m} d_i\theta_{ik}^3\right) + \sum_{i,j=1}^{m} d_id_j\xi_{ik}^{ij}(\alpha),$$

$$c_{kl}(\alpha) = d_k^2\gamma_{kl}(\alpha) + 2d_kd_l\theta_{kl}^2 + \sum_{\substack{i,j=1\\i\neq j}}^{m} d_id_j\xi_{kl}^{ij}(\alpha).$$

On the basis of the estimates (5.26) and (5.27) obtained for the hierarchical Lyapunov function, the conditions for the stability of solutions of the system (2.1) with respect to the moving invariant set

$$A(r) = A_1(r) \cup A_2(r) \cup \cdots \cup A_m(r)$$

are written in the form similar to the conditions of Lyapunov theorems of stability of motion.

Theorem 5.3 *Assume that the vector-function $f(t, x, \alpha)$ in the system (2.1) is continuous on $R_+ \times R^n \times R^d$ and the following conditions are satisfied:*

(1) *for each $\alpha \in S \subseteq R^d$ there exists a function $r = r(\alpha) > 0$ such that the set $A(r)$ is not empty at all $\alpha \in S$;*

(2) *for the function (5.24) locally Lipshitz with respect to x all the conditions of the Assumptions 5.1 – 5.3 are satisfied, as well as*

  (a) $\gamma_1(\|x\|) \leq V(t, x, d)$ *at* $\|x\| > r(\alpha)$,

  (b) $V(t, x, d) \leq \gamma_2(\|x\|)$ *at* $\|x\| < r(\alpha)$,

  (c) $D^+V(t, x, d) \leq \varkappa^{\mathrm{T}}(\|x\|)Q(\alpha)\varkappa(\|x\|)$ *at* $\|x\| > r(\alpha)$, $\alpha \in S$,

  (d) $D^+V(t, x, d) = 0$ *at* $\|x\| = r(\alpha)$, $\alpha \in S$,

  (e) $D^+V(t, x, d) \geq \varkappa^{\mathrm{T}}(\|x\|)R(\alpha)\varkappa(\|x\|)$ *at* $\|x\| < r(\alpha)$, $\alpha \in S$;

(3) *there exist constant $(m \times m)$-matrices θ_1, θ_2 such that:*

  (a) $\frac{1}{2}(Q(\alpha) + Q^{\mathrm{T}}(\alpha)) \leq \theta_1$ *at all* $\alpha \in S$,

  (b) $\frac{1}{2}(R(\alpha) + R^{\mathrm{T}}(\alpha)) \geq \theta_2$ *at all* $\alpha \in S$, θ_1 *is a negative-semidefinite matrix and θ_2 is a positive-semidefinite matrix;*

(4) *for the given function $r(\alpha) > 0$ and the functions $(\gamma_1, \gamma_2) \in KR$-class the relation $\gamma_1(r) = \gamma_2(r)$ is true.*

Then the set $A(r)$ is invariant for solutions of the system (2.1), and the solutions $x(t, \alpha)$ of the system (2.1) are stable with respect to the set $A(r)$.

Proof Under the conditions of the Theorem 5.3 its proof is analogous to that of the Theorems 2.3 and 5.1. Therefore we will give some fragments of the proof. In particular, under the conditions of the Assumption 5.1, one can find the functions $\gamma_1, \gamma_2 \in KR$-class in the estimate (5.25) of the function $V(t, x, e)$. Since

$$V(t, x, e) = \sum_{i=1}^{m} v_{ii}(t, x_i) + \sum_{\substack{i,j=1 \\ i \neq j}}^{m} v_{ij}(t, x_i, x_j),$$

under the conditions of the Assumption 5.1,

$$\sum_{i=1}^{m} \varphi_i(\|x_i\|) + \sum_{\substack{i,j=1 \\ i \neq j}}^{m} \varphi_{ij}(\|x_{ij}\|) \leq V(t, x, e) \leq \sum_{i=1}^{m} \psi_i(\|x_i\|) + \sum_{\substack{i,j=1 \\ i \neq j}}^{m} \psi_{ij}(\|x_{ij}\|)$$

at all $(t, x_i, x_j) \in R_+ \times R^{n_i} \times R^{n_j}$. From the condition (2) of the Assumption 5.1 it follows that there exist functions α_1, $\alpha_2 \in KR$-class, such that

$$\alpha_1(\|x\|) \leq \sum_{i=1}^{m} \varphi_i(\|x_i\|) \quad \text{and} \quad \alpha_2(\|x\|) \geq \sum_{i=1}^{m} \psi_i(\|x_i\|),$$

and functions β_1, $\beta_2 \in KR$-class, such that

$$\beta_1(\|x\|) \leq \sum_{\substack{i,j=1 \\ i \neq j}}^{m} \varphi_{ij}(\|x_{ij}\|) \quad \text{and} \quad \beta_2(\|x\|) \geq \sum_{\substack{i,j=1 \\ i \neq j}}^{m} \psi_{ij}(\|x_{ij}\|).$$

Then, since $\alpha_1, \alpha_2, \beta_1, \beta_2$ belong to the KR-class, one can find functions $\gamma_1, \gamma_2 \in KR$-class, for which

$$\gamma_1(\|x\|) \leq \alpha_1(\|x\|) + \beta_1(\|x\|) \quad \text{and} \quad \gamma_2(\|x\|) \geq \alpha_2(\|x\|) + \beta_2(\|x\|).$$

The constructed functions $\gamma_1(\|x\|)$ and $\gamma_2(\|x\|)$ secure the estimates 2(a), (b) from the Theorem 5.3.

From the conditions 2 (c), (e) and 3 (a), (b) of the Theorem 5.3 it follows that $\lambda_M(\theta_1) \leq 0$ and $\lambda_m(\theta_2) \geq 0$, where $\lambda_M(\cdot)$ and $\lambda_m(\cdot)$ are the maximal and the minimal eigenvalues of the matrices θ_1 and θ_2 respectively. Therefore for all $\alpha \in S \subseteq R^d$ we obtain

$$D^+V(t, x, d) \leq 0 \quad \text{at} \quad \|x\| > r(\alpha),$$
$$D^+V(t, x, d) = 0 \quad \text{at} \quad \|x\| = r(\alpha),$$
$$D^+V(t, x, d) \geq 0 \quad \text{at} \quad \|x\| < r(\alpha).$$

These inequalities along with the condition (4) of the Theorem 5.3 enable us to prove its statement.

5.4 Stability of a Class of Time Invariant Uncertain Systems

Consider the class of large-scale systems of the form

$$\frac{dx_i}{dt} = A_i x_i + \sum_{j=1}^{K} \alpha_{ij} \Phi_{ij}(x) + G_i(x), \quad i = 1, 2, \ldots, N, \qquad (5.28)$$

where $x_i \in R^{n_i}$, A_i are constant $(n_i \times n_i)$-matrices, $\Phi_{ij} \in C(R^n, R^n)$, $G_i \in C(R^n, R^n)$ and $\alpha_{ij} \in S$ are uncertainties parameters.

Assumption 5.4 The summands in the right-hand part of the system (5.28) are such that

(1) the equilibrium state $x = 0$, $x = (x_1, \ldots, x_N)^{\mathrm{T}}$, of the system (5.28) is unique at any $\alpha_{ij} \in \mathcal{S}$, $i = 1, 2, \ldots, N$, $j = 1, 2, \ldots, K$;

(2) the independent subsystems (5.28) are asymptotically stable, i. e., for the given positive-definite matrices P_i there exist solutions of the Lyapunov equations

$$A_i^{\mathrm{T}} Q_i + Q_i A_i = -P_i, \quad i = 1, 2, \ldots, N,$$

in the form of the positive-definite matrices Q_i, $i = 1, 2, \ldots, N$;

(3) in the open domain $E \subset R^n$ the following expressions are determined:

$$R_{ij}(x) = \begin{cases} \Phi_{ij}(x) x_i^{\mathrm{T}} \big/ \|x_i\|^2, & \text{if } x_i \neq 0, \\ 0, & \text{if } x_i = 0, \end{cases}$$

$$L_i(x) = \begin{cases} G_i(x) x_i^{\mathrm{T}} \big/ \|x_i\|^2, & \text{if } x_i \neq 0, \\ 0, & \text{if } x_i = 0; \end{cases}$$

(4) at all $x \in E$ the elements of $(m \times m)$-matrix $W = [\omega_{ij}(x)]$ are determined, where $m = \max\{K, N\}$ and $\omega_{ij}(x)$ have the form

(a) if $K = N$, then $\omega_{ij}(x) = \begin{cases} 1 - 2\xi_{ij} & \text{at } i = j, \\ -2\xi_{ij} & \text{at } i \neq j, \end{cases}$

(b) if $K > N$, then $\omega_{ij}(x) = \begin{cases} 1 - 2\xi_{ij} & \text{at } i = j, \ i \leq N, \\ -2\xi_{ij} & \text{at } i \neq j, \ i \leq N, \\ 0 & \text{at } N < i \leq K, \end{cases}$

(c) if $K < N$, then $\omega_{ij}(x) = \begin{cases} 1 - 2\xi_{ij} & \text{at } i = j, \ j \leq K, \\ -2\xi_{ij} & \text{at } i \neq j, \ j \leq K, \\ \eta_i & \text{at } K < j \leq N, \end{cases}$

where

$$\xi_{ij} = \delta_M \left[\left(P_i^{-\frac{1}{2}} \right)^{\mathrm{T}} Q_i \left(\varepsilon_{ij} R_{ij}(x) + K^{-1} L_i(x) \right) P_i^{-\frac{1}{2}} \right],$$

$$\eta_i = K^{-1} \delta_M \left[\left(P_i^{-\frac{1}{2}} \right)^{\mathrm{T}} Q_i L_i(x) P_i^{-\frac{1}{2}} \right],$$

$\delta_M(\cdot)$ is the maximal eigenvalue of the matrix $(\cdot)$ and $P_i^{-\frac{1}{2}}$ denotes the positive-definite symmetrical matrix Q_i such that $Q_i^2 = P_i$, $i = 1, 2, \ldots, N$.

We now prove the following statement.

Theorem 5.4 *Assume that the vector-functions*

$$f_i(x, \alpha_{ij}) = A_i x_i + \sum_{j=1}^{K} \alpha_{ij} \Phi_{ij}(x) + G_i(x), \quad i = 1, 2, \ldots, N,$$

are continuous on $R^n \times S$, all the conditions of the Assumption 5.4 are satisfied, and in addition, in the domain $E \subset R^n$ there exists a diagonal matrix $D = \mathrm{diag}(d_1, \ldots, d_m)$ with constant components $d_i > 0$, $i = 1, 2, \ldots, m$, such that the matrix $W^T D + DW$ is positive-definite in the domain E.

Then the equilibrium state $x = 0$ of the system (5.28) is uniformly asymptotically stable.

Proof Apply the scalar approach in the method of vector Lyapunov functions. Let $K = N$ and $V(x) = \sum\limits_{i=1}^{N} d_i V_i(x_i)$, where $V_i(x_i) = x_i^T Q_i x_i$, $d_i > 0$ are some constants. Under the condition (3) of the Assumption 5.4 obtain the relations

$$
\begin{aligned}
\Phi_{ij}(x) &= R_{ij}(x) x_i, \\
G_i(x) &= L_i(x) x_i, \quad i, j = 1, 2, \ldots, N,
\end{aligned}
\tag{5.29}
$$

in the domain $E \subset R^n$.

Taking into account the relations (5.29), obtain

$$
\begin{aligned}
\left.\frac{dV}{dt}\right|_{(5.28)} &= \sum_{i=1}^{N} d_i \left[\left(A_i x_i + \sum_{j=1}^{N} \varepsilon_{ij} \Phi_{ij}(x) + G_i(x) \right)^T Q_i x_i \right. \\
&\quad \left. + x_i^T Q_i \left(A_i x_i + \sum_{j=1}^{N} \varepsilon_{ij} \Phi_{ij}(x) + G_i(x) \right) \right] \\
&= \sum_{i=1}^{N} d_i \left[-x_i^T P_i x_i + \left(\sum_{j=1}^{N} \varepsilon_{ij} \Phi_{ij}^T(x) + G_i^T(x) \right) Q_i x_i \right. \\
&\quad \left. + x_i^T Q_i \left(\sum_{j=1}^{N} \varepsilon_{ij} \Phi_{ij}(x) + G_i(x) \right) \right] \sum_{i=1}^{N} d_i \left[-\|P_i^{\frac{1}{2}} x_i\|^2 \right. \\
&\quad \left. + 2 x_i^T (P_i^{-\frac{1}{2}})^T (P_i^{\frac{1}{2}})^T Q_i \left(\sum_{j=1}^{N} \varepsilon_{ij} R_{ij}(x) + L_i(x) \right) P_i^{-\frac{1}{2}} P_i^{\frac{1}{2}} x_i \right] \\
&\leq -\sum_{i=1}^{N} d_i \left\{ \|P_i^{\frac{1}{2}} x_i\|^2 - 2 \|P_i^{\frac{1}{2}} x_i\|^2 \delta_M \right. \\
&\quad \left. \times \left[(P_i^{-\frac{1}{2}})^T Q_i \left(\sum_{j=1}^{N} \varepsilon_{ij} R_{ij}(x) + L_i(x) \right) P_i^{-\frac{1}{2}} \right] \right\} \\
&\leq -\sum_{i=1}^{N} \left[d_i \|P_i^{\frac{1}{2}} x_i\|^2 \left(1 - 2 \sum_{j=1}^{N} \varepsilon_{ij} \right) \right] \\
&= -\left[\|P_1^{\frac{1}{2}} x_1\|, \ldots, \|P_N^{\frac{1}{2}} x_N\| \right] (W^T D + DW) \\
&\quad \times \left[\|P_1^{\frac{1}{2}} x_1\|, \ldots, \|P_N^{\frac{1}{2}} x_N\| \right]^T = Y^T (W^T D + DW) Y,
\end{aligned}
$$

where $Y = \left[\|P_1^{\frac{1}{2}}x_1\|, \ldots, \|P_N^{\frac{1}{2}}x_N\|\right]^{\mathrm{T}}$.

Note that the vector $Y = 0$ if and only if $x_i = 0$, $i = 1, 2, \ldots, N$, since P_i are positive-definite matrices according to the condition (2) of the Assumption 5.4.

Since by the condition of the Theorem 5.4 the matrix $W^{\mathrm{T}}D + DW$ is positive-definite, the expression $\frac{dV}{dt}(x)\big|_{(5.28)}$ is negatively definite in the domain E. Therefore the equilibrium state $x = 0$ of the system (5.28) is uniformly asymptotically stable in the domain $E \subset R^n$.

Now consider the case $K > N$. For the system (5.28) we construct the system

$$\frac{dx_i}{dt} = A_i x_i + \sum_{j=1}^{K} \varepsilon_{ij} \Phi_{ij}(x) + G_i(x), \quad i = 1, 2, \ldots, N,$$

$$\frac{dx_{N+1}}{dt} = -x_{N+1},$$

(5.30)

$$\cdots \cdots \cdots \cdots$$

$$\frac{dx_K}{dt} = -x_K,$$

where $\mathrm{col}(x_{N+1}, \ldots, x_K) \in R^{K-N}$. Repeating the above reasoning written for the case $K = N$, for the system (5.30) we obtain the statement of the Theorem 5.4.

$K < N$ represents the system (5.28) in the form

$$\frac{dx_i}{dt} = A_i x_i + \sum_{j=1}^{N} \varepsilon_{ij} \Phi_{ij}(x) + G_i(x), \quad i = 1, 2, \ldots, N,$$

where $\Phi_{ij}(x) = 0$, at $K < j \le N$. Therefore $R_{ij}(x) = 0$ at $i = 1, 2, \ldots, N$ and $j = K, \ldots, N$. Applying the reasoning written for the case $K = N$, to the system, we obtain the statement of the Theorem 5.4. The theorem is proved.

Remark 5.2 ξ_{ij} and η_i are functions of the uncertainties $\varepsilon_{ij} \in S$ and the vector $x \in E$. The conditions of the Theorem 5.4 can be simplified, if instead of ξ_{ij} and η_i one applies

$$\xi_{ij}^*(\alpha_{ij}) = \sup_{x \in E} \xi_{ij}(\alpha_{ij}, x) \quad \text{and} \quad \eta_i^* = \sup_{x \in E} \eta_i(x).$$

Remark 5.3 The Theorem 5.4 will remain valid if ξ_{ij} and ξ_{ij}^* are replaced by

$$\overline{\xi}_{ij}(x) = \sup_{\alpha_{ij} \in S} \xi_{ij}(\alpha_{ij}, x) \quad \text{or} \quad \overline{\overline{\xi}}_{ij} = \sup_{\alpha_{ij} \in S} \xi_{ij}^*(\alpha_{ij}),$$

$$i = 1, 2, \ldots, K, \quad j = 1, 2, \ldots, N.$$

Remark 5.4 The presence of uncertainties parameters in the expression of the matrix $W^{\mathrm{T}} + W$ provides the possibility to facilitate the process of choosing the values d_i, $i = 1, 2, \ldots, N$, at which the matrix $W^{\mathrm{T}} D + DW$ is positive-definite.

Chapter 6

Interval and Parametric Stability of Uncertain Systems

The concept of vector Lyapunov functions is adapted to many dynamic problems of nonlinear systems. In this chapter vector Lyapunov functions are used, both canonical ones and those whose components are quadratic forms. Along with differential inequalities, vector Lyapunov functions present efficient tools for the analysis of stability of uncertain systems.

6.1 Conditions for the Stability of a Quasilinear System (Continued)

Consider the equations of perturbed motion

$$\frac{dx_1}{dt} = P_1 x_1 + Q_1(x_1, x_2, \alpha),$$

$$\frac{dx_2}{dt} = P_2 x_2 + Q_2(x_1, x_2, \alpha),$$

(6.1)

where $x_1 \in R^{n_1}$, $x_2 \in R^{n_2}$, P_k are constant real $(n_k \times n_k)$-matrices, $Q_k \colon R^{n_1} \times R^{n_2} \times R^d \to R^{n_k}$ and $k = 1, 2$, $\mathcal{S} \subseteq R^d$ is a compact set in R^d.

Using the linear nonsingular transformations $z_k = A_k x_k$ $(\det A_k \neq 0)$, $k = 1, 2$, one can reduce the linear part of the system (6.1) to the Jordan form

$$\frac{dz_k}{dt} = \Lambda_k z_k + Z_k(z_1, z_2, \alpha),$$

(6.2)

where $\Lambda_k = A_k^{-1} P_k A_k = [\delta_{sr} \lambda_r^k + \delta_{s-1,r} \mu_r^k]$, $s, r = 1, 2, \ldots, n_k$, $k = 1, 2$.

It is known (see Tikhonov [1965a]) that in the equations (6.2) the variables z_k^s, $s = 1, 2, \ldots, n_k$, of the vectors z_k, $k = 1, 2$, corresponding to the real proper numbers λ_k^s, will be real, and those corresponding to the complex numbers will be complex. Assuming $z_k^s = r_k^s \exp(i\theta_s)$, for real variables obtain $z_k^s = \pm r_k^s$, $\theta_s \in (0, \pi)$.

Determine the norm of the vector z_k by the formula

$$\|z_k\| = \sum_{s=1}^{n_k} |z_k^s|, \quad k = 1, 2,$$

and the norm of the vector $z = (z_1^T, z_2^T)^T$ as the sum of the norms

$$\|z\| = \|z_1\| + \|z_2\|.$$

Determine the moving set $\mathcal{A}^*(\varkappa)$ as follows:

$$\mathcal{A}^*(\varkappa) = \{z \in R^{n_1+n_2}: \ \|z\| = \varkappa(\alpha)\}, \tag{6.3}$$

where $\varkappa(\alpha) > 0$ at all $\alpha \in \mathcal{S} \subseteq R^d$, in addition, $\varkappa(\alpha) \to \varkappa_0 > 0$ at $\|\alpha\| \to 0$ and $\varkappa(\alpha) \to \infty$ at $\|\alpha\| \to +\infty$.

The objective of the further analysis is to obtain the conditions under which the set $\mathcal{A}^*(\varkappa)$ will be invariant for solutions of the system (6.1) and uniformly asymptotically stable.

To analyze the behavior of solutions of the system (6.2) with respect to the set (6.3), use the vector function $V = (v_1, v_2)^T$ with the components

$$\begin{aligned}
v_1 &= \|z_1\| = \sum_{s=1}^{n_1} |z_1^s| = \sum_{s=1}^{n_1} r_1^s, \\
v_2 &= \|z_2\| = \sum_{s=1}^{n_2} |z_1^s| = \sum_{s=1}^{n_2} r_2^s.
\end{aligned} \tag{6.4}$$

Transform the equations (6.2) to the form

$$\begin{aligned}
\frac{dr_k^s}{dt} &= \sigma_k^s r_k^s + \frac{1}{2} \sum_{s=1}^{n_k} \mu_k^{s-1} (r_k^{s-1} e^{-i\theta_s} + \bar{r}_k^{s-1} e^{i\theta_s}) \\
&\quad + \frac{1}{2} \big(Z_k^s(r_1^s, r_2^s, \alpha) e^{-i\theta_s} + \bar{Z}_k^s(r_1^s, r_2^s, \alpha) e^{i\theta_s} \big), \\
&\quad s = 1, \ldots, n_k, \quad k = 1, 2,
\end{aligned} \tag{6.5}$$

where $\sigma_k^s = \operatorname{Re} \lambda_s^k$.

For the full derivatives of the components (6.4) of the vector function $V = (v_1, v_2)^T$ obtain the expressions

$$\begin{aligned}
\frac{dv_k}{dt} &= \sum_{s=1}^{n_k} \frac{dr_k^s}{dt} = \sum_{s=1}^{n_k} \sigma_k^s r_k^s + \frac{1}{2} \sum_{s=1}^{n_k} \mu_k^{s-1} (r_k^{s-1} e^{-i\theta_s} + \bar{r}_k^{s-1} e^{i\theta_s}) \\
&\quad + \frac{1}{2} \sum_{s=1}^{n_k} (Z_k^s(r_1^s, r_2^s, \alpha) e^{-i\theta_s} + \bar{Z}_k^s(r_1^s, r_2^s, \alpha) e^{i\theta_s}), \quad k = 1, 2.
\end{aligned} \tag{6.6}$$

Note that

$$\sum_{s=1}^{n_k} \sigma_k^s r_k^s \geq \sigma_m^k v_k, \quad \sigma_m^k = \min_s \{\sigma_k^s\}, \tag{6.7}$$

$$\frac{1}{2} \sum_{s=1}^{n_k} \mu_k^{s-1} (r_k^{s-1} e^{-i\theta_s} + \bar{r}_k^{s-1} e^{i\theta_s}) \geq \mu_m^k v_k, \quad \mu_m^k \leq \mu_k^{s-1}, \tag{6.8}$$

$$\sum_{s=1}^{n_k} \sigma_k^s r_k^s \leq \sigma_M^k v_k, \quad \sigma_M^k = \max_s \{\sigma_k^s\}, \tag{6.9}$$

$$\frac{1}{2} \sum_{s=1}^{n_k} \mu_k^{s-1} (r_k^{s-1} e^{-i\theta_s} + \bar{r}_k^{s-1} e^{i\theta_s}) \leq \mu_M^k v_k, \quad \mu_M^k \geq \mu_k^{s-1}. \tag{6.10}$$

The estimates $(6.7)-(6.10)$ will further be applied for the estimation of the full derivatives of the functions v_k, $k = 1,2$ along solutions of the full system (6.1).

We now indicate the conditions under which the solutions of the system (6.1) uniformly asymptotically tend to the moving invariant set (6.3).

Theorem 6.1 *Let the equations of perturbed motion (6.1) be such that under the given function $\varkappa(\alpha)$ there exists a vector $\beta = (\beta_1, \beta_2)^{\mathrm{T}}$, $\beta_k \neq 0$, and*

(1) *for the function $V(r_1, r_2, \beta) = \sum_{k=1}^{2} \beta_k v_k(r_k)$ the following estimates are true*

$$a(\|r\|) \leq V(r_1, r_2, \beta) \quad at \quad \|r_1\| + \|r_2\| > \varkappa(\alpha),$$
$$V(r_1, r_2, \beta) \leq b(\|r\|) \quad at \quad \|r_1\| + \|r_2\| \leq \varkappa(\alpha),$$

where the functions a, b belong to K-class and $\|r\| = \|r_1\| + \|r_2\|$;

(2) *for the given function $\varkappa(\alpha)$ and functions v_1, v_2 there exist constants $\bar{a}_{ij} \in R$ and $\underline{a}_{ij} \in R$ such that:*

(a) *at $\|r\| < \varkappa(\alpha)$*

$$\frac{1}{2} \sum_{s=1}^{n_k} \left(Z_k^s(r_1^s, r_2^s, \alpha) e^{-i\theta_s} + \bar{Z}_k^s(r_1^s, r_2^s, \alpha) e^{i\theta_s} \right) > [v_k]^{1/2} \sum_{j=1}^{2} \underline{a}_{ij} [v_j]^{1/2}$$

$$at \; all \quad \alpha \in \mathcal{S} \subseteq R^d,$$

(b) *at $\|r\| = \varkappa(\alpha)$*

$$\sum_{k=1}^{2} \beta_k \left[\sum_{s=1}^{n_k} \sigma_k^s r_k^s + \frac{1}{2} \sum_{s=1}^{n_k} \mu_k^{s-1} \left(r_k^{s-1} e^{-i\theta_s} + \bar{r}_k^{s-1} e^{i\theta_s} \right) \right.$$

$$\left. + \frac{1}{2} \sum_{s=1}^{n_k} \left(Z_k^s(r_1^s, r_2^s, \alpha) e^{-i\theta_s} + \bar{Z}_k^s(r_1^s, r_2^s, \alpha) e^{i\theta_s} \right) \right] = 0$$

$$at \; all \quad \alpha \in \mathcal{S} \subseteq R^d,$$

(c) *at $\|r\| > \varkappa(\alpha)$*

$$\frac{1}{2} \sum_{s=1}^{n_k} \left(Z_k^s(r_1^s, r_2^s, \alpha) e^{-i\theta_s} + \bar{Z}_k^s(r_1^s, r_2^s, \alpha) e^{i\theta_s} \right) < [v_k]^{1/2} \sum_{j=1}^{2} \bar{a}_{ij} [v_j]^{1/2}$$

$$at \; all \quad \alpha \in \mathcal{S} \subseteq R^d;$$

(3) *the following conditions are satisfied:*

(a) *the matrix* $P = [p_{kj}]$ *with the elements*

$$
p_{kj} = \begin{cases} \beta_k[(\sigma_M^k + \mu_M^k) + \bar{a}_{kj}] & at \quad k = j, \\ (\beta_k \bar{a}_{kj} + \beta_j \bar{a}_{jk})/2 & at \quad k \neq j \end{cases}
$$

is negative-definite,

(b) *the matrix* $Q = [q_{kj}]$ *with the elements*

$$
q_{kj} = \begin{cases} \beta_k[(\sigma_m^k + \mu_m^k) + \underline{a}_{kj}] & at \quad k = j, \\ (\beta_k \underline{a}_{kj} + \beta_j \underline{a}_{jk})/2 & at \quad k \neq j \end{cases}
$$

is positive-definite;

(4) *under the given functions* $\varkappa(\alpha)$ *the relation*

$$
a(\varkappa(\alpha)) = b(\varkappa(\alpha))
$$

is true at all $\alpha \in \mathcal{S} \subseteq R^d$.

Then for the system (6.1) *the set* $\mathcal{A}^*(\varkappa)$ *is invariant and uniformly asymptotically stable.*

Proof From the components (6.4) of the vector-function $V = (v_1, v_2)$ construct a scalar Lyapunov function

$$
V(r_1, r_2, \beta) = \sum_{k=1}^{2} \beta_k v_k(r_k), \tag{6.11}
$$

where $\beta_k \neq 0$, $k = 1, 2$. Obviously, the function (6.11) is continuously differentiable, positive-definite, and radially unbounded.

Taking into account (6.6), for the expression of the full derivative of the function (6.11) find

$$
\left. \frac{dV}{dt} \right|_{(6.2)} = \sum_{k=1}^{2} \beta_k \left[\sum_{s=1}^{n_k} \sigma_k^s r_k^s + \frac{1}{2} \sum_{s=1}^{n_k} \mu_k^{s-1} (r_k^{s-1} e^{-i\theta_s} + \bar{r}_k^{s-1} e^{i\theta_s}) \right.
$$

$$
\left. + \frac{1}{2} \sum_{s=1}^{n_k} (Z_k^s(r_1^s, r_2^s, \alpha) e^{-i\theta_s} + \bar{Z}_k^s(r_1^s, r_2^s, \alpha) e^{i\theta_s}) \right] \quad at \ all \quad \alpha \in \mathcal{S} \subseteq R^d.
$$

$$
\tag{6.12}
$$

Now it is necessary to determine the conditions under which the function (6.11) increases along solutions of the system (6.2) in the domain $\|z\| < \varkappa(\alpha)$ and decreases in the domain $\|z\| > \varkappa(\alpha)$ at all $\alpha \in \mathcal{S} \subseteq R^d$.

Let $\|z\| < \varkappa(\alpha)$ at all $\alpha \in \mathcal{S} \subseteq R^d$. According to the estimates (6.7), (6.8) and the condition (2)(a) of the theorem, obtain

$$\left.\frac{dV}{dt}\right|_{(6.2)} > \sum_{k=1}^{2}\left\{\beta_k(\alpha_m^k + \mu_m^k)v_k + \beta_k[v_k]^{1/2}\sum_{j=1}^{2}\underline{a}_{ij}|v_j|^{1/2}\right\}$$

at all $\alpha \in \mathcal{S} \subseteq R^d$.

Denote $w^{\mathrm{T}} = (v_1^{1/2},\, v_2^{1/2})$ and $Q^* = [q_{kj}^*]$, where

$$q_{kj}^* = \begin{cases} \beta_k[(\sigma_m^k + \mu_m^k) + \underline{a}_{kj}] & \text{at} \quad k = j, \\ \beta_k\underline{a}_{kj} & \text{at} \quad k \neq j. \end{cases}$$

For the estimate (6.12) obtain the following inequality:

$$\left.\frac{dV}{dt}\right|_{(6.2)} > w^{\mathrm{T}}Q^*w = w^{\mathrm{T}}\left[\frac{1}{2}\left(Q^* + Q^{*T}\right)\right]w = w^{\mathrm{T}}Qw$$

at all $\alpha \in \mathcal{S} \subseteq R^d$.

By the condition (3)(a) of the Theorem 6.1 the matrix Q is positive-definite and symmetrical, therefore all its eigenvalues are real and $\lambda_m(Q) > 0$. Hence

$$\left.\frac{dV}{dt}\right|_{(6.2)} > \lambda_m(Q)w^{\mathrm{T}}w = \lambda_m(Q)\xi_M^{-1}V, \tag{6.13}$$

since $w^{\mathrm{T}}w = (v_1 + v_2) \geq \sigma_M^{-1}V$ and $\xi_M = \max_k(\beta_k)$.

Thus, in the domain $\|r\| < \varkappa(\alpha)$ the function (6.11) along solutions of the system (6.2) increases at all $\alpha \in \mathcal{S} \subseteq R^d$.

The inequalities (6.9), (6.10) are true and with the condition (2)(b) of the Theorem 6.1 satisfied, the reasoning similar to the above results in the following expression for the estimation of the function dV/dt in the domain $\|z\| > \varkappa(\alpha)$:

$$\left.\frac{dV}{dt}\right|_{(6.2)} < \lambda_M(P)w^{\mathrm{T}}w = \lambda_M(P)\xi_m^{-1}V, \tag{6.14}$$

where $\xi_m = \min_k(\beta_k)$, $\lambda_M(P) < 0$.

Hence, in the domain $\|r\| > \varkappa(\alpha)$ the function (6.11) along solutions of the system (6.2) decreases at all $\alpha \in \mathcal{S} \subseteq R^d$.

Taking into account the conditions (6.13) and (6.14) together with the other conditions of the Theorem 6.1, one can make the conclusion about the uniform asymptotic stability of the moving invariant set (6.3).

Note that the inequalities (2)(a), (c) and the relation (2)(b) contain the uncertainties parameter $\alpha \in \mathcal{S} \subseteq R^d$. The actual check of these conditions when solving specific problems is made by the method corresponding to the specification of constraints for uncertainties parameters. The development of new efficient procedures of checking those conditions, parallel with the existing ones, is of separate interest.

6.2   Interval Stability of a Linear Mechanical System

In this section a new approach to the analysis of the integral stability is proposed, which is based on the idea of maximal expansion of the initial system in combination with the method of vector Lyapunov functions.

Consider the linear system of differential equations

$$\frac{dx}{dt} = Ax, \quad x(t_0) = x_0, \tag{6.15}$$

where $x \in R^n$, $t \in R_+ = [0, \infty)$ and A is a constant $(n \times n)$-matrix.

The system

$$\frac{dy}{dt} = \tilde{A}y, \quad y(t_0) = y_0, \tag{6.16}$$

where $y \in R^m$, $t \in R_+$, $\tilde{A}$ is a constant $(m \times m)$-matrix, $m \geq n$, is called an expansion of the system (6.15) if the connection between solutions of the systems (6.15) and (6.16) is determined by the relation

$$x(t; t_0, x_0) = T^I y(t; t_0, y_0) \quad \text{at all} \quad t \geq t_0, \tag{6.17}$$

as soon as $y_0 = Tx_0$. Here T is some $(m \times n)$-matrix and T^I is the generalized inverse transformation, $m \geq n$.

Note that in the case $n \leq m$ the sense of the relation (6.17) coincides with that in the Definition 2.4 of the article by Ikeda and Šiljak [1981], where the concept of the generalized decomposition of a dynamic system is introduced. The system (6.16) is decomposed, with crossing of the main diagonal blocks of the matrix A.

The system (6.16) is an expansion of the system (6.15) if, and only if, for the matrix $M = \tilde{A} - TAT^I$ one of the relations $MT = 0$ and $T^I M = 0$ holds true.

Now consider a system of equations of the form

$$\frac{dx}{dt} = (A_0 + pA_1)x, \quad x(t_0) = x_0, \tag{6.18}$$

where $x \in R^n$, A_0 and A_1 are constant $(n \times n)$-matrices, $p \in [0, \bar{p}] \subseteq R_+$ is a constant scalar parameter.

Along with the system (6.18), consider the system

$$\frac{dy_1}{dt} = A_0 y_1 + pA_1 y_2, \quad y_1(t_0) = y_{10},$$

$$\frac{dy_2}{dt} = (p - \bar{p})A_1 y_1 + (A_0 + \bar{p}A_1)y_2, \quad y_2(t_0) = y_{20}, \tag{6.19}$$

where $y_1 \in R^n$ and $y_2 \in R^n$, $\bar{p} > 0$ are some fixed values of parameter.

It is easy to check that the system (6.19) is an expansion of the system (6.18), since the condition $MT = 0$ is satisfied, where $T = (I, I)^\mathrm{T}$.

Assume that in the system (6.19) the matrices $A_0 + \bar{p}A_1$ are stable, i.e., there exist symmetrical positive-definite matrices B_1 and B_2 which satisfy the matrix Lyapunov equations

$$A_0^\mathrm{T} B_1 + B_1 A_0 = -2Q_1, \qquad (6.20)$$

$$(A_0 + \bar{p}A_1)^\mathrm{T} B_2 + B_2(A_0 + \bar{p}A_1) = -2Q_2 \qquad (6.21)$$

for some symmetrical positive-definite matrices Q_1, Q_2.

Definition 6.1 The system (6.18) is *intervally stable*, if the equilibrium state $x = 0$ is asymptotically stable in the sense of Lyapunov at any $p \in [0, \bar{p}]$.

The conditions for the interval stability of the system (6.18) are contained in the following statement.

Theorem 6.2 *Let the system* (6.18) *be such that:*

(1) *the matrices A_0 and $A_0 + \bar{p}A_1$ at some $\bar{p} \in S$ are stable;*

(2) *the value $\bar{p}$ satisfies the inequality*

$$\bar{p}^2 < \frac{4\lambda_m(Q_1)\lambda_m(Q_2)}{\|B_1 A_1\| \|B_2 A_1\|}, \qquad (6.22)$$

where $\lambda_m(\cdot)$ are minimal proper numbers of the matrices Q_1 and Q_2 from the equations (6.20) *and* (6.21) *respectively.*

Then the system (6.18) *is intervally stable.*

Proof For the expanded system (6.19) construct the vector Lyapunov function

$$V(y) = (v_1(y_1), v_2(y_2))^\mathrm{T},$$

where $v_1(y_1) = y_1^\mathrm{T} B_1 y_1$ and $v_2(y_2) = y_2^\mathrm{T} B_2 y_2$.

Consider the full derivatives of the components of the vector Lyapunov function along solutions of the system (6.19):

$$\left.\frac{dv_1}{dt}\right|_{(6.19)} = 2y_1^\mathrm{T} B_1(A_0 y_1 + pA_1 y_2) \le -2\lambda_m(Q_1)\|y_1\|^2$$

$$+ 2|p| \|B_1 A_1\| \|y_1\| \|y_2\|,$$

$$\left.\frac{dv_2}{dt}\right|_{(6.19)} = 2y_2^\mathrm{T} B_2((A_0 + pA_1)y_2 + (p - \bar{p})A_1 y_1) \le -2\lambda_m(Q_2)\|y_2\|^2$$

$$+ 2|p - \bar{p}| \|B_2 A_1\| \|y_1\| \|y_2\|.$$

According to the results of the monograph by Šiljak [1978], for the asymptotic stability of the system (6.19) it is sufficient that the matrix

$$S = \begin{pmatrix} -\lambda_m(Q_1) & |p|\,\|B_1 A_1\| \\ |p - \overline{p}|\,\|B_2 A_1\| & -\lambda_m(Q_2) \end{pmatrix}$$

should be an M-matrix, i.e., the following condition should be satisfied:

$$\lambda_m(Q_1)\lambda_m(Q_2) - |p|\,|\overline{p} - p|\,\|B_1 A_1\|\,\|B_2 A_1\| > 0.$$

Note that for $p \in [0, \overline{p}]$ the inequality $|p|\,|\overline{p}-p| \le \dfrac{1}{4}\overline{p}^2$ is true, therefore, taking into account the condition (2) of the Theorem 6.2, we obtain the inequality (6.22). This completes the proof of the Theorem 6.2, since the stability of the system (6.19) implies the stability of the system (6.18).

Consider a mechanical system with a finite number of degrees of freedom, which is described by the equations (see Cao and Shu [1999])

$$(R + p\widetilde{R})\frac{d^2 x}{dt^2} + (D + p\widetilde{D})\frac{dx}{dt} + (K + p\widetilde{K})x = 0, \qquad (6.23)$$

where $x \in R^n$, R, D, K, $\widetilde{R}$, $\widetilde{D}$, $\widetilde{K}$ are constant $(n \times n)$-matrices.

Assume that $\det R \neq 0$ and $|p| < \|R^{-1}\widetilde{R}\|^{-1}$, and transform the system (6.23) to the form

$$\frac{dy}{dt} = \left(A_0 + pA_1 + \ldots + p^k A_k + \ldots\right) y,$$

where $y = (x, \dot{x})^{\mathrm{T}}$, $y \in R^{2n}$, and A_k are constant $(2n \times 2n)$-matrices of the form

$$A_0 = \begin{pmatrix} 0 & I \\ -R^{-1}K & -R^{-1}D \end{pmatrix},$$

$$A_k = (-1)^k F \begin{pmatrix} (R^{-1}\widetilde{R})^{k-1} & 0 \\ 0 & (R^{-1}\widetilde{R})^{k-1} \end{pmatrix} G, \quad k \ge 1.$$

Here

$$F = \begin{pmatrix} 0 & 0 \\ 0 & R^{-1} \end{pmatrix}, \quad G = \begin{pmatrix} 0 & 0 \\ \widetilde{K} - R^{-1}\widetilde{R}K & \widetilde{D} - R^{-1}\widetilde{R}D \end{pmatrix}.$$

Now assume that the matrices A_0 and $\overline{A}$ are stable, and

$$\overline{A} = \begin{pmatrix} 0 & I \\ -\overline{R}^{-1}\overline{K} & -\overline{R}^{-1}\overline{D} \end{pmatrix},$$

$$\overline{R} = R + \overline{p}\widetilde{R}, \quad \overline{K} = K + \overline{p}\widetilde{K}, \quad \overline{D} = D + \overline{p}\widetilde{D}.$$

In this case, for the given positive-definite matrices Q_1 and Q_2 there exist solutions of the Lyapunov equations

$$A_0^{\mathrm{T}} P_1 + P_1 A_0 = -2Q_1,$$

$$\overline{A}^{\mathrm{T}} P_2 + P_2 \overline{A} = -2Q_2,$$

in the form of positive-definite symmetrical matrices P_1 and P_2.

Theorem 6.3 *Let the interval system (6.23) be such that:*

(1) *the matrices A_0 and $\overline{A}$ are stable;*

(2) *the value $\overline{p}$ satisfies the inequality*

$$\overline{p}^2 < \frac{4\lambda_m(Q_1)\lambda_m(Q_2)(1 - \overline{p}\|R^{-1}\widetilde{R}\|)^3}{\|P_1F\|\,\|P_2F\|\,\|G\|^2}.$$

Then the system (6.23) is intervally stable.

Proof Consider a system which is an expansion of the system (6.23):

$$\frac{dz_1}{dt} = A_0 z_1 + \sum_{k=1}^{\infty} p^k A_k z_2,$$

$$\frac{dz_2}{dt} = \overline{A} z_2 + \sum_{k=1}^{\infty} (p^k - \overline{p}^k) A_k z_1. \tag{6.24}$$

For the system (6.24) construct the vector Lyapunov function $V(z) = (v_1(z_1), v_2(z_2))^{\mathrm{T}}$, where $v_1(z_1) = z_1^{\mathrm{T}} P_1 z_1$, $v_2(z_2) = z_2^{\mathrm{T}} P_2 z_2$, and estimate the full derivatives of the components of the vector function along solutions of this system:

$$\left.\frac{dv_1}{dt}\right|_{(6.24)} = 2z_1^{\mathrm{T}} P_1 \Big(A_0 z_1 + \sum_{k=1}^{\infty} p^k A_k z_2\Big) \le -2\lambda_m(Q_1)\|z_1\|^2$$

$$+ 2|p|\,\|P_1F\|\,\|G\| \sum_{k=1}^{\infty} \overline{p}^{k-1}\|R^{-1}\widetilde{R}\|^{k-1}\|z_1\|\,\|z_2\|$$

$$= -2\lambda_m(Q_2)\|z_1\|^2 + \frac{2|p|\,\|P_1F\|\,\|G\|}{1 - \overline{p}\|R^{-1}\widetilde{R}\|}\|z_1\|\,\|z_2\|,$$

$$\left.\frac{dv_2}{dt}\right|_{(6.24)} = 2z_1^{\mathrm{T}} P_1 \Big(\overline{A} z_2 + \sum_{k=1}^{\infty} (p^k - \overline{p}^k) A_k z_1\Big) \le -2\lambda_m(Q_2)\|z_2\|^2$$

$$+ 2\|P_2F\|\,\|G\| \sum_{k=1}^{\infty} \|R^{-1}\widetilde{R}\|^{k-1}|p^k - \overline{p}^k|\,\|z_1\|\,\|z_2\|$$

$$\le -2\lambda_m(Q_2)\|z_2\|^2$$

$$+ 2\|P_2F\|\,\|G\|\,|p - \overline{p}| \sum_{k=1}^{\infty} k\overline{p}^{k-1}\|R^{-1}\widetilde{R}\|^{k-1}\|z_1\|\,\|z_2\|$$

$$= -2\lambda_m(Q_2)\|z_2\|^2 + 2\|P_2F\|\,\|G\|\frac{|p - \overline{p}|}{(1 - \overline{p}\|R^{-1}\widetilde{R}\|)^2}\|z_1\|\,\|z_2\|.$$

The system (6.24) is stable if the matrix

$$
S_1 = \begin{pmatrix} \lambda_m(Q_1) & \dfrac{\|P_1 F\|\,\|G\|\,|p|}{1 - \overline{p}\|M^{-1}\widetilde{M}\|} \\[2ex] \dfrac{\|P_2 F\|\,\|G\|\,|p - \overline{p}|}{(1 - \overline{p}\|M^{-1}\widetilde{M}\|)^2} & \lambda_m(Q_2) \end{pmatrix}
$$

is an M-matrix. For that it is sufficient that the following inequality be true:

$$
\det S_1 = \lambda_m(Q_1)\lambda_m(Q_2) - \frac{\overline{p}^2\,\|P_1 F\|\,\|P_2 F\|\,\|G\|^2}{4(1 - \overline{p}\|M^{-1}\widetilde{M}\|)^3} > 0.
$$

The further proof of the Theorem 6.3 is similar to that of the Theorem 6.2.

6.3   Parametric Stability of an Uncertain Time Invariant System

Consider the large-scale system

$$
\frac{dx}{dt} = f(x, p), \tag{6.25}
$$

where $x = (x_1, \ldots, x_n)^{\mathrm{T}}$, $p = (p_1, \ldots, p_m)^{\mathrm{T}}$, $f = (f_1, \ldots, f_n)^{\mathrm{T}}$, $f_i \colon \mathbb{R}^n \times \mathbb{R}^m \to \mathbb{R}$, and $i = n$. Let the function $f(x, p)$ be sufficiently smooth and let the conditions for the existence and uniqueness of the initial problem for the system (6.25) exist. Represent the system (6.25) in the form

$$
\frac{dx_i}{dt} = g_i(x_i, p) + h_i(x, p), \quad i = 1, \ldots, s. \tag{6.26}
$$

The equations

$$
\frac{dx_i}{dt} = g_i(x_i, p), \quad i = 1, \ldots, s,
$$

where $x = (x_1^{\mathrm{T}}, \ldots, x_s^{\mathrm{T}})^{\mathrm{T}}$, the functions $g_i \colon \mathbb{R}^{n_i} \times \mathbb{R}^m \to \mathbb{R}^{n_i}$, describe the dynamics of the independent subsystems of the system (6.26).

Let for some value of the parameter p^* the system (6.25) have the equilibrium state $x^* = ((x_1^*)^{\mathrm{T}}, \ldots, (x_s^*)^{\mathrm{T}})^{\mathrm{T}}$, $x_i^* \in \mathbb{R}^{n_i}$, $i = 1, \ldots, s$, so that

$$
g_i(x_i^*, p^*) + h_i(x^*, p^*) = 0 \quad \text{at all} \quad i = 1, \ldots, s. \tag{6.27}
$$

Note that the condition (6.27) does not imply that $g_i(x_i^*, p^*) = 0$, i.e., x_i^* is not necessarily the equilibrium state of the i-th subsystem. Determine the functions $\tilde{g}_i \colon \mathbb{R}^{n_i} \times \mathbb{R}^m \to \mathbb{R}^{n_i}$, $\tilde{h}_i \colon \mathbb{R}^n \times \mathbb{R}^m \to \mathbb{R}^{n_i}$, $i = 1, \ldots, s$ in the form

$$
\tilde{g}_i(x_i, p) = g_i(x_i, p) - g_i(x_i^e(p), p),
$$
$$
\tilde{h}_i(x, p) = h_i(x, p) - h_i(x^e(p), p),
$$

where $x^e(p) = ((x_1^e(p))^{\mathrm{T}}, \ldots, (x_s^e(p))^{\mathrm{T}})^{\mathrm{T}}$ is the equilibrium state of the system (6.26), corresponding to the value of the parameter p. Instead of the system (6.26) consider the system

$$\frac{dx_i}{dt} = \tilde{g}_i(x_i, p) + \tilde{h}_i(x, p), \quad i = 1, \ldots, s, \tag{6.28}$$

for which $\tilde{g}_i(x_i^e(p), p) = 0$ and $\tilde{h}_i(x^e(p), p) = 0$ at all $i = 1, \ldots, s$ and which obviously has the same solutions as the system (6.26) under the same initial conditions. Thus, along with the system (6.26), we will analyze the system (6.28) which consists of s related subsystems

$$\frac{dx_i}{dt} = \tilde{g}_i(x_i, p), \quad i = 1, \ldots, s. \tag{6.29}$$

We make the following assumptions regarding the systems (6.25) and (6.28).

Assumption 6.1 The systems of equations (6.25), (6.28) are such that:

(1) the functions

$$f_i(x, p), \quad i = 1, \ldots, n,$$

are defined and continuous on some open set $\Gamma \subset \mathbb{R}^n \times \mathbb{R}^m$;

(2) the functions $\tilde{g}_i(x_i, p)$, $i = 1, \ldots, s$, are defined and continuous on some open set $\Gamma_i \subset \mathbb{R}^{n_i} \times \mathbb{R}^m$, $i = 1, \ldots, s$, $\mathbb{R}^{n_1} \times \ldots \times \mathbb{R}^{n_s} = \mathbb{R}^n$, together with elements of the matrices

$$\frac{\partial \tilde{g}_i(x_i, p)}{\partial x_i}, \quad \left(\left| \frac{\partial^2 g_{n_1 + \cdots + n_{i-1} + k}(x_i, p)}{\partial x_{n_1 + \cdots + n_{i-1} + l} \partial x_i} \right| \right)_{k,l=1}^{n_i},$$

$$\left(\left| \frac{\partial^2 g_{n_1 + \cdots + n_{i-1} + k}(x_i, p)}{\partial x_{n_1 + \cdots + n_{i-1} + l} \partial R_j} \right| \right)_{k,l=1}^{n_i},$$

$$i = 1, \ldots, s, \quad j = 1, \ldots, b, \quad p = (R_1^{\mathrm{T}}, \ldots, R_b^{\mathrm{T}})^{\mathrm{T}};$$

(3) the functions $\tilde{h}_i(x, p)$ are defined and continuous on some open set $\Gamma \subset \mathbb{R}^n \times \mathbb{R}^m$ together with the partial derivatives

$$\frac{\partial \tilde{h}_i(x, p)}{\partial x_j}, \quad i, j = 1, \ldots, s;$$

(4) for some value of the parameter vector p^* there exists an equilibrium state $x^* = x^e(p^*) = ((x_1^*)^{\mathrm{T}}, \ldots, (x_s^*)^{\mathrm{T}})^{\mathrm{T}}$, $x_i^* \in \mathbb{R}^{n_i}$, $i = 1, \ldots, s$, of the system (6.28), so that $f(x^*, p^*) = 0$, $\tilde{g}_i(x_i^*, p^*) = 0$, $\tilde{h}_i(x^*, p^*) = 0$, $(x^*, p^*) \in \Gamma$;

(5) the following inequality is true:

$$\det \left(\left. \frac{\partial f(x, p)}{\partial x} \right|_{(x^*, p^*)} \right) \neq 0;$$

(6) the matrices

$$\frac{\partial g_i(x_i, p^*)}{\partial x_i} + \frac{\partial h_i(x, p^*)}{\partial x_i}$$

are stable in the point x^*.

If in the space $\mathbb{R}^m$ a domain Ω_p is defined such that for each p from Ω_p there exists an equililbrium state of the system (6.28), then the question of the parametric asymptotic stability of the system (6.28) with respect to the domain Ω_p comes to the question of the existence of an appropriate Lyapunov function.

To apply the comparison method for the solution of the problem, see below some auxiliary results.

Definition 6.2 The function $f \in C(\mathbb{R}^n, \mathbb{R}^n)$ is called *weakly quasimonotone nondecreasing with respect to the cone* $K \subset \mathbb{R}^n$, if for any positive functional $\varphi \in K^*$ the conditions $0 \xrightarrow{K} \leq x \xrightarrow{K} \leq y$ and $\varphi(y - x) = 0$ imply the inequality $\varphi(f(y) - f(x)) \geq 0$.

We now prove the following statement.

Lemma 6.1 *Let the function* $g \in C(\mathbb{R}_+ \times K, \mathbb{R}^m)$ *and* $g(t, u)$ *be a function, weakly quasimonotone nondecreasing with respect to* u *relative to the cone* $K \subset \mathbb{R}^n$ *for each* $t \in \mathbb{R}_+$. *Let* $r(t)$ *be the maximal solution of the system*

$$\frac{du}{dt} = g(t, u), \quad u(t_0) = u_0,$$

with respect to the cone $K \subset \mathbb{R}^n$, *which exists on the interval* $[t_0, +\infty)$, *and for* $t \geq t_0$

$$\frac{dV(t)}{dt} \xrightarrow{K} \leq g(t, V(t)),$$

where $V(t) \in C(\mathbb{R}_+, \mathbb{R}^m)$.
Then if $V(t)$ *is a function positive with respect to the cone* $K \subset \mathbb{R}^n$, $V(t) \xrightarrow{K} \geq 0$, $t \geq t_0$, *then the inequality* $V(t_0) \xrightarrow{K} \leq u_0$ *implies the estimate*

$$V(t) \xrightarrow{K} \leq r(t) \quad \text{at all} \quad t \geq t_0.$$

Proof Let $t_0 < T < +\infty$. Since $r(t) = \lim_{\varepsilon \to 0} u(t, \varepsilon)$ is uniform on $[t_0, T]$ at all sufficiently small $\varepsilon > 0$, where $u(t, \varepsilon)$ is a solution of the auxiliary system of equations

$$\frac{du}{dt} = g(t, u) + \varepsilon\psi, \quad u(t_0) = u_0 + \varepsilon\psi,$$

and ψ is some fixed element from int K, it is sufficient to prove that

$$V(t) \xrightarrow{K} < u(t, \varepsilon) \quad \text{at all} \quad t \in [t_0, T].$$

Assume that this inequality is not true. Then there exists $t_1 \in (t_0, T]$ such that

$$u(t_1, \varepsilon) - V(t_1) \in \partial K, \quad u(t, \varepsilon) \overset{K}{\to} > V(t) \overset{K}{\to} \geq 0, \quad t \in [t_0, t_1).$$

It means that there exists a functional $\varphi \in K_0^*$, $K_0 = K - \{0\}$, such that

$$\varphi(u(t_1, \varepsilon) - V(t_1)) = 0.$$

From the weak quasimonotony of the function g it follows that

$$\varphi(g(t_1, u(t_1, \varepsilon)) - g(t_1, V(t_1))) \geq 0.$$

Assuming $w(t) = \varphi(u(t, \varepsilon) - V(t))$, $t \in [t_0, t_1]$, obtain $w(t) > 0$, $t \in [t_0, t_1)$ and $w(t_1) = 0$,

$$\frac{dw(t_1)}{dt} = \lim_{h \to 0} \frac{w(t_1 + h) - w(t_1)}{h}$$

irrespective of the way in which h tends to zero. If $h \to 0-$, obtain $\dfrac{dw(t_1)}{dt} \leq 0$. On the other hand,

$$\frac{dw(t_1)}{dt} = \varphi\left(\frac{du(t_1, \varepsilon)}{dt} - \frac{dV(t_1)}{dt}\right) = \varphi\left(g(t_1, u(t_1, \varepsilon)) + \varepsilon\psi - \frac{dV(t_1)}{dt}\right)$$
$$> \varphi(g(t_1, u(t_1, \varepsilon)) - g(t_1, V(t_1))) \geq 0.$$

The obtained contradiction completes the proof of the Lemma 6.1.

Thus, if as a result of the estimation of the derivative of the Lyapunov function $V(t, x)$ one obtains the inequality

$$\frac{dV(t)}{dt} \leq g(t, V(t)),$$

then, as a comparison system, one can use the system

$$\frac{du}{dt} = g(t, u),$$

where $g(t, u)$ is a function weakly quasimonotone nondecreasing in u with respect to some cone, and $V(t)$ is a positive function with respect to the same cone.

Let us formulate the criterion of Sevastianov-Kotelanskiy (see Matrosov [2001]), we will need it for the solution of the main theorem.

Proposition 6.1 For the equilibrium state $x = 0$ of the system $\dot{x} = Ax$, where the function $f(x) = Ax$ satisfies Wazevsky's condition with respect to the cone $\mathbb{R}_+^n$, to be asymptotically stable, it is necessary and sufficient that $\text{sign}\,\Delta_i(A) = (-1)^i$, $i = 1, \ldots, n$, where $\Delta_i(A)$ are the principal minors of the matrix A.

The boundary of the possible domain of the parametric asymptotic stability of the system (6.25)

$$\Pi_{r,q} = \{(x,p) \mid \Omega_{r_i}: \ \|x_i - x_i^*\| < r_i, \quad \Omega_{q_i}: \ \|R_j - R_j^*\| < q_j,$$
$$i = 1, \ldots, s, \quad j = 1, \ldots, b\}$$

can be estimated by using the approach indicated in Appendix. Note that later for matrices the Hilbert-Schmidt norm will be used.

Introduce the notation

$$\Phi_i = r_i \max_{\overline{\Pi}_{r_i,q}} \|A_i(x_i, p)\| + \sum_{c=1}^{s} \max_{\overline{\Pi}_{r,q}} \|A_{ic}(x, p)\|$$

$$+ \sum_{j=1}^{b} q_j \left(\max_{\overline{\Pi}_{r_i,q}} \|B_{ij}(x_i, p)\| + \max_{\overline{\Pi}_{r,q}} \|B_{ij}(x, p)\| \right),$$

$$A_i(x_i, p) = \left(\left| \frac{\partial^2 g_{n_1 + \ldots + n_{i-1} + k}(x_i, p)}{\partial x_{n_1 + \ldots + n_{i-1} + l} \partial x_i} \right| \right)^{n_i}_{k,l=1},$$

$$B_{ij}(x_i, p) = \left(\left| \frac{\partial^2 g_{n_1 + \ldots + n_{i-1} + k}(x_i, p)}{\partial x_{n_1 + \ldots + n_{i-1} + l} \partial R_j} \right| \right)^{n_i}_{k,l=1},$$

$$A_{ic}(x, p) = \left(\left| \frac{\partial^2 h_{n_1 + \ldots + n_{i-1} + k}(x, p)}{\partial x_{n_1 + \ldots + n_{i-1} + l} \partial x_c} \right| \right)^{n_i}_{k,l=1},$$

$$B_{ij}(x, p) = \left(\left| \frac{\partial^2 h_{n_1 + \ldots + n_{i-1} + k}(x, p)}{\partial x_{n_1 + \ldots + n_{i-1} + l} \partial R_j} \right| \right)^{n_i}_{k,l=1},$$

$$\Pi_{r_i,q} = \{(x_i, p) \mid: \ \|x_i - x_i^*\| < r_i, \quad \|R_j - R_j^*\| < q_j, \quad j = 1, \ldots, b\},$$

$\lambda_m(\cdot)$ is the minimum eigenvalue of the respective matrix, Q_i, $i = 1, \ldots, s$, is an arbitrary symmetrical positive-definite matrix with the dimensions $n_i \times n_i$, P_i, and $i = 1, \ldots, s$ is a symmetrical positive-definite matrix which is a solution of the matrix equation

$$\left(\frac{\partial g_i(x_i, p^*)}{\partial x_i} \bigg|_{x_i = x_i^*} + \frac{\partial h_i(x, p^*)}{\partial x_i} \bigg|_{x = x^*} \right)^{\mathrm{T}} P_i$$

$$+ P_i \left(\frac{\partial g_i(x_i, p^*)}{\partial x_i} \bigg|_{x_i = x_i^*} + \frac{\partial h_i(x, p^*)}{\partial x_i} \bigg|_{x = x^*} \right) = -Q_i, \quad i = 1, \ldots, s.$$

$$(6.30)$$

Having used the Assumption 6.1, determine the sufficient conditions for the parametric asymptotic stability of the system (6.28) with respect to the specified domain, we can prove the following theorem.

Theorem 6.4 *Let for the vector functions* $g_i(x_i, p)$, $h_i(x, p)$, $i = 1, \ldots, s$, *the system* (6.28) *and the domain* $\Pi_{r,q}$ *the matrix* $A = (a_{ij})_{i,j=1}^{s}$ *with the* elements

$$a_{ii} = \frac{-\lambda_m(Q_i) + 2\|P_i\|\Phi_i}{2\lambda_m(P_i)},$$

$$a_{ij} = \frac{2\|P_i\|^2 \sum\limits_{\substack{k=1 \\ k \neq i}}^{s} \left(\max\limits_{\Pi_{r,q}} \left\| \frac{\partial h_i(x,p)}{\partial x_k} \right\| \right)^2}{\lambda_m(P_j)\left(\lambda_m(Q_i) - 2\|P_i\|\Phi_i\right)} \geq 0, \quad i \neq j,$$

satisfy the condition

$$\operatorname{sign}\Delta_i(A) = (-1)^i, \quad i = 1, \ldots, s,$$

where $\Delta_i(A)$ are the principal minors of the matrix A. Then the system (6.28) is parametrically asymptotically stable with respect to the system Ω_q.

Proof Choose an arbitrary value of the parameter vector p from the domain Ω_q. According to the definition of the domain $\Pi_{r,q}$ there exists an equilibrium state $x^e(p)$ from the domain Ω_r. By substitution of the variable $z = x - x^e(p)$ reduce the system (6.28) to the form

$$\frac{dz_i}{dt} = \tilde{g}_i(z_i + x_i^e(p), p) + \tilde{h}_i(z + x^e(p), p), \quad i = 1, \ldots, s, \qquad (6.31)$$

and the subsystems (6.29) to the form

$$\frac{dz_i}{dt} = \tilde{g}_i(z_i + x_i^e(p), p). \qquad (6.32)$$

As components of the vector Lyapunov function, choose the functions

$$v_i(z_i) = z_i^{\mathrm{T}} P_i z_i, \quad i = 1, \ldots, s,$$

where the matrices P_i, $i = 1, \ldots, s$ are determined from the equations (6.30). Estimate the derivatives of the components of the vector Lyapunov function along solutions of the system (6.31):

$$\left.\frac{dv_i(z_i)}{dt}\right|_{(6.31)} = (\tilde{g}_i(z_i + x_i^e(p), p))^{\mathrm{T}} P_i z_i + z_i^{\mathrm{T}} P_i(\tilde{g}_i(z_i + x_i^e(p), p))$$

$$+ (\tilde{h}_i(z + x^e(p), p))^{\mathrm{T}} P_i z_i + z_i^{\mathrm{T}} P_i(\tilde{h}_i(z + x^e(p), p))$$

$$= \left.\frac{dv_i(z_i)}{dt}\right|_{(6.32)} + z^{\mathrm{T}} \left(\left.\frac{\partial \tilde{h}_i(x,p)}{\partial x}\right|_{x=x^e(p)}\right)^{\mathrm{T}} P_i z_i$$

$$+ z_i^{\mathrm{T}} P_i \left(\left.\frac{\partial \tilde{h}_i(x,p)}{\partial x}\right|_{x=x^e(p)}\right) z + (o_i(z))^{\mathrm{T}} P_i z_i + z_i P_i(o_i(z)).$$

$$(6.33)$$

Here $o_i(z)$ is a vector with the dimension n_i and its module is an infinitesimal quantity compared with the module z in some neighborhood $z = 0$. It is known that the function $V_i(z_i)$ tolerates the following estimate of the time derivative in view of the system (6.32):

$$\left.\frac{dv_i(z_i)}{dt}\right|_{(6.32)} \leq z_i^{\mathrm{T}}\left(\left(\left.\frac{\partial g_i(x_i, p^*)}{\partial x_i}\right|_{x_i=x_i^*}\right)^{\mathrm{T}} P_i + P_i\left(\left.\frac{\partial g_i(x_i, p^*)}{\partial x_i}\right|_{x_i=x_i^*}\right)\right)z_i$$

$$+ 2\|P_i\|\left(r_i \max_{\Pi_{r_i,q}}\|A_i(x_i, p)\| + \sum_{j=1}^{b} q_j \max_{\Pi_{r_i,q}}\|B_{i,j}(x_i, p)\|\right)\|z_i\|^2$$

$$+ 2\|o(z_i)\|\|P_i\|\|z_i\|, \quad i = 1, \ldots, s,$$

where $o(z_i)$ is a vector with the dimension n_i whose module is an infinitesimal quantity compared with the module z_i in some neighborhood $z_i = 0$.

Continuing the estimate of (6.33), obtain

$$\left.\frac{dv_i(z_i)}{dt}\right|_{(6.31)} \leq z_i^{\mathrm{T}}\left(\left.\frac{\partial g_i(x_i, p^*)}{\partial x_i}\right|_{x_i=x_i^*} + \left.\frac{\partial h_i(x, p^*)}{\partial x_i}\right|_{x=x^*}\right)^{\mathrm{T}} P_i z_i$$

$$+ z_i^{\mathrm{T}} P_i\left(\left.\frac{\partial g_i(x_i, p^*)}{\partial x_i}\right|_{x_i=x_i^*} + \left.\frac{\partial h_i(x, p^*)}{\partial x_i}\right|_{x=x^*}\right)z_i$$

$$+ z_i^{\mathrm{T}}\left(\left.\frac{\partial h_i(x, p)}{\partial x_i}\right|_{x=x^e(p)} - \left.\frac{\partial h_i(x, p^*)}{\partial x_i}\right|_{x=x^*}\right)^{\mathrm{T}} P_i z_i$$

$$+ z_i^{\mathrm{T}} P_i\left(\left.\frac{\partial h_i(x, p)}{\partial x_i}\right|_{x=x^e(p)} - \left.\frac{\partial h_i(x, p^*)}{\partial x_i}\right|_{x=x^*}\right)z_i$$

$$+ 2\|P_i\|\left(r_i \max_{\Pi_{r_i,q}}\|A_i(x_i, p)\| + \sum_{j=1}^{b} q_j \max_{\Pi_{r_i,q}}\|B_{ij}(x_i, p)\|\right)\|z_i\|^2$$

$$+ 2\left(\sum_{\substack{j=1 \\ j\neq i}}^{s} \max_{\Pi_{r,q}}\left\|\frac{\partial h_i(x, p)}{\partial x_j}\right\|\|P_i\|\|z_j\|\right)\|z_i\| + 2(\|o(z_i)\| + \|o_i(z)\|)\|P_i\|\|z_i\|$$

$$\leq (-\lambda_m(Q_i) + 2\|P_i\|\Phi_i)\|z_i\|^2 + 2\left(\sum_{\substack{j=1 \\ j\neq i}}^{s} \max_{\Pi_{r,q}}\left\|\frac{\partial h_i(x, p)}{\partial x_j}\right\|\|P_i\|\|z_j\|\right)\|z_i\|$$

$$+ 2(\|o(z_i)\| + \|o_i(z)\|)\|P_i\|\|z_i\|, \quad i = 1, \ldots, s.$$

$$(6.34)$$

Estimate the last summand in the formula (6.34). Choose arbitrarily some positive number ε. According to the definition of $o_i(z)$ there exists a neighborhood $\Omega_i(\varepsilon)$ of the point $z = 0$ such that in it $\|o_i(z)\| \leq c_i\|z\|^{1+2\varepsilon}$, where c_i, $i = 1, \ldots, s$ are some positive constants. Analogously, according to the definition of $o(z_i)$ there exists a neighborhood $\Omega_{z_i}(\varepsilon)$ of the point $z_i = 0$ such that in it $\|o(z_i)\| \leq \bar{c}_i\|z_i\|^{1+2\varepsilon}$, $\bar{c}_i > 0$.

The functions $v_i(z_i) = z_i^{\mathrm{T}} P_i z_i$, $i = 1, \ldots, s$, satisfy the estimate

$$\lambda_m(P_i)\|z_i\|^2 \leq v_i(z_i) \leq \lambda_M(P_i)\|z_i\|^2. \qquad (6.35)$$

Then for all z from $\Omega_i(\varepsilon) \cap \Omega_{z_i}(\varepsilon)$ the following estimate is true:

$$2(\|o(z_i)\| + \|o_i(z)\|)\|P_i\| \|z_i\| \leq 2(c_i\|z\|^{1+2\varepsilon} + \bar{c}_i\|z_i\|^{1+2\varepsilon})\|P_i\| \|z\|$$

$$\leq 2(c_i + \bar{c}_i)\|P_i\| \|z\|^{2+2\varepsilon} = 2(c_i + \bar{c}_i)\|P_i\| \left(\sum_{j=1}^{s} \frac{v_j(z_j)}{\lambda_m(P_j)} \right)^{1+\varepsilon}$$

$$\leq 2(c_i + \bar{c}_i)\|P_i\| \|V(z)\|^{1+\varepsilon}\|\lambda\|^{1+\varepsilon} = M_i\|P_i\| \|V(z)\|^{1+\varepsilon},$$

$$i = 1, \ldots, s,$$

$$(6.36)$$

where $\lambda = \left(\dfrac{1}{\lambda_m(P_1)}, \ldots, \dfrac{1}{\lambda_m(P_s)} \right)^{\mathrm{T}}$ and $M_i = 2(c_i + \bar{c}_i)\|\lambda\|^{1+\varepsilon}$. Denote $-\lambda_m(Q_i) + 2\|P_i\|\Phi_i = -\Delta$, $\Delta > 0$, and estimate the sum of the first two summands in the formula (6.34). Taking into account that

$$-az^2 + bz \leq -\frac{a}{2}z^2 + \frac{b^2}{2a}$$

at $a > 0$, $z \in \mathbb{R}$, and considering the estimates (6.35), we obtain

$$-\Delta\|z_i\|^2 + \left(2\|P_i\| \sum_{\substack{j=1 \\ j\neq i}}^{s} \max_{\overline{\Pi}_{r,q}} \left\| \frac{\partial h_i(x,p)}{\partial x_j} \right\| \|z_j\| \right) \|z_i\|$$

$$\leq \frac{-\Delta\|z_i\|^2}{2} + \frac{\left(2\|P_i\| \sum\limits_{\substack{j=1 \\ j\neq i}}^{s} \max\limits_{\overline{\Pi}_{r,q}} \left\| \frac{\partial h_i(x,p)}{\partial x_j} \right\| \|z_j\| \right)^2}{2\Delta}$$

$$\leq \frac{-\Delta\|z_i\|^2}{2} + \frac{2\|P_i\|^2 \sum\limits_{\substack{j=1 \\ j\neq i}}^{s} \left(\max_{\overline{\Pi}_{r,q}} \left\| \frac{\partial h_i(x,p)}{\partial x_j} \right\| \right)^2 \sum\limits_{\substack{j=1 \\ j\neq i}}^{s} \|z_j\|^2}{\Delta}$$

$$\leq \frac{-\Delta}{2\lambda_m(P_i)}v_i(z_i) + \frac{\sum\limits_{\substack{j=1 \\ j\neq i}}^{s} \left(2\|P_i\|^2 \sum\limits_{\substack{k=1 \\ k\neq i}}^{s} \left(\max_{\overline{\Pi}_{r,q}} \left\| \frac{\partial h_i(x,p)}{\partial x_k} \right\| \right)^2 \frac{1}{\lambda_m(P_j)} \right) v_j(z_j)}{\Delta}.$$

$$(6.37)$$

Thus, from the formula (6.34), taking into account (6.36) and (6.37), for all z from $\Omega_i(\varepsilon) \cap \Omega_{z_i}(\varepsilon)$ the following estimates are true:

$$\left. \frac{dv_i(z_i)}{dt} \right|_{(6.31)} \leq a_{ii}v_i(z_i) + \sum_{\substack{j=1 \\ j\neq i}}^{s} a_{ij}v_j(z_j) + M_i\|V(z)\|^{1+\varepsilon}\|P_i\|, \quad i = 1, \ldots, s.$$

Hence in some neighborhood of the point $z = 0$

$$\left. \frac{dV(z)}{dt} \right|_{(6.31)} \leq AV(z) + \|V(z)\|^{1+\varepsilon}(M_1\|P_1\|, \ldots, M_s\|P_s\|)^{\mathrm{T}},$$

where the matrix $A = (a_{ij})_{i,j=1}^s$, and the elements a_{ij} are specified in the condition of the theorem.

The function

$$f(u) = Au + \|u\|^{1+\varepsilon}(M_1\|P_1\|, \ldots, M_s\|P_s\|)^{\mathrm{T}}$$

weakly quasimonotone nondecreasing in u with respect to the cone $\mathbb{R}_+^s$, and $V(z)$ is positive with respect to the same cone. Then, according to the lemma 6.1, as a comparison system for the large-scale system (6.31) one can use the system

$$\frac{du}{dt} = Au + \|u\|^{1+\varepsilon}(M_1\|P_1\|, \ldots, M_s\|P_s\|)^{\mathrm{T}}. \tag{6.38}$$

Taking into account that the matrix A satisfies the Sevastianov-Kotelyansky condition, and applying the theorem of stability by linear approximation, the equilibrium state $u = 0$ of the system (6.38) is asymptotically stable in the cone $\mathbb{R}_+^s$. According to the comparison method, the equilibrium state $z = 0$ of the system (6.31), i. e., the equilibrium state $x = x^e(p)$ of the system (6.28) is asymptotically stable. Since p was chosen arbitrarily from Ω_q, the system (6.28) is parametrically asymptotically stable with respect to the domain Ω_q.

The theorem is proved.

Example 6.1 Consider the system of Lotka-Volterra differential equations (see Freedman [1987]) in the form

$$\begin{aligned}
\dot{X}_1 &= \bar{x}_1(A_1X_1 + a_1 + A_{11}X_1 + A_{12}X_2), \\
\dot{X}_2 &= \bar{x}_2(A_2X_2 + a_2 + A_{21}X_1 + A_{22}X_2),
\end{aligned} \tag{6.39}$$

where $X_1 = (x_1, x_2)^{\mathrm{T}}$, $X_2 = (x_3)$, $a_1 = (p^4+2, -p^6-p^2-1)^{\mathrm{T}}$, $a_2 = (p^2+1)$,

$$A_1 = \begin{pmatrix} -1 & -1 \\ 1 & 0 \end{pmatrix}, \quad A_2 = (-1), \quad A_{11} = \begin{pmatrix} 0 & -(p^4-p^2) \\ p^4-p^2 & 0 \end{pmatrix},$$

$$A_{12} = \begin{pmatrix} 0 \\ p^2 \end{pmatrix}, \quad A_{21} = (0 \ -p^2), \quad A_{22} = (0), \quad \bar{x}_1 = \begin{pmatrix} x_1 & 0 \\ 0 & x_2 \end{pmatrix}, \quad \bar{x}_2 = (x_3).$$

Obviously, in addition to the fixed equilibrium state of the system (6.39) in the origin of coordinates for all values of the parameter p there exists another equilibrium state

$$X_1^e(p) = (p^2 + 1, \ 1), \quad X_2^e(p) = (1),$$

whose stability will be analyzed below.

Represent the system (6.39) in the form

$$\begin{aligned}
\frac{dX_1}{dt} &= \tilde{g}_1(X_1, p) + \tilde{h}_1(x, p), \\
\frac{dX_2}{dt} &= \tilde{g}_2(X_2, p) + \tilde{h}_2(x, p).
\end{aligned} \tag{6.40}$$

Here

$$\tilde{g}_1(X_1,p)= g_1(X_1,p)- g_1(X_1^e(p),p), \quad \tilde{g}_2(X_2,p)= g_2(X_2,p) - g_2(X_2^e(p),p),$$
$$\tilde{h}_1(x,p) = h_1(x,p) - h_1(x^e(p),p), \quad \tilde{h}_2(x,p) = h_2(x,p) - h_2(x^e(p),p),$$
$$g_1(X_1,p)= \bar{x}_1(A_1X_1 + a_1+ A_{11}X_1), \quad g_2(X_2,p)= \bar{x}_2(A_2X_2 + a_2 + A_{22}X_2),$$
$$h_1(x,p) = \bar{x}_1 A_{12}X_2, h_2(x,p) = \bar{x}_2 A_{21}X_1.$$

Construct the components of the vector Lyapunov functions in the form

$$v_i(X_i, X_i^e(p)) = (X_i - X_i^e(p))^{\mathrm{T}} P_i(X_i - X_i^e(p)), \quad i = 1,2,$$

where P_i are matrices determined from the matrix equations

$$\left(\frac{\partial g_i(X_i,p)}{\partial X_i}\bigg|_{(X_i^e(p),p)} + \frac{\partial h_i(x,p)}{\partial X_i}\bigg|_{(x^e(p),p)}\right)^{\mathrm{T}} P_i$$
$$+ P_i\left(\frac{\partial g_i(X_i,p)}{\partial X_i}\bigg|_{(X_i^e(p),p)} + \frac{\partial h_i(x,p)}{\partial x_i}\bigg|_{(X^e(p),p)}\right) = -I, \quad i = 1,2.$$

Here I is a unit matrix with the respective dimensions. Since the matrices $\left(\frac{\partial g_i(X_i,p)}{\partial X_i}\big|_{(X_i^e(p),p)} + \frac{\partial h_i(x,p)}{\partial X_i}\big|_{(x^e(p),p)}\right)$, $i = 1,2$, are supposed to be stable, the matrices P_i, $i = 1,2$, always exist, they are symmetrical and positive-definite.

Since

$$g_1(X_1,p) = \begin{pmatrix} (p^4 + 2)x_1 - x_1^2 - (p^4 - p^2 + 1)x_1x_2 \\ -(p^6 + p^2 + 1)x_2 + (p^4 - p^2 + 1)x_1x_2 \end{pmatrix},$$

$$h_1(x,p) = \begin{pmatrix} 0 \\ p^2 x_2 x_3 \end{pmatrix},$$

then the matrix

$$\frac{\partial g_1(X_1,p)}{\partial X_1}\bigg|_{(X_1^e(p),p)} + \frac{\partial h_1(x,p)}{\partial X_1}\bigg|_{(x^e(p),p)}$$
$$= \begin{pmatrix} -1 - p^2 & -1 - p^6 \\ 1 - p^2 + p^4 & 0 \end{pmatrix} = \begin{pmatrix} a_{11} & a_{12} \\ a_{21} & a_{22} \end{pmatrix}$$

is stable at all p.

Owing to the fact that $g_2(X_2,p) = (p^2 + 1)x_3 - x_3^2$, $h_2(x,p) = -p^2 x_2 x_3$, the one-element matrix

$$\frac{\partial g_2(X_2,p)}{\partial X_2}\bigg|_{(X_2^e(p),p)} + \frac{\partial h_2(x,p)}{\partial X_2}\bigg|_{(x^e(p),p)} = -1$$

is stable at all p.

Find the matrices P_1 and P_2:

$$P_1 = \begin{pmatrix} \dfrac{a_{22}(a_{11}+a_{22})+a_{21}(a_{21}-a_{12})}{2(a_{11}+a_{22})(a_{12}a_{21}-a_{11}a_{22})} & \dfrac{-a_{22}a_{12}-a_{11}a_{21}}{2(a_{11}+a_{22})(a_{12}a_{21}-a_{11}a_{22})} \\ \dfrac{-a_{22}a_{12}-a_{11}a_{21}}{2(a_{11}+a_{22})(a_{12}a_{21}-a_{11}a_{22})} & \dfrac{a_{11}(a_{11}+a_{22})+a_{12}(a_{12}-a_{21})}{2(a_{11}+a_{22})(a_{12}a_{21}-a_{11}a_{22})} \end{pmatrix},$$

$$P_2 = \frac{1}{2}.$$

For the system (6.40) construct the vector Lyapunov function

$$V(x, x^e(p)) = (v_1(X_1, X_1^e(p)), \; v_2(X_2, X_2^e(p))).$$

Estimate the derivatives of its time components in view of the solutions of the system (6.40):

$$\left.\frac{dv_1(X_1, X_1^e(p))}{dt}\right|_{(6.40)} \leq \frac{-v_1(X_1, X_1^e(p))}{2\lambda_m(P_1)}$$

$$+ 2\left\|\frac{\partial h_1(x,p)}{\partial X_2}\right\|_{(x^e(p),p)}^2 \|P_1\|^2 \frac{V_2(X_2, X_2^e(p))}{\lambda_m(P_2)}$$

$$+ 2\|P_1\|(c_1 + \bar{c}_1)\|\lambda\|^{1+\varepsilon}\|V\|^{1+\varepsilon},$$

$$\left.\frac{dv_2(X_2, X_2^e(p))}{dt}\right|_{(6.40)} \leq \frac{-v_2(X_2, X_2^e(p))}{2\lambda_m(P_2)}$$

$$+ 2\left\|\frac{\partial h_2(x,p)}{\partial X_1}\right\|_{(x^e(p),p)}^2 \|P_2\|^2 \frac{V_1(X_1, X_1^e(p))}{\lambda_m(P_1)}$$

$$+ 2\|P_2\|(c_2 + \bar{c}_2)\|\lambda\|^{1+\varepsilon}\|V\|^{1+\varepsilon},$$

where c_1, $\bar{c}_1$, c_2, $\bar{c}_2$ are some positive constants, ε is an arbitrary positive number, $\lambda = \left(\frac{1}{\lambda_m(P_1)}, \frac{1}{\lambda_m(P_2)}\right)$. Calculate

$$\left\|\frac{\partial h_1(x,p)}{\partial X_2}\right\|_{(x^e(p),p)} = p^2, \quad \left\|\frac{\partial h_2(x,p)}{\partial X_1}\right\|_{(x^e(p),p)} = p^2.$$

Thus, for the system (6.40) obtain the comparison system

$$\frac{du}{dt} = Au + \|u\|^{1+\varepsilon}(2\|P_1\|(c_1+\bar{c}_1), \; 2\|P_2\|(c_2+\bar{c}_2))\|\lambda\|^{1+\varepsilon}, \qquad (6.41)$$

where $u = (u_1, u_2)$,

$$A = \begin{pmatrix} \dfrac{-1}{2\lambda_m(P_1)} & 4p^4\|P_1\|^2 \\ \dfrac{p^4}{2\lambda_m(P_1)} & -1 \end{pmatrix}.$$

For the zero equilibrium state of the system (6.41) with respect to the cone $\mathbb{R}^2_+$ to be asymptotically stable it is sufficient that the matrix A should have positive extradiagonal elements and satisfy the Sevastianov-Kotelyansky condition. So, provided that the inequality

$$1 - 4p^8 \|P_1\|^2 > 0 \tag{6.42}$$

is true, the equilibrium state of the system (6.40), hence the system (6.39), is asymptotically stable. From the inequality (6.42) one can easily obtain that the system (6.39) is parametrically asymptotically stable with respect to the domain

$$\Omega_q = \{p \in \mathbb{R} \mid |p| < 0.7197\}.$$

Chapter 7

Stability of Solutions of Uncertain Impulsive Systems

One of the directions of the development of the general Lyapunov theory of stability of motion is the application of this theory to discontinuous systems. Among systems of this type, which are applied in practice, are fuzzy, hybrid, and impulsive systems.

The objective of this chapter is to obtain the conditions for the stability of the motion of impulsive systems with uncertainly known parameters. For this purpose we propose to use the block-diagonal Lyapunov function and the comparison principle.

7.1   Problem Setting

Let $0 < \tau_1 < \tau_2 < \ldots < \tau_k < \ldots$ and $\tau_k \to \infty$ for $k \to \infty$. The vector-function w belongs to $PC(R_+ \times R^n, R^m)$, if $w\colon (\tau_{k-1}, \tau_k] \times R^n \to R^m$ is continuous on $(\tau_{k-1}, \tau_k] \times R^n$ and for any $x \in R^n$ there exists a limit

$$\lim_{(t,y) \to (\tau_k^+, x)} w(t,y) = w(\tau_k^+, x), \quad k = 1, 2, \ldots .$$

The function V belongs to the V_0-class if

(1) $V \in PC(R_+ \times D(\rho), R_+)$, $V(t, 0) = 0$;

(2) $V(t, x)$ is locally Lipschitz with respect to x at each $(t, x) \in (\tau_{k-1}, \tau_k] \times D(\rho)$, where $D(\rho) = \{x \in R^n \colon \|x\| < \rho\}$, $\rho = \mathrm{const} > 0$;

(3) for each $k = 1, 2, \ldots$ there exist and are finite the limits

$$V(t_0 - 0, x_0) = \lim_{\substack{(t,x) \to (t_0, x_0) \\ (t,x) \in G_k}} V(t, x), \quad V(t_0 + 0, x_0) = \lim_{\substack{(t,x) \to (t_0, x_0) \\ (t,x) \in G_{k+1}}} V(t, x),$$

$$G_k = \{(t, x) \in R_+ \times D(\rho) \colon \tau_{k-1} < t < \tau_k\}.$$

The system of equations

$$\frac{dx}{dt} = f(t, x, \alpha), \qquad t \neq \tau_k,$$

$$\Delta x(t) = I_k(x(t), \alpha), \qquad t = \tau_k, \tag{7.1}$$

$$x(t_0) = x_0$$

will be called an uncertain impulsive system of the general form.

Here $x(t) \in R^n$ is the state vector of the system at the point $t \in R_+$, $f \in PC(R_+ \times R^n \times R^d, R^n)$, $I_k \colon R^n \times R^d \to R^n$, $k = 1, 2, \ldots$ denotes the instantaneous value of the state vector of the system at the points of impulsive action τ_k and $\alpha \in S \subseteq R^d$ ($d \geq 1$, S is a compact set) represents the "uncertainties" parameter of the system (7.1).

The impulsive system (7.1) without the uncertainties parameter will be called the nominal impulsive system.

The set of all points of impulsive action upon the continuous component of the system (7.1) forms the set $E = \{\tau_1, \tau_2, \ldots : \tau_1 < \tau_2 < \ldots\} \subset R_+$, which is unbounded and closed on R_+.

Regarding the system (7.1) we assume that a motion subordinate to the equations (7.1) is described by the function $x(t, t_0, x_0, \alpha) \overset{\Delta}{=} x(t, \alpha)$ with the following properties in any open neighbourhood D of the state $x = 0$, $D \subseteq R^n$:

(a) the motion $x(t, \alpha)$ of the system (7.1) is continuous from the left on $[t_0, \infty)$ for any $t_0 \geq 0$ and such that $x(t_0, \alpha) = x_0$ for any $(t_0, x_0, \alpha) \in \mathcal{T}_i \times D \times S$;

(b) the motion $x(t, \alpha)$ of the system (7.1) is differentiable with respect to t and

$$\frac{dx(t, \alpha)}{dt} = f(t, x(t, \alpha), \alpha)$$

almost everywhere on (t_0, ∞), apart from the range $E \subset R_+$, for any $\alpha \in S$;

(c) for any $t = \tau_k \in E$ the condition

$$x(t^+, \alpha) = \lim_{t^* \to t,\, t^* > t} x(t^*, \alpha) = x(t, \alpha) + I_k(x(t, \alpha), \alpha)$$

is satisfied for any $\alpha \in S$ and $k = 1, 2, \ldots$.

It is assumed that for any $\alpha \in S$ the order of the system (7.1) remains unchanged throughout the time of its functioning and for all $t \in \mathcal{T}$ the equilibrium state $x = 0$ is unique for all values $\alpha \in S$, i.e. $f(t, 0, \alpha) = I_k(0, \alpha) = 0$.

Then the analysis of the strict stability of the system (7.1) is realized under different assumptions about the dynamic behavior of the continuous and the discrete components of the nominal impulsive system.

Represent the system (7.1) in the form

$$\frac{dx}{dt} = f(t, x, 0) + \Delta f(t, x, \alpha), \qquad t \neq \tau_k,$$
$$\Delta x(t) = I_k(x(t),\, 0) + \Delta I_k(x(t),\, \alpha), \quad t = \tau_k, \qquad (7.2)$$
$$x(t_0) = x_0,$$

where the notations Δf and ΔI_k are obvious. Then the nominal impulsive system

$$\frac{dx}{dt} = f(t, x, 0), \qquad t \neq \tau_k,$$
$$\Delta x(t) = I_k(x(t),\, 0), \quad t = \tau_k, \qquad (7.3)$$
$$x(t_0) = x_0,$$

is decomposed into m interrelated subsystems

$$\frac{dx_i}{dt} = f_i(t, x_i, 0) + r_i(t, \hat{x}, 0), \qquad t \neq \tau_k,$$
$$\Delta x_i(t) = I_{ik}(x_i(t),\, 0) + a_{ik}(\hat{x}(t),\, 0), \quad t = \tau_k, \qquad (7.4)$$
$$x_i(t_0) = x_{i0}, \quad i = 1, 2, \ldots, m.$$

Obviously,

$$x(t) = (x_1^{\mathrm{T}}(t),\, \ldots,\, x_m^{\mathrm{T}}(t))^{\mathrm{T}} \in R^n, \quad n = \sum_{i=1}^{m} n_i,$$
$$r_i(t, \hat{x}, 0) = f_i(t, x_1, \ldots, x_m, 0) - f_i(t, x_i, 0),$$
$$a_{ik}(\hat{x}(t), 0) = I_{ik}(x_1(t), \ldots, x_m(t), 0) - I_{ik}(x_i(t), 0),$$
$$f_i\colon R_+ \times R^{n_i} \to R^{n_i}, \quad r_i\colon R_+ \times R^{n-n_i} \to R^{n_i},$$
$$I_{ik}\colon R^{n_i} \to R^{n_i}, \quad a_{ik}\colon R^{n-n_i} \to R^{n_i}.$$

Make some assumption regarding the dynamic behavior of the independent nominal subsystems

$$\frac{dx_i}{dt} = f_i(t, x_i, 0), \qquad t \neq \tau_k,$$
$$\Delta x_i(t) = I_{ik}(x_i(t),\, 0), \qquad t = \tau_k, \qquad (7.5)$$
$$x_i(t_0) = x_{i0}, \quad i = 1, 2, \ldots, m,$$

of the system (7.4).

Assumption 7.1 Among m independent continuous components (7.4) of the impulsive system there exist r asymptotically stable and $m - r \geq 1$ unstable ones.

Assumption 7.2 Among m independent discrete components of the impulsive system (7.4) there exist r unstable and $m - r \geq 1$ asymptotically stable ones.

Note that the analysis of the stability of the impulsive systems (7.1) with the described dynamic properties of the continuous and the discrete components has never been conducted before in the theory of impulsive systems (see Lakshmikantham, Bainov and Simeonov [1989], Samoilenko, Perestyuk [1995], Martynyuk [2000e], and others).

7.2   Principle of Comparison with a Block-Diagonal Matrix Function

For the analysis of the system (7.1), construct the matrix function $U(t, x)$ in the form

$$U(t, x) = \mathrm{diag}\{U_1(t, x),\, U_2(\tau_k, x)\}, \qquad (7.6)$$

where $U_1 \in PC(R_+ \times R^n, R^{m' \times m'})$, $m' = \sum_{i=1}^{r} n_i$, $U_2 \colon E \times R^{n-m'} \to R^{(n-m') \times (n-m')}$.

Auxiliary functions of the form (7.6) will be called matrix block-diagonal functions. Particular cases are two-component vector functions; one of their components characterizes the continuous component, the other one characterizes the discrete component.

The matrix-values function $U_1(t, x)$ is constructed on the basis of r stable subsystems of the continuous part of the system (7.4), and $U_2(\tau_k, x)$ on the basis of $m - r$ asymptotically stable components of the discrete part of the system (7.4).

Assumption 7.3 The elements $v_{ij}(t, \cdot)$ of the matrix-valued function $U_1(t, x)$ satisfy the following conditions:

(a) v_{ij} are continuous on $(\tau_{k-1}, \tau_k] \times R^{m'}$ for all $x \in R^{m'}$ and there exist

$$\lim_{(t,y) \to (\tau_k^+, x)} v_{ij}(t, y) = v_{ij}(\tau_k^+, x), \quad i, j = 1, 2, \ldots, r;$$

(b) v_{ij} are locally Lipschitz with respect to x;

(c) for any $(t, x) \in (\tau_{k-1}, \tau_k] \times R^n$ the derivatives

$$D_- v_{ij}(t, x) = \liminf\{[v_{ij}(t + \theta,\, x + \theta f(t, x, \alpha)) - \\ - v_{ij}(t, x)]\, \theta^{-1} \colon \ \theta \to 0^-\}, \quad i, j = 1, 2, \ldots, m',$$

are defined along the solutions of the respective subsystems (of the system (7.1));

(d) for $(t, x_p) \in E \times R^{n-m'}$ the first differences are defined:

$$\Delta v_{pq}(\tau_k, x_p) = v_{pq}(\tau_{k+1}, x(\tau_{k+1})) -$$
$$- v_{pq}(\tau_k, x(\tau_k)), \quad p, q = 1, 2, \ldots, m', \quad k = 1, 2, \ldots .$$

On the basis of the matrix-valued function $U(t, x)$ construct the scalar auxiliary function

$$V(t, x, \eta) = \eta^T U(t, x) \eta, \quad \eta \in R_+^m.$$

Consider the comparison equations

$$\frac{du}{dt} = g_1(t, u, \mu(\alpha)), \qquad t \neq \tau_k,$$
$$u(\tau_k^+) = \psi_k(u(\tau_k), \mu(\alpha)), \qquad \qquad (7.7)$$
$$u(t_0) = u_0 \geq 0,$$

and

$$\frac{dv}{dt} = g_2(t, v, \mu(\alpha)), \qquad t \neq \tau_k,$$
$$v(\tau_k^+) = \phi_k(v(\tau_k), \mu(\alpha)), \qquad \qquad (7.8)$$
$$v(t_0) = v_0 \geq 0.$$

Here $g_1, g_2 \in PC(R_+^3, R)$, $g_2(t, v, \mu(\alpha)) \leq g_1(t, u, \mu(\alpha)))$ for all $\alpha \in \mathcal{S}$, $\mu(\alpha) \geq 0$, $\psi_k, \phi_k \colon R_+^2 \to R$ are not decreasing in their first arguments, and in addition, $\phi_k(u, \mu(\alpha)) \leq \psi_k(u, \mu(\alpha))$ for each k, $\alpha \in \mathcal{S}$ and $\mu(\alpha) \geq 0$.

Let $u^+(t, t_0, u_0)$ and $v^-(t, t_0, v_0)$ be the upper and the lower solutions of the equations (7.7) and (7.8) respectively, defined for all $t \geq t_0$. Then from the theory of differential equations (see Lakshmikantham, Leela, and Martynyuk [1988b]) it follows that

$$v^-(t, t_0, v_0) \leq u^+(t, t_0, u_0)$$

for all $t \geq t_0$, $\mu(\alpha) \geq 0$, as soon as

$$v_0 \leq u_0.$$

Theorem 7.1 *Assume that:*

(1) *all the conditions of the Assumptions 7.1–7.3 are satisfied;*

(2) *there exist functions $g_1, g_2 \in PC(R_+^3, R)$, $\psi_k, \phi_k \colon R_+^2 \to R$ such that the following two-sided inequalities hold true:*

    (a) *$g_2(t, V(t, x, \eta), \mu(\alpha)) \leq D_- V(t, x, \eta) \leq g_1(t, V(t, x, \eta), \mu(\alpha))$ for $t \neq \tau_k$ and for all $(t, x) \in R_+ \times D(\rho)$,*

    (b) *$\phi_k(V(t, x, \eta), \mu(\alpha)) \leq V(t, x + I_k(x, \alpha), \eta) \leq \psi_k(V(t, x, \eta), \mu(\alpha))$ for $t = \tau_k$ for some $\rho_0 = \rho_0(\rho) > 0$ such that the condition $x \in D(\rho_0)$ implies the inclusion $x + I_k(x, \alpha) \in D(\rho)$ for all k and $\mu(\alpha) \geq 0$;*

(3) the vector $\eta \in R_+^m$ and the initial values (x_0, u_0, v_0) are chosen so that $v_0 \leq V(t, x_0, \eta) \leq u_0$ for $t = t_0$;

(4) under the initial conditions specified in the condition (3) the motions $x(t, \alpha)$ of the system (7.1) are defined for all $t \geq t_0$.

Then the two-way estimate

$$v^-(t, t_0, v_0) \leq V(t, x(t, \alpha), \eta) \leq u^+(t, t_0, u_0) \qquad (7.9)$$

holds for all $t \geq t_0$ for any $\alpha \in \mathcal{S}$ and $\mu(\alpha) \geq 0$.

Proof The statement of the Theorem 7.1 follows from the Theorem 4.9.1 from the monograph by Lakshmikantham, Leela, and Martynyuk [1988b] (also see Lakshmikantham and Devi [1993]).

7.3   Conditions for Strict Stability

See the following definitions necessary for further reasoning.

Definition 7.1 The equilibrium state $x = 0$ of the impulsive system (7.1):

(a) is *strictly equistable*, if for the given $t_0 \in R_+$, $\varepsilon_1 > 0$ and $\Delta > 0$ it is possible to indicate positive functions $\delta_1 = \delta_1(t_0, \varepsilon_1, \Delta)$, $\delta_2 = \delta_2(t_0, \varepsilon_1, \Delta)$ and $\varepsilon_2 = \varepsilon_2(t_0, \varepsilon_1, \Delta)$, c t_0 at fixed (ε_1, Δ) and ordered by the inequalities $\varepsilon_2 < \delta_2 < \delta_1 < \varepsilon_1$, such that $\varepsilon_2 < \|x(t, \alpha)\| < \varepsilon_1$ for all $t \geq t_0$ and any $\alpha \in \mathcal{S}$, as soon as $\delta_2 < \|x_0\| < \delta_1$ and $|t - \tau_k| > \Delta$;

(b) is *strictly uniformly stable*, if in the definition 7.1(a) the functions δ_1, δ_2 and ε_2 do not depend on t_0.

Definition 7.2 The zero solution of the comparison equations (7.7) and (7.8) is *strictly equistable* , if for the given $t_0 \in R_+$, $\varepsilon_1 > 0$, $\widetilde{\Delta} > 0$ it is possible to indicate positive functions $\tilde{\delta}_1 = \tilde{\delta}_1(t_0, \tilde{\varepsilon}_1, \widetilde{\Delta})$, $\tilde{\delta}_2 = \tilde{\delta}_2(t_0, \tilde{\varepsilon}_1, \widetilde{\Delta})$ and $\tilde{\varepsilon}_2 = \tilde{\varepsilon}_2(t_0, \tilde{\varepsilon}_1, \widetilde{\Delta})$, continuous with respect to t_0 and ordered by the inequalities $\tilde{\varepsilon}_2 < \tilde{\delta}_2 < \tilde{\delta}_1 < \tilde{\varepsilon}_1$, such that

$$\tilde{\varepsilon}_2 < v^-(t, t_0, v_0) \leq u^+(t, t_0, u_0) < \tilde{\varepsilon}_1$$

for all $t \geq t_0$ and $\mu(\alpha) \geq 0$, as soon as $\tilde{\delta}_2 < v_0 \leq u_0 < \tilde{\delta}_1$ and $|t - \tau_k| > \widetilde{\Delta}$.

The sufficient conditions for the strict stability of the equilibrium state $x = 0$ of the system (7.1) are contained in the following statement.

Theorem 7.2 *Assume that:*

(1) *all the conditions of the Theorem 7.1 are satisfied;*

(2) *for all $t \geq t_0$ the vector-functions $f(t, x, \alpha) = 0$ and $I_k(x, \alpha) = 0$, if, and only if, $x = 0$ for each k and for any $\alpha \in S$;*

(3) *there exist functions $a, b \in K$-class such that for the function $V(t, x, \eta) = \eta^T U(t, x)\eta$ the estimates*

$$b(\|x\|) \leq V(t, x, \eta) \leq a(\|x\|)$$

hold for all $(t, x) \in R_+ \times D(\rho)$.

Then the equilibrium state $x = 0$ of the uncertain impulsive system (7.1) has the same type of strict stability as the zero solution of the comparison equations (7.7) and (7.8).

Proof Consider the property of strict equistability of the motion $x = 0$ of the system (7.1). Owing to the condition (2) of the Theorem 7.2, the system (7.1) has the unique equilibrium state $x = 0$ for any change of $\alpha \in S$ and for all $k = 1, 2, \ldots$. Assume that $t_0 \in R_+$ and $0 < \varepsilon_1 < \rho^* = \min(\rho_0, \rho)$ are given.

Let the zero solution of the comparison equations (7.7), (7.8) be strictly equistable. In this case, for the given $\varepsilon_1 > 0$ choose $\tilde{\varepsilon}_1$ from the relation $\tilde{\varepsilon}_1 = b(\varepsilon_1)$. For $\varepsilon_1 > 0$ choose $\tilde{\delta}_1 = \tilde{\delta}_1(t_0, \varepsilon_1, \Delta)$, $\tilde{\delta}_2 = \tilde{\delta}_2(t_0, \varepsilon_1, \Delta)$ and $\tilde{\varepsilon}_2 = \tilde{\varepsilon}_2(t_0, \varepsilon_1, \Delta)$, continuous with respect to t_0 and satisfying the inequalities

$$\tilde{\varepsilon}_2 < \tilde{\delta}_2 < \tilde{\delta}_1 < b(\varepsilon_1),$$

$$\tilde{\varepsilon}_2 < v^-(t, t_0, v_0) < u^+(t, t_0, u_0) < b(\varepsilon_1), \quad t \geq t_0, \quad \mu(\alpha) \geq 0,$$

as soon as $\tilde{\delta}_2 < v_0 \leq u_0 < \tilde{\delta}_1$, $|t - \tau_k| > \Delta$. Under the condition (3) of the Theorem 7.1 (see the condition (1) of the Theorem 7.2), choose the vector $\eta \in R_+^m$ and the values (x_0, u_0, v_0) so that

$$v_0 = V(t, x_0, \eta) = u_0 \quad \text{at} \quad t = t_0.$$

From the fact that the function $V(t, x, \eta)$ is continuous in the neighborhood of the point (t_0, x_0) and $V(t, 0, \eta) = 0$ for all $t \in R_+$, it follows that under the given $\tilde{\delta}_1, \tilde{\delta}_2, \tilde{\delta}_2 < \tilde{\delta}_1$, there exist $\delta_1, \delta_2, \delta_2 < \delta_1 < \varepsilon_1$, such that

$$\tilde{\delta}_2 < V(t_0, x_0, \eta) < \tilde{\delta}_1, \tag{7.10}$$

as soon as $\delta_2 < \|x_0\| < \delta_1$, where $\delta_2 = \delta_2(t_0, \varepsilon_1, \Delta)$, $\delta_1 = \delta_1(t_0, \varepsilon_1, \Delta)$.

Now choose the value $\varepsilon_2 = \varepsilon_2(t_0, \varepsilon_1, \Delta) > 0$ so that $a(\varepsilon_2) \leq \tilde{\varepsilon}_2$ and $\varepsilon_2 < \delta_2$. As a result, obtain the sequence of inequalities $\varepsilon_2 < \delta_2 < \delta_1 < \varepsilon_1$ (see the Definition 7.1). Show that under the above mentioned choice of the values ε_2, δ_2 and δ_1 the equilibrium state $x = 0$ of the system (7.1) is strictly equistable as soon as all the conditions of the Theorem 7.2 are satisfied. This means that the motion $x(t, \alpha)$ of the system (7.1) for any $\alpha \in S$ satisfies the estimate

$$\varepsilon_2 < \|x(t, \alpha)\| < \varepsilon_1 \quad \text{for all} \quad t \geq t_0,$$

as soon as $\delta_2 < \|x_0\| < \delta_1$.

If this statement is incorrect, then there exists an alternative for the two cases of the motion $x(t, \alpha)$ of the system (7.1).

Case 1. For some k there exists $t_2 \in (\tau_k, \tau_{k+1}]$ such that $\varepsilon_2 \geq \|x(t_2, \alpha)\|$ for any $\alpha \in \mathcal{S}$. Obviously, for $t_0 \leq t \leq t_2$ the motion $x(t, \alpha)$ remains in the domain $D(\rho)$ and according to the conditions of the Theorem 7.1, the estimate (7.9) holds. Hence, taking into account the condition (3) of the Theorem 7.2 and the estimate (7.9), we obtain

$$a(\varepsilon_2) \geq a(\|x(t_2, \alpha)\|) \geq V(t, x(t, \alpha), \eta) \geq v^-(t_0, t_0, v_0) > \tilde{\varepsilon}_2 \geq a(\varepsilon_2).$$

The obtained contradiction proves that $t_2 \bar{\in} (\tau_k, \tau_{k+1}]$ and $\|x(t, \alpha)\| > \varepsilon_2$ for any $t \geq t_0$ and $\alpha \in \mathcal{S}$.

Case 2. For some p there exists $t_1 \in (\tau_p, \tau_{p+1}]$ such that $\varepsilon_1 < \|x(t_1, \alpha)\|$ and $\|x(t, \alpha)\| < \varepsilon_1$ for any $t_0 \leq t < t_1$. According to the condition 2(b) of the Theorem 7.2 $\|x(\tau_p, \alpha)\| < \varepsilon_1 < \rho_0$ and hence

$$\|x(\tau_p^+, \alpha)\| = \|x(\tau_p, \alpha) + I_k(x(\tau_p, \alpha), \alpha)\| < \rho. \qquad (7.11)$$

Then there must exist $\tilde{t} \in (\tau_p, t_1]$ such that

$$\varepsilon_1 \leq \|x(\tilde{t}, \alpha)\| < \rho. \qquad (7.12)$$

This means that for $t_0 \leq t \leq \tilde{t}$ the estimate $\|x(t, \alpha)\| < \rho$ holds true, and therefore, according to the Theorem 7.1, the estimate (7.9) holds true as well. According to the condition (3) of the Theorem 7.2 and the relations (7.10) and (7.11),

$$b(\varepsilon_1) \leq b(\|x(\tilde{t}, \alpha)\|) \leq V(\tilde{t}, x(\tilde{t}, \alpha), \eta) \leq u^+(\tilde{t}, t_0, u_0) < b(\varepsilon). \qquad (7.13)$$

The contradiction (7.13) proves that $\tilde{t} \bar{\in} (\tau_p, t_1]$ therefore $\|x(t, \alpha)\| < \varepsilon_1$ for all $t \geq t_0$ and for any $\alpha \in \mathcal{S}$.

The Theorem 7.2 is proved.

Now we will obtain the sufficient conditions for the stability of an uncertain impulsive system on the basis of a vector auxiliary function.

7.4   Application of the Vector Approach

Let for the system (7.1) a matrix function $U(t, x)$ be constructed in the form

$$U(t, x) = \text{diag}\{U_1(t, x), U_2(\tau_k, x)\}, \qquad (7.14)$$

where $U_1 \in PC(R_+ \times R^n, R^{m' \times m'})$ and $m' = \sum_{i=1}^{r} n_i$, $U_2 \colon E \times R^{n-m'} \to R^{(n-m') \times (n-m')}$.

Using the matrix-valued function $U(t, x)$, construct the vector auxiliary function

$$L(t, x, \eta) = AU(t, x)\eta, \quad \eta \in R_+^m, \tag{7.15}$$

where $A = \mathrm{diag}(A_1, A_2)$, A_i are constant matrices with their dimensions corresponding to those of the blocks of the matrix function (7.14).

For the system (7.1) and the function (7.15) consider the comparison system

$$\frac{du}{dt} = g(t, u, \mu(\alpha)), \quad t \neq \tau_k,$$
$$u(\tau_k^+) = \psi_k(u(\tau_k), \mu(\alpha)), \tag{7.16}$$
$$u(t_0) = u_0 \geq 0.$$

Here $g \in PC(R_+^3, R^m)$, $g(t, u, \mu(\alpha)))$ is quasimonotone with respect to u for all $\alpha \in S$, $\mu(\alpha) \geq 0$, ψ_k, ϕ_k and $R_+^2 \to R$ are not decreasing in their first arguments.

Let $u^+(t, t_0, u_0)$ be the upper solution of the system of equations (7.16) determined for all $t \geq t_0$. The theory of differential equations implies the following statement.

Theorem 7.3 *Assume that:*

(1) *all the conditions of the Assumptions 7.1 – 7.3 are satisfied;*

(2) *there exist functions $g \in PC(R_+^3, R)$, $\psi_k, \phi_k \colon R_+^2 \to R$ such that the following estimates hold true:*

   (a) $D^+L(t, x, \eta) \leq g(t, L(t, x, \eta), \mu(\alpha))$ *for $t \neq \tau_k$ and for all $(t, x) \in R_+ \times D(\rho)$,*

   (b) $L(t, x + I_k(x, \alpha), \eta) \leq \psi_k(L(t, x, \eta), \mu(\alpha))$ *for $t = \tau_k$ for some $\rho_0 = \rho_0(\rho) > 0$ such that the condition $x \in D(\rho_0)$ implies the inclusion $x + I_k(x, \alpha) \in D(\rho)$ for all k and $\mu(\alpha) \geq 0$;*

(3) *the matrix A, the vector $\eta \in R_+^m$, and the initial values (x_0, u_0, v_0) are chosen so that $L(t, x_0, \eta) \leq u_0$ at $t = t_0$;*

(4) *under the initial conditions specified in the condition (3), the motions $x(t, \alpha)$ of the system (7.1) are defined for all $t \geq t_0$.*

Then the change of the components of the vector function (7.15) along solutions of the system (7.1) is estimated by the inequality

$$L(t, x(t, \alpha), \eta) \leq u^+(t, t_0, u_0) \tag{7.17}$$

at all $t \geq t_0$ for any $\alpha \in S$ and $\mu(\alpha) \geq 0$.

The proof is similar to that of the comparison Theorem 3.4.1 in the monograph by Lakshmikantham, Leela, and Martynyuk [1988b].

Recall the definition of stability in the sense of Lyapunov, adapted to the uncertain impulsive system (7.1).

Definition 7.3 The equilibrium state $x = 0$ of the impulsive system (7.1) is:

(a) *equistable* if for the given $t_0 \in R_+$, $\varepsilon > 0$ and $\Delta > 0$ it is possible to indicate a positive function $\delta = \delta(t_0, \varepsilon, \Delta)$ continuous with respect to t_0 for fixed (ε, Δ), such that $\|x(t, \alpha)\| < \varepsilon$ for all $t \geq t_0$ and for any $\alpha \in S$, as soon as $\|x_0\| < \delta$ and $|t - \tau_k| > \Delta$;

(b) *uniformly stable* if in the definition 7.1(a) the function δ does not depend on t_0.

Definition 7.4 The zero solution of the comparison system (7.16) is *equistable* if for the given $t_0 \in R_+$, $\varepsilon_1 > 0$, $\Delta > 0$ it is possible to indicate a positive function $\delta_1 = \delta_1(t_0, \varepsilon_1, \Delta)$, $\delta_1 < \varepsilon_1$, continuous with respect to t_0, such that

$$u^+(t, t_0, u_0) < \varepsilon_1$$

for all $t \geq t_0$ and $\mu(\alpha) \geq 0$, as soon as $u_0 < \delta_1$ and $|t - \tau_k| > \Delta$.

The conditions for the stability of the equilibrium state $x = 0$ of the system (7.1) are contained in the following statement.

Theorem 7.4 *Let the equations of perturbed motion (1.1) be such that:*

(1) *all the conditions of the Theorem 7.3 are satisfied;*

(2) *for all $t \geq t_0$ the vector-functions $f(t, x, \alpha) = 0$ and $I_k(x, \alpha) = 0$, if, and only if, $x = 0$ for any k and for any $\alpha \in S$;*

(3) *there exist functions a, $b \in K$-class and a vector $c \in R^m$ such that for the function $L(t, x, \eta)$ the estimates*

$$b(\|x\|) \leq c^T L(t, x, \eta) \leq a(\|x\|)$$

hold true for all $(t, x) \in R_+ \times D(\rho)$.

Then the equilibrium state $x = 0$ of the system (7.1) has the same type of stability as the zero solution of the comparison system (7.16).

Proof Let the following quantities be specified: $t_0 \in R_+$ and $0 < \varepsilon < \rho^* = \min(\rho_0, \rho)$. Assume that the zero solution of the comparison system (7.16) is equistable. Then for the given $t_0 \in R_+$ and $b(\cdot) > 0$ there exists $\delta_1(t_0, \varepsilon) > 0$ such that

$$\|u(t; t_0, u_0)\| < b(\varepsilon) \quad \text{for all} \quad t \geq t_0,$$

as soon as $\|u_0\| < \delta_1$, where $u(t; t_0, u_0)$ is any solution of the comparison system (7.16). Under the specified function $L(t, x, b)$ according to the condition (3) of the Theorem 7.3, for $t = \tau_0^+$ we obtain the estimate $e^T L(t, x_0, b) \leq u_0$, where $u_0 = a(\|x_0\|)$ is admissible. Choose $\delta_2 = \delta_2(\varepsilon)$ so that $a(\delta_2) < b(\varepsilon)$, and define $\delta = \min(\delta_1, \delta_2) > 0$.

For the given $\varepsilon > 0$ and δ defined above, show that $\|x(t, \alpha)\| < \varepsilon$ for all $t \geq t_0$ and $\alpha \in \mathcal{S}$, as soon as $\|x_0\| < \delta$, where $x(t, \alpha)$ is a solution of the system (7.1), determined for all $t \geq t_0$. If that is not so, there must exist a solution $x(t, \alpha)$ at least for one $\alpha \in \mathcal{S}$ such that for $\|x_0\| < \delta$ and $t^* > t_0$ ($\tau_k < t^* < \tau_{k+1}$ for some k) the following conditions will be satisfied:

$$\|x(t^*, \alpha)\| \geq \varepsilon \quad \text{and} \quad \|x(t, \alpha)\| < \varepsilon \quad \text{at} \quad t_0 \leq t \leq \tau_k.$$

Since $0 < \varepsilon < \rho^*$, from the condition (2)(b) of the Theorem 7.3 it follows that $\|x_k^+\| = \|x_k + I_k(x_k, \alpha)\| < \rho$, where $x_k = x(\tau_k, \alpha)$ and $\|x_k\| < \varepsilon$. Hence one can find t^0, $\tau_k < t^0 < t^*$, such that $\varepsilon \leq \|x(t^0, \alpha)\| < \rho^*$ for all $\alpha \in \mathcal{S}$.

Denoting $m(t) = c^T L(t, x(t), b)$ for $t_0 \leq t \leq t^0$ and taking into account the conditions $(2) - (4)$ of the Theorem 7.3,

$$e^T L(t, x(t), b) \leq e^T u(t; t_0, u_0), \quad t_0 \leq t \leq t^0, \tag{7.18}$$

where $u(t; t_0, u_0)$ is the maximal solution of the comparison system (7.15). From the inequality (7.18) we obtain

$$b(\varepsilon) \leq b(\|x(t^0, \alpha)\|) \leq e^T L(t^0, x(t^0, \alpha), b) \leq e^T u(t^0, t_0, a(\|x_0\|)) < b(\varepsilon).$$

Hence $\|x(t, \alpha)\| < \varepsilon$ for all $t \geq t_0$ and any $\alpha \in \mathcal{S}$. The stability of the condition $x = 0$ of the system (7.1) is proved.

7.5 Robust Stability of Impulsive Systems

Consider the linear impulsive system

$$\frac{dx}{dt} = Ax(t), \quad t \neq \tau_i,$$
$$\Delta x(t) = Bx(t), \quad t = \tau_i, \tag{7.19}$$
$$x(t_0) = x_0.$$

Here $A, B \in R^{n \times n}$ are the interval matrices from the compacts $\mathcal{A} = \{A_1 \leq A \leq A_2\}$ and $\mathcal{B} = \{B_1 \leq B \leq B_2\}$, $n \times n$-matrices A_s and B_s, and $s = 1, 2$, have the given elements, $\Delta x(t) = x(t_i^+) - x(t_i)$, $i = 1, 2, \ldots$.

Definition 7.5 An uncertain robust system (7.19) is robust exponentially stable in the whole if for any values of the parameters $A \in \mathcal{A}$ and $B \in \mathcal{D}$, a zero solution of system (7.19) is exponentially stable in the whole.

We transform system (7.19) to the form

$$\frac{dx}{dt} = (A_0 + \Delta A)x(t), \quad t \neq \tau_i,$$

$$\Delta x(t) = (B_0 + \Delta B)x(t), \quad t = \tau_i, \tag{7.20}$$

where $A_0 = \frac{1}{2}(A_1 + A_2)$, $B_0 = \frac{1}{2}(B_1 + B_2)$, $\Delta A = E_A \Phi_A F_A$, $\Delta B = E_B \Phi_B F_B$, $\Phi_A \in \Phi^* = \{\Phi \in R^{n^2 \times n^2} : \Phi = \text{diag}\{E_{11}, \dots, E_{nn}\}\}$ with the elements $|\varepsilon_{kl}| \leq 1$, $k, l = 1, 2, \dots, n$.

The matrices E_A and F_A are determined in terms of the elements of the matrix $C = \frac{1}{2}(A_2 - A_1)$ by the formulas

$$E_A E_A^{\mathrm{T}} = \text{diag}\left\{ \sum_{j=1}^{n} c_{1j}, \sum_{j=1}^{n} c_{2j}, \dots, \sum_{j=1}^{n} c_{nj} \right\} \in R^{n \times n},$$

$$F_A F_A^{\mathrm{T}} = \text{diag}\left\{ \sum_{j=1}^{n} c_{j1}, \sum_{j=1}^{n} c_{j2}, \dots, \sum_{j=1}^{n} c_{jn} \right\} \in R^{n \times n}.$$

The matrices E_B and F_B are determined in the same way in terms of the matrix $D = \frac{1}{2}(B_2 - B_1)$.

Further we need the following estimate.

Lemma 7.1 *If the matrix $\Phi \in \Phi^*$, then for any constant $\lambda > 0$ and any vectors $\xi \in R^{n^2}$ and $\eta \in R^{n^2}$ the inequality*

$$2\xi^{\mathrm{T}} \Phi \eta \leq \lambda^{-1} \xi^{\mathrm{T}} \xi + \lambda \eta^{\mathrm{T}} \eta \tag{7.21}$$

holds true.

Proof Let us show that inequality (7.21) is valid for one block of the matrix Φ_A, i.e.,

$$2\xi^{\mathrm{T}} \Phi \eta \leq \lambda^{-1} \xi^{\mathrm{T}} \xi + \lambda \eta^{\mathrm{T}} \eta, \quad (\xi, \eta) \in R^n.$$

In fact, $2\xi^{\mathrm{T}} \Phi \eta \leq 2|\xi^{\mathrm{T}} \eta|$ by virtue of the condition $|\varepsilon_{kl}| \leq 1$ for the elements of matrix Φ. Then, consider the inequality

$$\left(\frac{1}{\sqrt{\lambda}} \xi^{\mathrm{T}} - \sqrt{\lambda} \eta^{\mathrm{T}} \right) \left(\frac{1}{\sqrt{\lambda}} \xi - \sqrt{\lambda} \eta \right) \geq 0,$$

which yields

$$2|\xi^{\mathrm{T}} \eta| \leq \frac{1}{\lambda} \xi^{\mathrm{T}} \xi + \lambda \eta^{\mathrm{T}} \eta.$$

Since $\Phi_A \Phi_A^{\mathrm{T}} = \Phi_A^{\mathrm{T}} \Phi_A \leq I^*$, where I^* is the $n^2 \times n^2$-identity matrix, extending the latter inequality to all blocks of the matrix Φ_A we arrive at estimate (7.21).

For system (7.20) assume that the function $V(x) = x^{\mathrm{T}} P x$ is constructed, where P is an $n \times n$-constant symmetric positive definite matrix. The function $V(x)$ satisfies the inequality

$$\lambda_m(P)\|x\|^2 \leq V(x) \leq \lambda_M(P)\|x\|^2$$

for all $x \in R^n$, where $\lambda_m(\cdot)$ and $\lambda_M(\cdot)$ are the minimal and maximal eigenvalues of the matrix P.

Assume that the matrix P is the solution of the Lyapunov equations

$$A_0^{\mathrm{T}} P + P A_0 = -Q_1,$$
$$B_0^{\mathrm{T}} P B_0 - P = Q_2, \tag{7.22}$$

where Q_1 and Q_2 are some symmetric positive definite matrices.

Theorem 7.5 *For system (7.20) let the function $V(x)$ be constructed with the matrix P which solves equations (7.22). If there exist constants $\mu_1 > 0$, $\mu_2 > 0$ and $\varkappa > 0$ such that*

(a) $-Q_1 + \mu_1 P E_A E_A^{\mathrm{T}} P + \mu_1^{-1} F_A^{\mathrm{T}} F_A \leq 0,$

(b) $(I + B_0)^{\mathrm{T}} P (I + B_0) + \mu_2 (I + B_0)^{\mathrm{T}} P E_B E_B^{\mathrm{T}} P (I + B_0) + (\mu_2^{-1} + \lambda_M(P)\|E_B\|^2) F_B^{\mathrm{T}} F_B - Q_2 \leq 0,$

(c) $\dfrac{\lambda_M(M)}{\lambda_M(P)} + \dfrac{\ln(\lambda_M(Q_2)/\lambda_m(P))}{\Delta t_k} \leq -\varkappa,$

where $M = -Q_1 + \mu_1 P E_A E_A^{\mathrm{T}} P + \mu_1^{-1} F_A^{\mathrm{T}} F_A$, I is an $n \times n$-identity matrix and the state $x = 0$ of system (7.19) is robust exponentially stable in the whole.

Proof For the function $V(x)$ for $t \neq \tau_i$, $i = 1, 2, \ldots$, we have

$$\begin{aligned}
\frac{dV}{dt} &= x^{\mathrm{T}} (A_0^{\mathrm{T}} P + P A_0) x + 2 x^{\mathrm{T}} P E_A \Phi_A E_A^{\mathrm{T}} x \\
&\leq -x^{\mathrm{T}} Q_1 x + \mu_1 x^{\mathrm{T}} P E_A E_A^{\mathrm{T}} P x + \mu_1^{-1} x^{\mathrm{T}} F_A^{\mathrm{T}} F_A x \\
&= x^{\mathrm{T}} M X \leq \lambda_M(M) x^{\mathrm{T}} x.
\end{aligned} \tag{7.23}$$

Since $V(x)$ is positive definite, inequality (7.23) provides

$$\frac{dV}{dt} \leq \frac{\lambda_M(M)}{\lambda_M(P)} V(x).$$

This estimate follows from inequality (7.21) and condition (a) of Theorem 7.5.

Then for $t = \tau_i$, $i = 1, 2, \ldots$, we have

$$\begin{aligned}
V(x_i^+) &= V(x_i + \Delta x_i) = x_i^{\mathrm{T}} (I + B_0 + \Delta B)^{\mathrm{T}} P (I + B_0 + \Delta B) x_i \\
&= x_i^{\mathrm{T}} (I + B_0)^{\mathrm{T}} P (I + B_0) x_i + 2 x_i^{\mathrm{T}} (I + B_0)^{\mathrm{T}} P E_B \Phi_B F_B x_i \\
&\quad + x_i^{\mathrm{T}} F_B^{\mathrm{T}} \Phi_B^{\mathrm{T}} E_B^{\mathrm{T}} P E_B \Phi_B F_B x_i \\
&\leq x_i^{\mathrm{T}} \{ (I + B_0)^{\mathrm{T}} P (I + B_0) + \mu_2^{-1} (I + B_0)^{\mathrm{T}} P E_B E_B^{\mathrm{T}} P (I + B_0) \\
&\quad + \mu_2^{-1} F_B^{\mathrm{T}} F_B \} x_i + x_i^{\mathrm{T}} F_B^{\mathrm{T}} \Phi_B^{\mathrm{T}} E_B^{\mathrm{T}} P E_B \Phi_B F_B x_i.
\end{aligned} \tag{7.24}$$

Since the matrix P is positive definite, there exists a nonsingular matrix K such that $P = K^{\mathrm{T}}K$. Taking this into account we estimate the second additive in estimate (7.24) as

$$
\begin{aligned}
x_i^{\mathrm{T}} F_B^{\mathrm{T}} \Phi_B^{\mathrm{T}} E_B^{\mathrm{T}} P E_B \Phi_B F_B x_i &= \|K E_B \Phi_B F_B x_i\|^2 \\
&\leq \|K\|^2 \|E_B\|^2 x_i^{\mathrm{T}} F_B^{\mathrm{T}} F_B x_i = \lambda_M(P)\|E_B\|^2 x_i^{\mathrm{T}} F_B^{\mathrm{T}} F_B x_i,
\end{aligned}
\tag{7.25}
$$

where $\|\cdot\|$ is spectral norm. In view of estimate (7.25) and condition (b) of Theorem 7.5 we transform inequality (7.24) to the form

$$
\begin{aligned}
V(x_i^+) &\leq x_i^{\mathrm{T}}\big\{(I + B_0)^{\mathrm{T}} P(I + B_0) + \mu_2(I + B_0)^{\mathrm{T}} P E_B E_B^{\mathrm{T}} P(I + B_0) \\
&\quad + (\mu_2^{-1} + \lambda_M(P)\|E_B\|^2)x_i \leq x_i^{\mathrm{T}} Q_2 x_i \\
&\leq \lambda_M(Q_2) x_i^{\mathrm{T}} x_i \leq \frac{\lambda_M(Q_2)}{\lambda_m(P)} V(x_i), \quad i = 1, 2, \dots.
\end{aligned}
\tag{7.26}
$$

It is easy to see that under condition (c) of Theorem 7.5 estimates (7.23) and (7.26) lead to the conclusion about robust exponential stability in the whole of the state $x = 0$ of system (7.19).

Further we consider a nonlinear robust impulsive system

$$
\begin{aligned}
\frac{dx}{dt} &= f(x) + g_1(t, x, \alpha), \quad t \neq \tau_i, \\
\Delta x(t) &= I_i(x) + g_2(t, x, \alpha), \quad t = \tau_i, \\
x(t_0) &= x_0, \quad i = 1, 2, \dots.
\end{aligned}
\tag{7.27}
$$

Here the vector-function $f \in C(R^n, R^n)$, $I_i\colon R^n \to R^n$, $g_k\colon R_+ \times R^n \times \mathcal{S} \to R^n$, $k = 1, 2$, are the functions characterizing uncertainties in system (7.27). Consider the function g_k of the class

$$
\mathcal{S}_g = \{g\colon g(t, x, \alpha) = Q(t, x)r(t, x) \text{ for all } \alpha \in \mathcal{S}\}.
$$

In this relation $Q\colon R_+ \times R^n \to R^{n \times m}$ is the matrix with complete information about its elements and $r\colon R_+ \times R^n \to R^m$ is the vector-function whose elements are not known exactly but satisfy the conditions $\|r(t, x)\| \leq \|m(t, x)\|$, where $m\colon R_+ \times R^n \to R^n$, $m(t, 0) = 0$ for all $t \in R_+$, is a known vector-function.

Let

$$
\begin{aligned}
g_1(t, x, \alpha) &= Q_1(t, x)r_1(t, x) \quad \forall \alpha \in \mathcal{S}, \\
g_2(t, x, \alpha) &= Q_2(t, x)r_2(t, x) \quad \forall \alpha \in \mathcal{S}, \\
\|r_1(t, x)\| &\leq \|m_1(t, x)\|, \quad \|r_2(t, x)\| \leq \|m_2(t, x)\|.
\end{aligned}
$$

The following statement holds.

Theorem 7.6 *For some nominal system (7.27) assume that for $g_1 = g_2 = 0$ function $V(x)\colon R^n \to R_+$ is constructed in the form*

$$
V(x) = V_0(x) + P_1(x)y + y^{\mathrm{T}} P_2(x)y, \quad y \in R^n,
$$

where $P_1(x) \in R^{1 \times n}$, $P_2 \in R^{n \times n}$, $P_2(x) \geq 0$ for all $x \in R^n$ and the following conditions are satisfied:

(1) *there exist constants $a_1, a_2 > 0$ and $p > 1$ such that*

$$a_1 \|x\|^p \leq V(x) \leq a_2 \|x\|^p \quad \forall x \in R^n;$$

(2) *function $V(x)$ is differentiable for $t \neq \tau_i$, $i \in \mathbb{N}$, and there exist constants c_i and functions $\mu_i(t) > 0$ such that for $t \in (t_i, t_{i+1}]$, $i \in \mathbb{N}$, the estimate*

$$\frac{\partial V(x)}{\partial x} f(x) + \frac{1}{2} \mu_i(t) \frac{\partial V}{\partial x} Q_1 Q_1^T \left(\frac{\partial V}{\partial x} \right)^T + \frac{1}{2} \mu_i^{-1}(t) m_1^T m_1 + c_i V(x) \leq 0$$

is valid;

(3) *there exist constants $M > 1$, $0 < \beta_i \leq M$ and $\sigma_i > 0$, $i \in \mathbb{N}$ such that*

$$V_0(x_i + I_i(x)) - \beta_i V(x_i) + \frac{1}{2} \sigma_i P_1(x_i) Q_2(t_i, x_i) Q_2^T(t_i, x_i) P_1^T(x_i)$$

$$+ \left\{ \frac{1}{2} \sigma_i^{-1} + \lambda_M (Q_2^T(t_i, x_i) P_2(x_i) Q_2(t_i, x_i)) \right\} m_2^T(t_i, x_i) m_2(t_i, x_i) \leq 0;$$

(4) *there exists a constant $\varkappa > 0$ such that for all $i \in \mathbb{N}$ the estimate*

$$-c_i + \frac{\ln \beta_i}{\tau_{i+1} - \tau_i} \leq -\varkappa$$

holds true.

Then the equilibrium state $x = 0$ of system (7.27) is robust exponentially stable in the whole.

Proof　For the values $t \neq \tau_i$ due to condition (2) we have for $D^+ V(x(t))$

$$D^+ V(x(t)) = \frac{\partial v}{\partial x}(f(x) + g_1(t, x, \alpha)) \leq c_i V(x(t)), \quad i = 1, 2, \dots. \quad (7.28)$$

For the values $t = \tau_i$, $i = 1, 2, \dots$, due to condition (3) of Theorem 7.6 we have

$$V(x_i^+) = V(x_i + I_i(x_i) + g_2(t_i, x_i, \alpha)) = V_0(x_i + I_i(x_i))$$
$$+ P_1(x_i) g_2(\tau_i, x_i, \alpha) + g_2^T(\tau_i, x_i, \alpha) P_2(x_i) g_2(\tau_i, x_i, \alpha) \leq \beta_i V_i(x_i),$$
$$i = 1, 2, \dots.$$

$$(7.29)$$

Estimates (7.28) and (7.29) imply that for solutions of system (7.27)) the estimate

$$\|x(t, \alpha)\| \leq \begin{cases} \left(\dfrac{a_2}{a_1} \right)^{1/p} \exp \left(\displaystyle\sum_{i=0}^{k} \gamma_i \right) p^{-1}, & \text{if } c_i \geq 0, \\[3mm] \left(\dfrac{M a_2}{a_1} \right)^{1/p} \exp \left(\displaystyle\sum_{i=0}^{k-1} \gamma_i \right) p^{-1}, & \text{if } c_i < 0 \end{cases}$$

is valid for all $t \in (\tau_i, \tau_{i+1}]$, where $\gamma_i = c_i(\tau_{i+1} - \tau_i) + \ln \beta_i$.

Hence it follows that under condition (4) of Theorem 7.6 the state $x = 0$ of system (7.27) is robust exponentially stable in the whole.

Corollary 7.1 Assume that in system (7.27) the vector-function $g_2(t, x, \alpha) = 0$ and there exists a function $V_0(x): R^n \to R_+$, for which conditions (1) and (2) of Theorem 7.6 are fulfilled and condition (3) is of the form

$$V_0(x_i + I_i(x_i)) - \beta_i V(x_i) \leq 0.$$

Then, if condition (4) of Theorem 7.6 is satisfied, the state $x = 0$ of system (7.27) is robust exponentially stable in the whole.

Example 7.1 Consider system (7.27) with the following components

$$f(x) = \begin{pmatrix} -2x_1 + 2x_2(x_1^2 + x_2^2) \\ -2x_2 - x_1(x_1^2 + x_2^2) \end{pmatrix},$$

$$I_i(x) = \begin{pmatrix} -1 + \exp\left(-1 + \frac{1}{2(i+3)}\right) & 0 \\ 0 & -1 + \exp\left(-1 + \exp\left(-1 + \frac{1}{2(i+3)}\right)\right) \end{pmatrix} x_i,$$

$$\tag{7.30}$$

$\tau_{i+1} - \tau_i = 1$, $g_1(t, x, \alpha) = Q_1(x)r_1(x)$, $Q_1(x) = I$, $\|r_1(x)\| \leq \|m_1(x)\|$, $m_1(x) = (\sqrt{2}(x_1 + x_2), \sqrt{2}x_2)^{\mathrm{T}}$, $g_2(t, x, \alpha) = Q_2(x)r_2(x)$, $Q_2(x) = I$, $\|r_2(x)\| \leq \|m_2(x)\|$, $m_2(x) = (e^{-3}x_1, e^{-3}x_2)^{\mathrm{T}}$.

The function $V(x)$ is taken with the terms $V_0(x) = \dfrac{1}{2}(x_1^2 + 2x_2^2)$, $P_1(x) = (x_1, 2x_2)$, $P_2(x) = \begin{pmatrix} 1/2 & 0 \\ 0 & 1 \end{pmatrix}$. Let $\mu_i(t) = 1$ and $\sigma_i = \exp\left(-3 + \dfrac{1}{i+3}\right)$, $i \in \mathbb{N}$. Then condition (1) of Theorem 7.6 becomes

$$D^+V(x(t))\big|_{(7.30)} = \frac{\partial V}{\partial x}(f(x) + g_1(t, x, \alpha)) \leq V(x(t)).$$

For $t = \tau_i$, $i = 1, 2, \dots$, we have

$$V_0(x_i + I_i(x_i)) + \frac{1}{2}\sigma_i P_1(x_i)Q_2(x_i)Q_2^{\mathrm{T}}(x_i)P_1^{\mathrm{T}}(x_i)$$

$$+ \left\{ \frac{1}{2}\sigma_i^{-1} + \lambda_M(Q_2^{\mathrm{T}}(x_i)P_2(x_i)Q_2(x_i)) \right\} m_2^{\mathrm{T}}(x_i)m_2(x_i)$$

$$\leq \exp\left(-2 + \frac{1}{i+3}\right)\left(1 + \frac{5}{2e} + \exp\left(-2 + \frac{1}{i+3}\right)\right)V(x_i)$$

$$= \beta_i V(x_i).$$

Since $c_1 = -1$, we have

$$-c_i + \frac{\ln \beta_i}{\tau_{i+1} - \tau_i} = -1 + \frac{1}{i+3} + \ln\left(1 + \frac{5}{2e} + e^{-2 - \frac{1}{i+2}}\right) \leq -0.0296.$$

By Theorem 7.6 the state $x = 0$ of system (7.19) with components (7.30) is robust exponentially stable in the whole.

7.6   Concluding Remarks

The proposed conditions for the stability of motion of impulsive systems with uncertain parameter values are certain developments of the known sufficient conditions for the stability of nominal impulsive systems, i. e., impulsive systems without "uncertainties" (see Lakshmikantham, Bainov, and Simeonov [1989], Samoilenko, Perestyuk [1995] and others). From the Theorem 7.4 one can obtain some corollaries which are interesting in themselves. For example,

Corollary 7.2 Assume that in the matrix blocks $U_1(t, x)$ and $U_2(\tau_k, x)$ of the function (7.2) the elements $v_{ij}(\cdot) = 0$ for all $i \neq j$ and the vector function $V(t, x) = U(t, x)b$, $V \colon R_+ \times S(\rho) \to R_+^m$, satisfy all the conditions of the Theorem 7.4 for $\alpha \equiv 0$. Then the properties of stability of the zero solutions of the comparison system (7.16) for $\mu(\alpha) \equiv 0$ imply the respective properties of stability of the zero solution of the system (7.1) at $\alpha \equiv 0$.

Note that if in the conditions of the Theorem 7.4 the Assumptions 7.1 and 7.2 are omitted and an ordinary vector function is considered, then the Theorem 7.4 will turn into the known Theorem 3.4.2 from the monograph by Lakshmikantham, Matrosov, and Sivasundaram [1991].

The system (7.1) includes two classes of uncertain impulsive systems which are of interest for applications. Class 1 of uncertain impulsive systems (IIS-1) includes all systems of the form

$$\frac{dx}{dt} = f(t, x), \quad t \neq \tau_k,$$
$$\Delta x(\tau_k) = I_k(x(\tau_k), \alpha), \tag{7.31}$$

where $x \in R^n$, $f \in C(R_+ \times R^n, R^n)$, $I_k \colon R^n \times R^d \to R^n$, $d \geq 1$, $k = 1, 2, \ldots$. It is obvious that the system (7.19) describes the evolution of the impulsive system, when the value of the state vectors at the moments of impulsive action is known uncertainly.

Class 2 of uncertain impulsive systems (IIS-2) includes all systems of the form

$$\frac{dx}{dt} = f(t, x, \alpha), \quad t \neq \tau_k,$$
$$\Delta x(\tau_k) = I_k(x(\tau_k)), \tag{7.32}$$

where $x \in R^n$, $f \in C(R_+ \times R^n \times R^d, R^n)$ and $I_k \colon R^n \to R^n$, $k = 1, 2, \ldots$. The system (7.20) describes the situation when the discrete component of a system does not depend on "uncertainties", while on the continuity interval the uncertainties of the state vector of the system are taken into consideration.

The problem of the analysis of dynamic properties of the system (7.20) presents an open field for research in the theory of impulsive dynamic systems. For example, for the above mentioned classes of systems it would be interesting to obtain the results described in Chapter 2 for systems of ordinary differential equations with uncertain parameter values.

Chapter 8

Stability of Solutions of Uncertain Dynamic Equations on a Time Scale

Dynamic equations on a time scale are adequate models, for the description of continuous- and discrete-time processes occurring in many fields of natural science and technology. It is interesting to analyze the stability of solutions of this class of equations (systems of equations), because here such conditions are defined under which a certain dynamic property of solutions not only occurs in the continuous and discrete-time cases, but also in the time domain "between" those states.

In this chapter the reader will find the results of the analysis of the stability of uncertain dynamic equations on the basis of the generalized direct Lyapunov method.

8.1   Elements of the Analysis on a Time Scale

The necessary information from the mathematical analysis on time scales is given here in accordance with the article by Bohner and Martynyuk [2007]. Further information can be found in Bohner and Peterson [2001] with an extensive list of references.

A time scale $\mathbb{T}$ is an arbitrary nonempty closed subset of a set of real numbers R. Examples of a time scale are: the set $\mathbb{R}$, integers $\mathbb{Z}$, natural numbers $\mathbb{N}$, and nonnegative natural numbers $\mathbb{N}_0$. The most common time scales are $\mathbb{T} = \mathbb{R}$ for a continuous system s, $\mathbb{T} = \mathbb{Z}$ for discrete systems, and $\mathbb{T} = q^{\mathbb{N}_0} = \{q^n \colon n \in \mathbb{N}_0\}$, where $q > 1$, for quantum analysis.

For any $t \in \mathbb{T}$ the function of a jump forward (backward) is determined by the relations

$$\sigma(t) = \inf\{s \in \mathbb{T} \colon s > t\}$$

(respectively $\rho(t) = \sup\{s \in \mathbb{T}; \ s < t\}$).

Using the operators $\sigma \colon \mathbb{T} \to \mathbb{T}$ and $\rho \colon \mathbb{T} \to \mathbb{T}$ the current values of $\{t\}$ on a time scale $\mathbb{T}$ are classified as follows: if $\sigma(t) = t$ $(\rho(t) = t)$, then the point $t \in \mathbb{T}$ is called right-(left-)dense; if $\sigma(t) > t$ $(\rho(t) < t)$, then the point $t \in \mathbb{T}$ is called right-(left-)scattered. It is assumed here that $\inf \varnothing = \sup \mathbb{T}$

159

(i.e. $\sigma(t) = t$, if $\mathbb{T}$ contains the maximal element t) and $\sup \varnothing = \inf \mathbb{T}$ (i.e., $\rho(t) = t$, if $\mathbb{T}$ contains the minimum element t). Along with the set $\mathbb{T}$ the set $\mathbb{T}^\kappa$ is applied.

If $\mathbb{T}$ contains the maximal left-scattered m, then $\mathbb{T}^\kappa = \mathbb{T} \setminus \{m\}$, in the other case $\mathbb{T}^\kappa = \mathbb{T}$. Therefore

$$\mathbb{T}^\kappa = \begin{cases} \mathbb{T} \setminus (\rho(\sup \mathbb{T}), \sup \mathbb{T}], & \text{if } \sup \mathbb{T} < \infty, \\ \mathbb{T}, & \text{if } \sup \mathbb{T} = \infty. \end{cases}$$

The distance from an arbitrary element $t \in \mathbb{T}$ to the element next to it is called the granularity of the time scale $\mathbb{T}$ and is determined by the formula

$$\mu(t) = \sigma(t) - t.$$

If $\mathbb{T} = \mathbb{R}$, then $\sigma(t) = t = \rho(t)$ and $\mu(t) = 0$; if $\mathbb{T} = \mathbb{Z}$, then $\sigma(t) = t + 1$, $\rho(t) = t - 1$ and $\mu(t) = 1$.

While considering equations on time scales, sometimes one applies the principle of induction on the time scale $\mathbb{T}$. According to the monograph by Bohner and Peterson [2001], this principle is stated as follows.

Theorem 8.1 *Assume that $t_0 \in \mathbb{T}$ and $\{S(t): t \in [t_0, \infty)\}$ is some family of statements satisfying the following conditions:*

(1) *the statement $S(t)$ holds at $t = t_0$;*

(2) *if $t \in [t_0, \infty)$ is right-scattered, and $S(t)$ holds, then $S(\sigma(t))$ holds, too;*

(3) *if $t \in [t_0, \infty)$ is right-dense, and $S(t)$ holds, then there exists a neighborhood W of the value t, such that $S(s)$ holds for all $s \in W \cap (t, \infty)$;*

(4) *if $t \in (t_0, \infty)$ is left-dense, and $S(s)$ holds for all $s \in [t_0, t)$, then $S(t)$ holds.*

Then $S(t)$ holds for all $t \in [t_0, \infty)$.

Now consider the function $f \colon \mathbb{T} \to \mathbb{R}$ and determine its Δ-derivative in the point $t \in \mathbb{T}^\kappa$.

Let $f \colon \mathbb{T} \to \mathbb{R}$ and the function f be determined for all $t \in \mathbb{T}^\kappa$. The function f is called Δ-differentiable in the point $t \in \mathbb{T}^\kappa$, if there exists such $\gamma \in \mathbb{R}$, that for any $\varepsilon > 0$ and W-neighborhood $t \in \mathbb{T}^\kappa$ the inequality

$$|[f(\sigma(t)) - f(s)] - \gamma[\sigma(t) - s]| < \varepsilon |\sigma(t) - s|$$

would hold for all $s \in W$. In this case denote $f^\Delta(t) = \gamma$.

If the function $f(t)$ is Δ-differentiable at any $t \in \mathbb{T}^\kappa$, then $f \colon \mathbb{T} \to \mathbb{R}$ is Δ-differentiable on $\mathbb{T}^\kappa$.

Some useful relations for the derivative of the function $f(t)$ are contained in the following statement.

Theorem 8.2 *Assume that* $f\colon \mathbb{T} \to \mathbb{R}$ *and* $t \in \mathbb{T}^\kappa$. *Then the following statements hold true:*

(1) *if f differentiable in the point t, then it is continuous in the point t;*

(2) *if f is continuous in t and t is right-scattered, then f is differentiable in the point t with the derivative*

$$f^\Delta(t) = \frac{f(\sigma(t)) - f(t)}{\mu(t)};$$

(3) *if t is right-dense, then the function f is differentiable in the point t if, and only if, there exists a limit*

$$\lim_{s \to t} \frac{f(t) - f(s)}{t - s},$$

which is a finite number, and in this case

$$f^\Delta(t) = \lim_{s \to t} \frac{f(t) - f(s)}{t - s};$$

(4) *if f is differentiable in the point $t \in \mathbb{T}^\kappa$, then $f(\sigma(t)) = f(t) + \mu(t)f^\Delta(t)$.*

Note that if $\mathbb{T} = \mathbb{R}$, then $f^\Delta(t) = f'(t)$, which is the Euler derivative of the function $f(t)$, and if $\mathbb{T} = \mathbb{Z}$, then $f^\Delta(t) = \Delta f(t) = f(t+1) - f(t)$, i.e., we obtain the first difference for the function $f(t)$.

Theorem 8.3 *Assume that the functions f, $g\colon \mathbb{T} \to \mathbb{R}$ are differentiable in the point $t \in \mathbb{T}^\kappa$. Then the following statements hold true:*

(1) *the sum $f + g\colon \mathbb{T} \to \mathbb{R}$ is differentiable in the point $t \in \mathbb{T}^\kappa$ and*

$$(f + g)^\Delta(t) = f^\Delta(t) + g^\Delta(t);$$

(2) *for any constant α the expression $\alpha f\colon \mathbb{T} \to \mathbb{R}$ is differentiable in the point $t \in \mathbb{T}^\kappa$ and $(\alpha f)^\Delta(t) = \alpha f^\Delta(t)$;*

(3) *the product of the two functions $fg\colon \mathbb{T} \to \mathbb{R}$ is differentiable in the point $t \in \mathbb{T}^\kappa$ and*

$$(fg)^\Delta(t) = f^\Delta(t)g(t) + f(\sigma(t))g^\Delta(t) = f(t)g^\Delta(t) + f^\Delta(t)g(\sigma(t));$$

(4) *if $f(t)f(\sigma(t)) \neq 0$ for all $t \in \mathbb{T}^\kappa$, then the function $1/f$ is differentiable in the point $t \in \mathbb{T}^\kappa$ and*

$$\left(\frac{1}{f}\right)^\Delta(t) = -\frac{f^\Delta(t)}{f(t)f(\sigma(t))};$$

(5) *if $g(t)g(\sigma(t)) \neq 0$ for all $t \in \mathbb{T}^\kappa$, then the expression f/g is differentiable in the point $t \in \mathbb{T}^\kappa$ and*

$$\left(\frac{f}{g}\right)^\Delta (t) = \frac{f^\Delta(t)g(t) - f(t)g^\Delta(t)}{g(t)g(\sigma(t))}.$$

Consider functions integrable on the time scale $\mathbb{T}$. The function $f\colon \mathbb{T} \to \mathbb{R}$ will be called ordered on $\mathbb{T}$ if, and only if, there exist right side and left side limits in all right-dense (rd) and left-dense (ld) points $\mathbb{T}$ respectively.

The function $f\colon \mathbb{T} \to \mathbb{R}$ is rd-continuous if it is continuous in the right-dense points of the scale $\mathbb{T}$ and there exists a left side limit in the left-dense points of the scale $\mathbb{T}$. The set of all rd-continuous functions $f\colon \mathbb{T} \to \mathbb{R}$ is denoted by $C_{\mathrm{rd}} = C_{\mathrm{rd}}(\mathbb{T}) = C_{\mathrm{rd}}(\mathbb{T}, \mathbb{R})$.

Theorem 8.4 *Assume that $f\colon \mathbb{T} \to \mathbb{R}$. Then the following statements hold:*

(1) *if f is continuous on $\mathbb{T}$, then it is rd-continuous on $\mathbb{T}$;*

(2) *if f is rd-continuous on $\mathbb{T}$, then it is ordered on $\mathbb{T}$;*

(3) *the function $\sigma\colon \mathbb{T} \to \mathbb{T}$ is rd-continuous;*

(4) *if f is ordered or rd-continuous on $\mathbb{T}$, then $f(\sigma(t))$ has the same properties on $\mathbb{T}$;*

(5) *if $f\colon \mathbb{T} \to \mathbb{R}$ is continuous and $g\colon \mathbb{T} \to \mathbb{R}$ is ordered and rd-continuous, then the function $f(g(t))$ is ordered or rd-continuous respectively.*

Some function $F\colon \mathbb{T} \to \mathbb{R}$ with the derivative $F^\Delta(t) = f(t)$ is Δ-antiderivative of the function $f(t)$. Then for any $a, b \in \mathbb{T}$ the following integral is determined:

$$\int_a^b f(t)\,\Delta t = F(b) - F(a).$$

Any rd-continuous function $f\colon \mathbb{T} \to \mathbb{R}$ has an Δ-antiderivative.

If $f^\Delta(t) \geq 0$ on $[a, b]$ and $s, t \in \mathbb{T}$, $a \leq s \leq t \leq b$, then

$$f(t) = f(s) + \int_s^t f^\Delta(\tau)\,\Delta\tau \geq f(s),$$

i.e., the function $f(t)$ is increasing on $\mathbb{T}$.

Some properties of the integration on $\mathbb{T}$ are contained in the following statement.

Theorem 8.5 *Let $a, b, c \in \mathbb{T}$, $\alpha \in \mathbb{R}$ and the functions $f, g \in C_{rd}(\mathbb{T})$. Then*

(i) $\int\limits_a^b [f(t) + g(t)]\Delta t = \int\limits_a^b f(t)\Delta t + \int\limits_a^b g(t)\Delta t;$

(ii) $\int\limits_a^b \alpha f(t)\Delta t = \alpha \int\limits_a^b f(t)\Delta t;$

(iii) $\int\limits_a^b f(t)\Delta t = -\int\limits_b^a f(t)\Delta t;$

(iv) $\int\limits_a^b f(t)\Delta t = \int\limits_a^c f(t)\Delta t + \int\limits_c^b f(t)\Delta t;$

(v) $\int\limits_a^b f(\sigma(t))g^\Delta(t)\Delta t = f(b)g(b) - f(a)g(a) - \int\limits_a^b f^\Delta(t)g(t)\Delta t;$

(vi) $\int\limits_a^a f(t)\Delta t = 0;$

(vii) $\int\limits_t^{\sigma(t)} f(\tau)\Delta\tau = \mu(t)f(t), \quad t \in \mathbb{T}^k;$

(viii) *if* $|f(t)| \le g(t)$ *on* $[a,b)$, *then* $\left|\int\limits_a^b f(t)\Delta t\right| \le \int\limits_a^b g(t)\Delta t;$

(ix) *if* $f(t) \ge 0$ *for all* $a \le t < b$, *then* $\int\limits_a^b f(t)\Delta t \ge 0.$

Now we advance to the rules of Δ-differentiation of a complex function. Remember that if $f, g\colon \mathbb{R} \to \mathbb{R}$, then

$$(f \circ g)'(t) = f'(g(t))g'(t).$$

Theorem 8.6 *Let* $g\colon \mathbb{R} \to \mathbb{R}$ *be continuous,* $g\colon \mathbb{T} \to \mathbb{R}$ Δ-*differentiable on* $\mathbb{T}^k$ *and* $f\colon \mathbb{R} \to \mathbb{R}$ *continuously differentiable. Then there exists c from the interval* $[t, \sigma(t)]$, *such that*

$$(f \circ g)^\Delta(t) = f'(g(c))g^\Delta(t).$$

Below find the rule of Δ-differentiation of a complex function $(f \circ g)^\Delta$, where $g\colon \mathbb{T} \to \mathbb{R}$, and $f\colon \mathbb{R} \to \mathbb{R}$.

Theorem 8.7 *Let the function* $f\colon \mathbb{R} \to \mathbb{R}$ *be continuously differentiable and the function* $g\colon \mathbb{T} \to \mathbb{R}$ *be* Δ-*differentiable. Then* $(f \circ g)\colon \mathbb{T} \to \mathbb{R}$ *is* Δ-*differentiable and the following representation holds true:*

$$(f \circ g)^\Delta(t) = \left\{ \int\limits_0^1 f'(g(t) + h\mu(t)g^\Delta(t))\, dh \right\} g^\Delta(t).$$

If $\sup \mathbb{T} = \infty$, then the improper integral is determined by the formula

$$\int_a^\infty f(t)\Delta t = \lim_{b \to \infty} F(t)\Big|_a^b$$

for all $a \in \mathbb{T}$.

The function $f \colon \mathbb{T} \to \mathbb{R}$ is called regressive if $1 + \mu(t)f(t) \neq 0$ for all $t \in \mathbb{T}^\kappa$, and positively regressive if $1 + \mu(t)f(t) > 0$ for all $t \in \mathbb{T}$.

For the operation $\oplus$ defined by the expression

$$(p \oplus q)(t) = p(t) + q(t) + \mu(t)p(t)q(t), \quad t \in \mathbb{T},$$

the pair $(\mathcal{R}, \oplus)$ is an Abelian group where $\mathcal{R}$ is a set of all rd-continuous and regressive functions $f \colon \mathbb{T} \to \mathbb{R}$.

If the operation $\ominus$ is defined by the relation

$$\ominus p(t) = -\frac{p(t)}{1 + \mu(t)p(t)},$$

then $p \ominus q = p \oplus (\ominus q)$ on $\mathcal{R}$.

Note that if $p, q \in \mathcal{R}$, then $\ominus p$, $\ominus q$, $p \oplus q$, $p \ominus q$, $q \ominus p \in \mathcal{R}$.

To determine an exponential function on a time scale $\mathbb{T}$ the following notions are required.

For some $h > 0$ consider the band (see Hilger [1990])

$$\mathbb{Z}_h = \left\{ z \in \mathbb{C} \colon \ -\frac{\pi}{h} < \mathrm{Im}(z) \leq \frac{\pi}{h} \right\}$$

and the set

$$\mathbb{C}_h = \left\{ z \in \mathbb{C} \colon \ z \neq -\frac{1}{h} \right\}.$$

Let $\mathbb{Z}_h = \mathbb{C}_0 = \mathbb{C}_h$, $h = 0$, be a set of complex numbers.

Then for $h \geq 0$ determine the cylindric transformation $\xi_h \colon \mathbb{C}_h \to \mathbb{Z}_h$ by the formula

$$\xi_h = \begin{cases} \dfrac{1}{h} \mathrm{Log}(1 + zh), & \text{if } h > 0, \\ z, & \text{if } h = 0, \end{cases}$$

where Log is the principal logarithmic function. The inverse cylindric transformation $\xi_h^{-1} \colon \mathbb{Z}_h \to \mathbb{C}_h$ is determined by the relation

$$\xi_h^{-1}(z) = \frac{e^{zh} - 1}{h} = (\exp zh - 1)h^{-1}.$$

For the function $p \in \mathcal{R}$ the exponential function $e_p(t, s)$ is determined by the expression

$$e_p(t, s) = \exp\left(\int_s^t \xi_{\mu(t)}(p(\tau))\Delta\tau \right), \quad (s, t) \in \mathbb{T} \times \mathbb{T}, \tag{8.1}$$

where $\xi_h(z)$ is the cylindric transformation of the function $p(t)$.

The following properties of the exponential function (8.1) are known (see Bohner and Peterson [2001]).

Let $p, q \in \mathcal{R}$ and $t, r, s \in \mathbb{T}$. Then:

(i) $e_0(t, s) \equiv 1$ and $e_p(t, t) \equiv 1$;

(ii) $e_p(\sigma(t), s) = (1 + \mu(t)p(t))e_p(t, s)$;

(iii) $\dfrac{1}{e_p(t, s)} = e_{\ominus p}(t, s)$;

(iv) $e_p(t, s) = \dfrac{1}{e_p(s, t)} = e_{\ominus p}(s, t)$;

(v) $e_p(t, s)e_p(s, r) = e_p(t, r)$;

(vi) $e_p(t, s)e_q(t, s) = e_{p \oplus q}(t, s)$;

(vii) $\dfrac{e_p(t, s)}{e_q(t, s)} = e_{p \ominus q}(t, s)$;

(viii) if $\mathbb{T} = \mathbb{R}$, then $e_p(t, s) = e^{\int_s^t p(\tau)d\tau}$, and if $p(t) = \text{const}$, then $e_p(t, s) = e^{p(t-s)}$;

(ix) if $\mathbb{T} = \mathbb{Z}$, then $e_p(t, s) = \displaystyle\prod_{\tau=s}^{t-1}(1 + p(\tau))$;

(x) if $\mathbb{T} = h\mathbb{Z}$, $h > 0$ and $p = \text{const}$, then $e_p(t, s) = (1 + hp)^{\frac{(t-s)}{h}}$.

On the basis of the exponential function (8.1), the formula of the variation of constants has the following form.

Let f be an rd-continuous function on $\mathbb{T}$ and $p \in \mathcal{R}$. Then the unique solution of the initial problem

$$x^{\Delta}(t) = -p(t)x(\sigma(t)) + f(t), \quad x(t_0) = x_0, \qquad (8.2)$$

where $t_0 \in \mathbb{T}$ and $x_0 \in \mathbb{R}$, is expressed by the formula

$$x(t) = e_{\ominus p}(t, t_0)x_0 + \int_{t_0}^{t} e_{\ominus p}(t, \tau)f(\tau)\Delta\tau.$$

For the functions f and p determined for the problem (8.2), the unique solution of the problem

$$x^{\Delta}(t) = p(t)x(t) + f(t), \quad x(t_0) = x_0,$$

where $t_0 \in \mathbb{T}$ and $x_0 \in \mathbb{R}$, are expressed by the formula

$$x(t) = e_p(t, t_0)x_0 + \int_{t_0}^{t} e_p(t, \sigma(\tau))f(\tau)\Delta\tau.$$

8.2   Theorems of the Direct Lyapunov Method

The generalized direct Lyapunov method for the analysis of the stability of the motion of uncertain systems with a finite number of degrees of freedom, which was developed in Chapter 2, applies to the class of uncertain systems of dynamic equations on time scales.

On a time scale $\mathbb{T}$ consider the following subsets:

$$\mathcal{A} = \{t \in \mathbb{T}: \ t \text{ left-dense and right-scattered}\};$$
$$\mathcal{C} = \{t \in \mathbb{T}: \ t \text{ left-scattered and right-dense}\}.$$

Denote the Euler derivative of the state vector of the system $x(t) \in \mathbb{R}^n$, $t \in \mathbb{T}$, by $\dot{x}(t)$, if it exists.

Consider the system of dynamic equations of perturbed motion

$$x^{\Delta}(t) = f(t, x(t), \alpha), \quad x(t_0) = x_0, \tag{8.3}$$

where

$$x^{\Delta}(t) = \begin{cases} \dfrac{x(\sigma(t)) - x(t)}{\mu(t)}, & \text{if } t \in \mathcal{A} \cup \mathcal{C}, \\ \dot{x}(t) & \text{in the other points,} \end{cases}$$

$x \in \mathbb{R}^n$, $f \colon \mathbb{T} \times \mathbb{R}^n \to \mathbb{R}^n$. Make the following assumptions regarding the system (4.1).

H_1. The vector-function $F(t) = f(t, x(t), \alpha)$ satisfies the condition $F \in C_{\mathrm{rd}}(\mathbb{T})$, if x is a differentiable function with its values in N, $N \subset \mathbb{R}^n$ is an open connected neighbourhood of the state $x = 0$.

H_2. The vector-function $f(t, x, \alpha)$ is component-wise regressive, i. e.,

$$e^{\mathrm{T}} + \mu(t)f(t, x, \alpha) \neq 0 \text{ for all } t \in [t_0, \infty), \text{ where } e^{\mathrm{T}} = (1, \dots, 1)^{\mathrm{T}} \in \mathbb{R}^n.$$

H_3. $f(t, x, \alpha) = 0$ for all $t \in [t_0, \infty)$ if, and only if, $x = 0$.

H_4. On the set $S \subset \mathbb{T} \times N$ the vector-function $f(t, x)$ is bounded and satisfies the Lipschitz condition.

Let a function $r(\alpha) > 0$ be specified, such that $r(\alpha) \to r_0$ $(r_0 = \text{const})$ at $\|\alpha\| \to 0$ and $r(\alpha) \to +\infty$ at $\|\alpha\| \to \infty$. In the normalized space $(\mathbb{R}^n, \|\cdot\|)$ consider the moving set

$$A(r) = \{x \in \mathbb{R}^n : \ \|x\| = r(\alpha)\} \tag{8.4}$$

and assume that at any $(\alpha \neq 0) \in \mathbb{R}^d$ the set $A(r)$ is nonempty.

For the system of dynamic equations (8.3) it makes sense to consider the stability of solutions with respect to the moving set (8.4) and/or to the equilibrium state $x = 0$.

Remember some definitions.

Definition 8.1 The equilibrium state $x = 0$ of the system (8.3):

(a) is *stable*, if for any solution $x(t, \alpha) = x(t; t_0, x_0, \alpha)$ there exists a comparison function of the K-class, such that

$$\|x(t; t_0, x_0, \alpha)\| \leq \psi(\|x_0\|)$$

for all $t \in \mathbb{T}$, $t > t_0$ and at any $\alpha \in \mathcal{S}$;

(b) is *asymptotically stable*, if it is stable in the sense of the definition 8.1(a) and there exists $\delta_0 > 0$ such that

$$\lim_{t \to \infty} \|x(t; t_0, x_0, \alpha)\| = 0$$

for $\|x_0\| < \delta_0$ and for all $\alpha \in \mathcal{S}$.

Definition 8.2 Motions of the system (8.3) are *stable with respect to the moving set* $A(r)$, if for the given $r(\alpha), \varepsilon > 0$ and $t_0 \in \mathbb{T}$ there exists $\delta(t_0, \varepsilon) > 0$ such that under the initial conditions

$$r(\alpha) - \delta < \|x_0\| < r(\alpha) + \delta$$

the solutions $x(t, \alpha)$ satisfy the two-way estimate

$$r(\alpha) - \varepsilon < \|x(t, \alpha)\| < r(\alpha) + \varepsilon$$

for all $t \in \mathbb{T}$ and any $\alpha \in \mathcal{S}$.

The definitions of the uniform stability, attraction, and asymptotic stability with respect to the moving set (8.4) are formulated, taking into account the Definitions 8.2 and 2.2–2.4.

As an auxiliary function for the analysis of the stability of the state $x = 0$ of the system (8.3) we will use the matrix-valued function

$$U(t, x) = [v_{ij}(t, x)], \quad i, j = 1, \dots, m, \tag{8.5}$$

where $v_{ii}(t, x) \colon \mathbb{T} \times \mathbb{R}^n \to \mathbb{R}_+$ at $i = 1, \dots, m$ and $v_{ij}(t, x) \colon \mathbb{T} \times \mathbb{R}^n \to \mathbb{R}$ at $i \neq j$, $i, j = 1, \dots, m$.

Assume that the elements $v_{ij}(t, x)$ of the function (8.5) satisfy the conditions:

(a) $v_{ij}(t, x)$ is continuously differentiable with respect to x for all $t \in \mathbb{T}$;

(b) $v_{ij}(t, x) = 0$ for all $t \in \mathbb{T}$, if only $x = 0$;

(c) $v_{ij}(t, x) = v_{ji}(t, x)$ for all $t \in \mathbb{T}$ and $i, j = 1, \dots, m$.

On the basis of the function (8.5) we construct a scalar function

$$V(t, x, \theta) = \theta^{\mathrm{T}} U(t, x)\theta, \quad \theta \in \mathbb{R}_+^m, \tag{8.6}$$

which is the principal one in the generalized Lyapunov method.

Definition 8.3 The matrix-valued function (8.5) is called:

(a) *positive- (negative-) semidefinite* on $\mathbb{T} \times N$, $N \subset \mathbb{R}^n$, if $V(t, x, \theta) \geq 0$
    $(V(t, x, \theta) \leq 0)$ for all $(t, x, \theta) \in \mathbb{T} \times N \times \mathbb{R}_+^m$ respectively;

(b) *positive-definite* on $\mathbb{T} \times N$, $N \subset \mathbb{R}^n$, if there exists a function $a \in K$-class, such that $V(t, x, \theta) \geq a(\|x\|)$ for all $(t, x, \theta) \in \mathbb{T} \times N \times \mathbb{R}_+^m$;

(c) *decreasing* on $\mathbb{T} \times N$, if there exists a function $b \in K$-class, such that $V(t, x, \theta) \leq b(\|x\|)$ for all $(t, x, \theta) \in \mathbb{T} \times N \times \mathbb{R}_+^m$;

(d) *radially unbounded* on $\mathbb{T} \times N$, if $V(t, x, \theta) \to +\infty$ for $\|x\| \to +\infty$, and $(t, x, \theta) \in \mathbb{T} \times N \times \mathbb{R}_+^m$.

Proposition 8.1 The matrix-valued function $U \colon \mathbb{T} \times \mathbb{R}^n \to \mathbb{R}^{m \times m}$ is positive-definite on $\mathbb{T}$ if, and only if, the function (8.5) can be represented in the form

$$\theta^T U(t, x)\theta = \theta^T U_+(t, x)\theta + a(\|x\|), \quad t \in \mathbb{T},$$

where $U_+(t, x)$ is a positive-semidefinite matrix-valued function and a belongs to the K-class.

Proposition 8.2 The matrix-valued function $U \colon \mathbb{T} \times \mathbb{R}^n \to \mathbb{R}^{m \times m}$ is decreasing on $\mathbb{T}$ if, and only if, the function (8.6) can be represented in the form

$$\theta^T U(t, x)\theta = \theta^T U_-(t, x)\theta + b(\|x\|), \quad t \in \mathbb{T},$$

where $U_-(t, x)$ is a negative-semidefinite matrix-valued function and b belongs to the K-class.

Now we will use the expression of the full Δ-derivative of the function (8.5) along solutions of the system (8.3). We will determine it as follows:

$$V^\Delta(t, x, \theta) = \theta^T U^\Delta(t, x)\theta, \quad \theta \in \mathbb{R}_+^m, \quad t \in \mathbb{T},$$

where $U^\Delta(t, x)$ is calculated element by element according to the formula

$$U^\Delta(t, x) = U^\Delta(t, x(t)) = U_t^\Delta(t, x(\sigma(t)))$$

$$+ f(t, x) \int_0^1 U_x(t, x(t) + h_\mu(t)f(t, x)) \, dh,$$

where U_t^Δ is a Δ-derivative of the matrix-valued function with respect to t, U_x' is an ordinary partial derivative with respect to x.

In a particular case, when

$$x^\Delta = A(\alpha)x, \quad \alpha \in \mathcal{S},$$

for the function $V(t, x, \theta) = x^T x$, $x \in \mathbb{R}^n$, we obtain

$$V^{\Delta}(t, x, \theta) = (x^T x)^{\Delta}(t) = x^T(t) x^{\Delta}(t) + x^{T\Delta}(t) x(\sigma(t))$$
$$= x^T(t) \left(A^T(\alpha) \oplus A(\alpha) \right) x(t),$$
$$A^T(\alpha) \oplus A(\alpha) = A^T(\alpha) + A(\alpha) + \mu(t) A^T(\alpha) A(\alpha).$$

Now formulate the statements about the stability of the equilibrium state $x = 0$ of the uncertain system (8.3).

Theorem 8.8 *Assume that the vector-function $f(t, x, \alpha)$ in the system (8.3) satisfies the assumptions H_1–H_4 on $\mathbb{T} \times N \times S$, and there exist:*

(1) *a matrix-valued function $U \colon \mathbb{T} \times N \to \mathbb{R}^{m \times m}$ and a vector $\theta \in \mathbb{R}_+^m$ such that the function $V(t, x, \theta) = \theta^T U(t, x) \theta$ is locally Lipschitz with respect to x for all $t \in \mathbb{T}$;*

(2) *comparision functions ψ_{i1}, ψ_{i2}, $\psi_{i3} \in K$-class and $(m \times m)$-matrices A_j, $j = 1, 2$, such that for all $(t, x) \in \mathbb{T} \times N$*

   (a) $\psi_1^T(\|x\|) A_1 \psi_1(\|x\|) \leq V(t, x, \theta)$,

   (b) $V(t, x, \theta) \leq \psi_2^T(\|x\|) A_2 \psi_2(\|x\|)$,

   (c) $(m \times m)$-*matrix* $A_3 = A_3(t, \alpha)$ *such that*

$$V^{\Delta}(t, x, \theta) \leq \psi_3^T(\|x\|) A_3 \psi_3(\|x\|)$$

   for all $(t, x) \in \mathbb{T} \times N$ and for all $\alpha \in S$,

   (d) $(m \times m)$-*matrix* $\bar{A}_3$ *such that*

$$\frac{1}{2} [A_3^T(t, \alpha) + A_3(t, \alpha)] \leq \bar{A}_3$$

   for all $\alpha \in S$.

Then, if the matrices A_1 and A_2 are positive-definite and the matrix $\bar{A}_3$ is negative-semidefinite, then the state $x = 0$ of the system (8.3) is stable under the conditions 2(a), 2(b), 2(d) and uniformly stable under the conditions 2(a)–2(d).

Proof From the fact that A_1 and A_2 are positive-definite matrices it follows that $\lambda_m(A_1) > 0$ and $\lambda_M(A_2) > 0$, where $\lambda_m(A_1)$ and $\lambda_M(A_2)$ are the minimum and the maximal eigenvalues of the matrices A_1 and A_2, respectively. Taking this into account, we represent the estimates (a), (b) from the condition 2 of the Theorem 8.1 in the form

$$\lambda_m(A_1) \bar{\psi}_1(\|x\|) \leq V(t, x, \theta) \leq \lambda_M(A_2) \bar{\psi}_2(\|x\|)$$

for all $(t, x) \in \mathbb{T} \times N$, where $\bar{\psi}_1, \bar{\psi}_2 \in K$-class, such that

$$\bar{\psi}_1(\|x\|) \leq \psi_1^T(\|x\|) \psi_1(\|x\|), \quad \bar{\psi}_2(\|x\|) \geq \psi_2^T(\|x\|) \psi_2(\|x\|)$$

for all $x \in N$.

Let $\varepsilon > 0$ and $S(t)$; there exists $\delta = \delta(\varepsilon) > 0$ such that the condition $\|x_0\| < \delta$ implies $\|x(t; t_0, x_0)\| < \varepsilon$.

Let

$$S^* = \{t \in [t_0, \infty): \quad S(t) \text{ not true}\}.$$

Let us show that under the conditions of the Theorem 8.1 the set $S^* = \varnothing$. Assume the opposite, i.e. $S^* \neq \varnothing$. From the fact that S^* is closed and nonempty it follows that $\inf S^* = t^* \in S^*$. Prove that the statement $S(t^*)$ is true. Let $t^* = t_0$, then $S(t_0)$ be true, since $\|x(t_0; t_0, x_0)\| < \varepsilon$ at $\|x_0\| < \varepsilon$ and $x(t_0; t_0, x_0) = x_0$ for all $\alpha \in S$.

Then, let $t^* \neq t_0$. Prove that $S(t^*)$ in this case, too. Indeed, choose $\delta_1 = \delta_1(\varepsilon)$ so that $\lambda_M(A_2)\bar{\psi}_2(\delta_1) < \lambda_m(A_1)\bar{\psi}_1(\varepsilon)$.

Now let $\delta = \min(\varepsilon, \delta_1)$ such that

$$\|x(t^*; t_0, x_0)\| \geq \varepsilon \quad \text{and} \quad \|x(t; t_0, x_0)\| < \varepsilon$$

at $t \in [t_0, t^*)$ and $\|x_0\| < \delta$. From the conditions 2(c), 2(d) of the Theorem 8.1 we obtain

$$V^{\Delta}(t, x, \theta) \leq \lambda_M(\bar{A}_3)\bar{\psi}_3(\|x\|) \leq 0$$

for all $(t, x, \theta) \in \mathbb{T} \times N \times \mathbb{R}_+^m$. Hence at $t = t^*$ we obtain

$$
\begin{aligned}
\lambda_m(A_1)\bar{\psi}_1(\varepsilon) = \lambda_m(A_1)\bar{\psi}_1(\|x(t^*; t_0, x_0)\|) \leq V(t^*, x(t^*), \theta) \leq \\
\leq V(t_0, x_0, \theta) < \lambda_M(A_2)\bar{\psi}_2(\delta)
\end{aligned}
\tag{8.7}
$$

at $\|x_0\| < \delta$. From (8.7) it follows that $S(t^*)$ is true, so that $t^* \notin S^*$. Therefore $S^* = \varnothing$.

The Theorem 8.8 is proved.

Corollary 8.1 *Let the vector function f in the system (8.3) satisfy the assumptions H_1–H_4 (Section 8.2) on $\mathbb{T} \times N \times S$ and let there exist at least one pair of indices $(p, q) \in [1, m]$ for which $(v_{pq}(t, x) \neq 0) \in U(t, x)$ and the function $V(t, x, \theta) = e^{\mathrm{T}} U(t, x)e = V(t, x)$ for all $(t, x) \in \mathbb{T} \times N$ satisfies the conditions:*

(a) $\psi_1(\|x\|) \leq V(t, x)$;

(b) $V(t, x) \leq \psi_2(\|x\|)$;

(c) $V^{\Delta}(t, x)|_{(8.3)} \leq 0$ *for all $\alpha \in S$,*

where ψ_1, ψ_2 are comparison functions of the K-class.

Then the state $x = 0$ of the uncertain system (8.3) is stable under the conditions (a) and (c) and uniformly stable under the conditions (a)–(c).

Theorem 8.9 *Assume that the vector-function $f(t, x, \alpha)$ in the system (8.3) satisfies the conditions H_1–H_4 on $\mathbb{T} \times N \times S$ and there exist:*

(1) *a matrix-valued function* $U\colon \mathbb{T} \times \mathbb{R}^n \to \mathbb{R}^{m \times m}$ *and a vector* $\theta \in \mathbb{R}_+^m$ *such that the function* $V(t, x, \theta) = \theta^{\mathrm{T}} U(t, x)\theta$ *is locally Lipschitz with respect to* x *for all* $t \in \mathbb{T}$;

(2) *the comparison functions* $\psi_{i1}, \psi_{i2}, \psi_{i3} \in K$-*class and* $(m \times m)$-*matrices* $B_j,\ j = 1, 2, 3,$ *such that:*

    (a) $\psi_1^{\mathrm{T}}(\|x\|)B_1\psi_1(\|x\|) \le V(t, x, \theta),$

    (b) $V(t, x, \theta) \le \psi_2^{\mathrm{T}}(\|x\|)B_2\psi_2(\|x\|)$ *for all* $(t, x, \theta) \in \mathbb{T} \times N \times \mathbb{R}_+^m$;

    (c) $(m \times m)$-*matrix* $B_3 = B_3(t, \alpha)$ *such that*

$$V^\Delta(t, x, \theta) \le \psi_3^{\mathrm{T}}(\|x\|)B_3\psi_3(\|x\|) + w(t, \psi_3(\|x\|), \alpha)$$

for all $(t, x, \alpha) \in \mathbb{T} \times N \times \mathcal{S}$, *where the function* $w(t, \cdot)$ *satisfies the condition*

$$\lim \frac{|w(t, \psi_3(\|x\|), \alpha)|}{\|\psi_3\|} = 0 \quad at \quad \|\psi_3\| \to 0 \qquad (8.8)$$

uniformly with respect to $t \in \mathbb{T}$ *for all* $\alpha \in \mathcal{S}$;

    (d) $(m \times m)$-*matrix* $\bar{B}_3$ *such that*

$$\frac{1}{2}\left[B_3^{\mathrm{T}}(t, \alpha) + B_3(t, \alpha)\right] \le \bar{B}_3$$

for all $\alpha \in \mathcal{S}$.

    Then if the matrices B_1 *and* B_2 *are positive-definite and the matrix* $\bar{B}_3$ *is negative-definite, then:*

(a) *under the conditions* 2(a), 2(c) *the state* $x = 0$ *of the system* (8.3) *is asymptotically stable on* $\mathbb{T}$;

(b) *under the conditions* 2(a)–2(c) *the state of the system* (8.3) *is uniformly asymptotically stable on* $\mathbb{T}$.

Proof Since the matrices B_1 and B_2 are positive-definite, we transform the inequalities 2(a), (b) of the Theorem 8.9 to the form

$$\lambda_m(B_1)\bar{\eta}_1(\|x\|) \le V(t, x, y) \le \lambda_M(B_2)\bar{\eta}_2(\|x\|).$$

This two-way inequality holds for all $(t, x) \in \mathbb{T} \times N$, where $\lambda_m(B_1) > 0$ and $\lambda_M(B_2) > 0$ with the comparison functions $\bar{\eta}_1, \bar{\eta}_2 \in K$-class, which satisfy the inequalities

$$\bar{\eta}_1(\|x\|) \le \eta_1^{\mathrm{T}}(\|x\|)\, \eta_1(\|x\|), \quad \bar{\eta}_2(\|x\|) \le \eta_2^{\mathrm{T}}(\|x\|)\, \eta_2(\|x\|)$$

for all $x \in N$.

Let $S_1(t)$ be given; the state $x = 0$ of the system (8.3) is uniformly stable and there exists $\varepsilon > \eta > 0$ such that for any $\delta_0 > 0$ there exists $\tau > 0$ such that from the conditions $t_0 \in \mathbb{T}$, $\|x_0\| < \eta$ it follows that $\|x(t; t_0, x_0)\| < \delta_0$ for all $t \geq t_0 + \tau$. Apply the reasoning similar to that used in the proof of the Theorem 8.8. Namely, let

$$S_1^* = \{ t \in [t_0, \infty) \colon S_1(t) \text{ not true} \}.$$

Show that under the conditions of the Theorem 8.9 the set S_1^* is empty. Assume the opposite, i. e. $S_1^* \neq \varnothing$. From the fact that the set S_1^* is closed and nonempty it follows that $\inf S_1^* = t^* \in S_1^*$. Note that at $t = t_0$ the statement $S_1^*(t)$ is true, i. e., $\|x(t_0; t_0, x_0)\| < \varepsilon$ for $\|x_0\| < \varepsilon$ and $\|x(t_0; t_0, x_0)\| < \eta$ at $\|x_0\| < \eta$, since $x(t_0; t_0, x_0) = x_0$. Let $t^* > t_0$. From the conditions 2(c), (d) of the Theorem 8.9 it follows that

$$V^\Delta(t, x, \theta) \leq 0$$

for all $(t, x) \in N$ and at any $\alpha \in S$. Here, according to the Theorem 8.8, the state $x = 0$ of the system (8.3) is uniformly stable. Thus, for the given $\varepsilon > 0$ there exists $\delta = \delta(\varepsilon) > 0$ such that for any solution $x(t; t_0, x_0)$ of the uncertain system (8.3) the condition $\|x(t^*)\| < \delta$, $t^* > t_0$, implies $\|x(t; t^*, x^*)\| < \varepsilon$ for $t > t^*$, where $x^* = x(t^*)$. Taking into account the condition (8.8), transform the condition 2 of the Theorem 8.9 to the form

$$V^\Delta(t, x, \theta) \leq (1 - \beta)\lambda_M(B_3^*)\overline{\eta}_3(\|x\|), \tag{8.9}$$

where $(t, x) \in \mathbb{T} \times N^*$, $N^* \subseteq N$, $0 < \beta < 1$, $\lambda_M(B_3^*) < 0$, $\lambda_M(B_3^*)$ is the maximal eigenvalue of the matrix B_3^* and the function $\overline{\eta}_3$ belongs to the K-class and satisfies the inequality $\overline{\eta}_3(\|x\|) \geq \eta_3^T(\|x\|)\eta_3(\|x\|)$ for all $x \in N^*$. From the inequality (8.9) we obtain

$$V^\Delta(t, x(t), \theta) \leq (1 - \beta)\lambda_M(B_3^*)\overline{\eta}_3(\|x(t)\|)$$

for all $(t, x(t)) \in \mathbb{T} \times N^*$. Integrating this inequality from t^* to t, we obtain

$$V(t, x(t), \theta) \leq v(t^*, x^*, \theta) - (1 - \beta)\lambda_M(B_3^*) \int_{t^*}^{t} \overline{\eta}_3(\|x(s)\|) \, \Delta s$$

$$\leq \lambda_M(B_2)\overline{\eta}_2(\|x^*\|) - (1 - \beta)\lambda_M(B_3^*) \int_{t^*}^{t} \overline{\eta}_3(\|x(s)\|) \, \Delta s. \tag{8.10}$$

For the given $0 < \eta < \varepsilon$ choose $\delta_0 = \delta(\varepsilon)$ and

$$\tau(\eta) = \lambda_M(B_2)\overline{\eta}_2(\delta_0) \big/ (1 - \beta)\lambda_M(B_3^*)\overline{\eta}_3(\delta(\eta)),$$

where $\delta(\eta)$ corresponds to δ in the definition of the uniform stability. Show that $\|x(t; t^*, x^*)\| < \delta(\eta)$ for all $t \in [t^*, t^* + \tau]$, as soon as $t^* > t_0$ and $\|x^*\| < \delta_0$.

Let this statement be not true. Then there should exist $t_1 \in [t^*, t^* + \tau]$ such that

$$\|x(t_1; t^*, x^*)\| \geq \delta(\eta). \tag{8.11}$$

From the estimate (8.10) and the inequality (8.11) for the values $t \in [t^*, t^* + \tau]$ we obtain

$$V(t, x(t), \theta) \leq \lambda_M(B_2)\overline{\eta}_2(\delta_0) - (1 - \beta)\lambda_M(B_3^*)\overline{\eta}_3(\delta(\eta))(t - t^*).$$

Hence at $t = t^* + \tau$ we obtain

$$0 < \lambda_m(B_1)\overline{\eta}_1(\delta(\eta)) \leq V(t^* + \tau, x(t^* + \tau), \theta)$$
$$\leq \lambda_M(B_2)\overline{\eta}_2(\delta_0) - (1 - \beta)\lambda_M(B_3^*)\overline{\eta}_3(\delta(\eta))\tau(\eta) = 0.$$

The obtained contradiction shows that there is no value $t_1 \in [t^*, t^* + \tau]$ at which the inequality (8.11) would hold. Therefore there exists $t_2 \colon t^* \leq t_2 \leq t^* + \tau$ such that $\|x(t_2; t^*, x^*)\| < \delta(\eta)$. From the uniform stability it follows that $\|x(t)\| < \eta$ for all $t \geq t_2$ and, in particular, at $t \geq t^* + \tau$. Therefore $\|x(t)\| < \eta$ at $t \geq t^* + \tau$, as soon as $t^* > t_0$ and $\|x(t^*)\| \leq \delta_0$. Hence $S_1(t^*)$ is true and t^* does not belong to S_1^*. This proves that the state $x = 0$ of the uncertain system of dynamic equations (8.3) is asymptotically stable.

Corollary 8.2 *Let the vector-function f in the system (8.3) satisfy the conditions H_1–H_4 on $\mathbb{T} \times N \times S$, let there exist at least one pair of indices $(p, q) \in [1, m]$, for which $(v_{pq}(t, x) \neq 0) \in U(t, x)$, and let the function $V(t, x, \theta) = e^T U(t, x)e = V(t, x)$ for all $(t, x) \in \mathbb{T} \times N$ satisfy the conditions:*

(a) $\psi_1(\|x\|) \leq V(t, x);$

(b) $V(t, x) \leq \psi_2(\|x\|);$

(c) *for all $\alpha \in S$ the following inequality holds*

$$V^\Delta(t, x)|_{(4.1)} \leq -\psi_3(\|x\|) + m(t, \psi_3(\|x\|), \alpha)$$

and

$$\lim \frac{|m(t, \psi_3(\|x\|), \alpha)|}{\psi_3(\|x\|)} \quad at \quad \psi_3 \to 0 \quad and \; for \; all \quad \alpha \in S$$

uniformly with respect to $t \in \mathbb{T}$, where ψ_1, ψ_2, ψ_3 are comparison functions of the class K.

Then under the conditions (a) and (c) the state $x = 0$ of the uncertain system (8.3) is asymptotically stable, and under the conditions (a)–(c) the state $x = 0$ of the uncertain system (8.3) is uniformly asymptotically stable.

Theorem 8.10 *Assume that the vector-function $f(t, x, \alpha)$ in the system (8.3) satisfies the conditions $H_1 - H_4$ on $\mathbb{T} \times \mathbb{N} \times S$ and there exist:*

(1) *a matrix-valued function* $U\colon\ \mathbb{T}\times\mathbb{R}^n\ \to\ \mathbb{R}^{m\times m}$ *and a vector* $\theta\in\mathbb{R}^m_+$ *such that the function* $V(t,x,\theta)=\theta^{\mathrm{T}}U(t,x)\theta$ *is locally Lipschitz with respect to* x *for all* $t\in\mathbb{T}$;

(2) $\psi_1\in K$ *and a symmetrical* $(m\times m)$-*matrix* A_1 *such that for all* $(t,x)\in\mathbb{T}\times N$:

(a) $\psi_1^{\mathrm{T}}(\|x\|)A_1\psi_1(\|x\|)\leq V(t,x,\theta)$ *and* $\lambda_m(A_1)$ *the minimum eigenvalue of the matrix* A_1,

(b) *there exists an* $(m\times m)$-*matrix* $C=C(t,\alpha)$ *such that*

$$V^\Delta(t,x,\theta)\geq\psi_1^{\mathrm{T}}(\|x\|)C(t,\alpha)\psi_1(\|x\|)+w(t,\psi_1(\|x\|))$$

for all $(t,x)\in\mathbb{T}\times N$ *and*

$$\lim_{\psi_1\to 0}\frac{w(t,\psi_1)}{\psi_1}=0\quad\text{uniformly with respect to }t\in\mathbb{T};$$

(c) *a constant* $(m\times m)$-*matrix* C^* *such that*

$$\frac{1}{2}[C^{\mathrm{T}}(t,\alpha)+C(t,\alpha)]\leq C^*$$

and at least at one value of $\alpha\in S$ *the following condition is satisfied* $\lambda_m(C^*)>0$;

(3) *the function* $\lambda_m(C^*)\lambda_m^{-1}(A_1)\ =\ p(t)\ \in\mathcal{R}$ *and* $\liminf p(t)>0$ *for* $t\to\infty$.

Then the state $x=0$ *of the uncertain system* (8.3) *is unstable.*

Proof Under the conditions 2(a), (c) of the Theorem 8.10 the estimate 2(b) can be represented in the form

$$V^\Delta(t,x,\theta)\geq p(t)v(t,x,\theta)+w(t,v(t,x,\theta)),\qquad(8.12)$$

where $p(t)\ =\ \lambda_m(C^*)\lambda_m^{-1}(A_1)$. The comparison equation for the inequality (8.12) has the form

$$u^\Delta(t)=p(t)u(t)+w(t,u(t))\qquad(8.13)$$

with the condition

$$\lim_{u\to 0}\frac{w(t,u)}{u}=0\quad\text{uniformly with respect to }t\in\mathbb{T}.$$

The instability of the state $u=0$ of the scalar equation (8.13) implies the instability of the state $x=0$ of the uncertain system (8.3), since $v(t)=v(t,x(t),\theta)\geq u(t)$ for all $t\in[t_0,\infty)\cap\mathbb{T}$ and at any $\alpha\in S$.

From the condition (3) of the Theorem 8.10 it follows that there exist

$p^* > 0$ and $t_1 \in [t_0, \infty) \cap \mathbb{T}$ such that $p(t) \geq p^*$ for all $t \in [t_1, \infty) \cap \mathbb{T}$, and for any $0 < \varepsilon < p^*$ there exists $\delta_1 > 0$ such that

$$|w(t, u)| \leq \varepsilon |u| \qquad (8.14)$$

for all $t \in [t_0, \infty) \cap \mathbb{T}$, as soon as $|u| < \delta_1$. Assume that the zero solution of the comparison equation (8.13) is stable on $[t_0, \infty) \cap \mathbb{T}$. Here for the given $\delta_1 > 0$ one can find $0 < \delta_2 < \delta_1$ such that $u(t) < \delta_1$ for all $t \in [t_1, \infty) \cap \mathbb{T}$, as soon as $u(t_1) < \delta_2$, where $u(t) = u(t; t_1, u_1)$ is the lower solution of the comparison equation (8.13). Show that $u(t) > 0$ on $[t_0, \infty) \cap \mathbb{T}$ under the conditions of the Theorem 8.10. If this is not so, then one can find $t_2 \in [t_1, \infty) \cap \mathbb{T}$ such that

$$u(t_2) \leq 0 \quad \text{and} \quad u(t) > 0 \quad \text{for all} \ t \in [t_1, t_2) \cap \mathbb{T}. \qquad (8.15)$$

From the equation (8.13) and the estimate (8.14) obtain

$$u^{\Delta}(t) \geq p(t)u(t) - \varepsilon|u(t)| \geq (p(t) - \varepsilon)u(t) \geq (p^* - \varepsilon)u(t) \qquad (8.16)$$

for all $t \in [t_1, t_2) \cap \mathbb{T}$. Hence follows that $u(t_2) > 0$, and this contradicts the assumption (8.15). Therefore $u(t) > 0$ on $[t_1, \infty) \cap \mathbb{T}$ and the estimate (8.16) is true for all $t \in [t_1, \infty) \cap \mathbb{T}$. In this case from (8.16) we obtain

$$u(t) = u(t; t_1, u_1) \geq e_{p^* - \varepsilon}(t, t_1)u(t_1) \qquad (8.17)$$

for all $t \in [t_1, \infty) \cap \mathbb{T}$, and this contradicts the fact that $u(t) \leq \delta_1$ for all $t \in [t_1, \infty) \cap \mathbb{T}$.

This proves the instability of the state $u = 0$ of the comparison equation (8.13), and according to the comparison principle, the state $x = 0$ of the uncertain system (8.3) is unstable.

Corollary 8.3 *Let the vector-function f in the system (8.3) satisfy the conditions H_1–H_4 on $\mathbb{T} \times N \times S$, $N \subseteq \mathbb{R}^n$, there exists at least one pair $(p, q) \in [1, m]$, for which $(v_{pq}(t, x) \neq 0) \in U(t, x)$, and the function $V(t, x, e) = e^{\mathrm{T}} U(t, x)e = V(t, x)$ for all $(t, x) \in \mathbb{T} \times N$ satisfies the conditions:*

(a) *$\psi_1(\|x\|) \leq V(t, x)$, $\psi_1 \in K$;*

(b) *for all $\alpha \in S$ the following inequality is true: $V^{\Delta}(t, x, \theta)|_{(8.3)} \geq \psi_3(\|x\|)$, $\psi_3 \in K$;*

(c) *the point $(x = 0) \in \partial L$;*

(d) *$V(t, x) = 0$ on $\mathbb{T} \times (\partial \mathbb{T} \cap B_{\Delta})$.*

Then the equilibrium state $x = 0$ of the system (8.3) is unstable.

8.3   Applications and the Discussion of the Results

Consider the uncertain system of dynamic equations

$$x^\Delta = Ax + g(\alpha, x),$$
$$\alpha \in \mathcal{S}, \qquad (8.18)$$

where, as above, α is a vector or a matrix of uncertain parameters from the compact set $\mathcal{S}$, and the vector-function $g \colon \mathcal{S} \times \mathbb{R}^n \to \mathbb{R}^n$ is known and rd-continuous on the solutions of the system (8.17), i.e., $g(\alpha, x(t)) = G(t) \in C_{\mathrm{rd}}(\mathbb{T}, \mathbb{R}^n)$ for all $\alpha \in \mathcal{S}$. Of the nominal system

$$x^\Delta = Ax$$

has a certain type of stability (asymptotic or exponential) and we can construct a Lyapunov function which will help to solve the problem of the stability of the uncertain system (8.18).

Note that for the system of dynamic equations (8.18) the known Lyapunov theorem of stability by first approximation is not valid.

In some cases it is acceptable to apply simple Lyapunov functions in the analysis of the stability of uncertain dynamic equations of the form (8.18).

Example 8.1 Consider the uncertain system

$$x^\Delta = f(x, \alpha), \quad \alpha \in \mathcal{S}, \qquad (8.19)$$

where $x(t) \in \mathbb{R}^n$ is the vector of the state of the system at a point $t \in \mathbb{T}$, the vector-function $f \colon \mathbb{R}^n \times \mathcal{S} \to \mathbb{R}^n$ and $f(0, \alpha) = 0$ for all $\alpha \in \mathcal{S}$.

The uncertain system (8.19) is quadratically stable if there exists a positive-definite matrix $Q \in \mathbb{R}^{n \times n}$ such that

$$f^{\mathrm{T}}(x, \alpha)x + x^{\mathrm{T}}f(x, \alpha) + \mu(t)f^{\mathrm{T}}(x, \alpha)f(x, \alpha) \le -x^{\mathrm{T}}Qx \qquad (8.20)$$

for all $\alpha \in \mathcal{S}$.

If $f(x, \alpha) = A(\alpha)x$, then the condition (8.20) takes the form

$$A^{\mathrm{T}}(\alpha) \oplus A(\alpha) \le -Q,$$

where $A^{\mathrm{T}}(\alpha) \oplus A(\alpha) = A^{\mathrm{T}}(\alpha) + A(\alpha) + \mu(t)A^{\mathrm{T}}(\alpha)A(\alpha)$. It is not difficult to check that the system

$$x^\Delta = A(\alpha)x, \quad \alpha \in \mathcal{S}$$

will be quadratically stable if

$$A^{\mathrm{T}}(\alpha) \oplus A(\alpha) < 0$$

for all $\alpha \in \mathcal{S}$. In this case for all $\alpha \in \mathcal{S}$ the simple function $V(x) = x^{\mathrm{T}}x$, $x \in \mathbb{R}^n$ is a Lyapunov function.

Example 8.2 Let the following uncertain system of pseudolinear dynamic equations be specified:

$$x^\Delta(t) = [A(t) + D(\alpha, x)] \, x(t), \quad \|D(\alpha, x)\| \le \beta, \tag{8.21}$$

for all $(\alpha, x) \in \mathcal{S} \times N$, where $\beta > 0$ is some constant. Assume that the nominal system

$$x^\Delta(t) = A(t)x(t), \quad x(t_0) = x_0,$$

is uniformly exponentially stable, i.e., its fundamental matrix satisfies the condition

$$\|\Phi(t, t_0)\| \le \gamma e_{-\lambda}(t, t_0), \tag{8.22}$$

where $\lambda, \gamma > 0$ $(-\lambda \in \mathcal{R}^+)$ and λ is uniformly regressive on $\mathbb{T}$.

Find the estimate of the quantity which bounds the norm of uncertainties and/or nonlinearity of the system (8.21), at which the property of the uniform exponential stability remains.

For any initial conditions $x(t_0) = x_0$ for $t_0 \in \mathbb{T}$ the solution of the system (8.21) can be represented in the form

$$x(t) = \Phi(t, t_0)x_0 + \int_{t_0}^t \Phi(t, \sigma(s))D(\alpha, x(s))x(s) \, \Delta s. \tag{8.23}$$

Taking into account the estimate (8.22), from the relation (8.23) we obtain

$$\|x(t)\| \le \gamma e_{-\lambda}(t, t_0) \, \|x_0\| + \int_{t_0}^t \gamma e_{-\lambda}(t, \sigma(s)) \|D(\alpha, x(s))\| \, \|x(s)\| \, \Delta s, \tag{8.24}$$

for all $t \ge t_0$. Taking into account the properties of exponential functions, transform the estimate (8.24) to the form

$$e_{-\lambda}(t_0, t)\|x(t)\| \le \gamma\|x_0\| + \int_{t_0}^t \gamma\|D(\alpha, x(s))\| e_{-\lambda}(t_0, s)e_{-\lambda}(s, \sigma(s))\|x(s)\|\Delta s$$

$$\le \gamma\|x_0\| + \int_{t_0}^t \gamma\beta(1 - \mu(s)\lambda)^{-1}e_{-\lambda}(t_0, s) \, \|x(s)\|\Delta s$$

$$\le \gamma\|x_0\| + \int_{t_0}^t \gamma\beta\delta e_{-\lambda}(t_0, s) \, \|x(s)\|\Delta s, \quad t \ge t_0,$$

$$\tag{8.25}$$

where δ is the constant of the uniform regressivity of λ: $0 < \delta^{-1} < (1 - \mu(t)\lambda)$.

Denote $z(t) = e_{-\lambda}(t_0, t)\|x(t)\|$ and rewrite the estimate (8.25) in the form of the inequality

$$z(t) \leq \gamma\|x_0\| + \int_{t_0}^{t} \gamma\beta\delta z(s)\Delta s, \quad t \geq t_0.$$

Applying Gronwall's lemma to this inequality, obtain

$$z(t) \leq \gamma\|x_0\| \, e_{\gamma\beta\delta}(t, t_0), \quad t \geq t_0,$$

and reverting to the estimate $\|x(t)\|$, find

$$\|x(t)\| \leq \gamma\|x_0\| \, e_{\gamma\beta\delta}(t, t_0)e_{-\lambda}(t, t_0), \quad t \geq t_0.$$

From this estimate it follows that the state $x = 0$ of the system (8.21) will be uniformly exponentially stable if $-\lambda \oplus \gamma\beta\delta \in \mathcal{R}^+$, and for all $t \in \mathbb{T}$ this expression will be negative. Note that the product $\gamma\beta\delta > 0$ is positively regressive, so that $\gamma\beta\delta \in \mathcal{R}^+$. Since $\mathcal{R}^+$ is a subgroup of $\mathcal{R}$, then $-\lambda \oplus \gamma\beta\delta \in \mathcal{R}^+$. Now find the estimate of β from the condition

$$-\lambda < -\lambda \oplus \gamma\beta\delta < 0 \quad \text{for all} \quad t \in \mathbb{T}.$$

Simple transformations of this inequality result in the estimate

$$0 < \beta < \frac{\lambda}{\gamma\delta(1 - \mu(t)\lambda)} \quad \text{for all} \quad t \in \mathbb{T}.$$

Note that from the condition of the uniform regressivity of λ it follows that the granularity $\mu(t)$ of the time scale $\mathbb{T}$ should be bounded, i. e., $0 < \mu(t) \leq \dfrac{\delta - 1}{\delta\lambda}$, $\delta > 1$.

Example 8.3 Consider the uncertain quasilinear system of dynamic equations

$$x^{\Delta}(t) = [A(t) + D_{\alpha}E] \, x(t) + g(\alpha, x(t)), \tag{8.26}$$

where $\alpha \in \mathbb{R}^{p \times q}$, D_{α}, E are constant matrices of the respective dimensions, $g \colon \mathcal{S} \times \mathbb{R}^n \to \mathbb{R}^n$.

Assume that the system (8.26) satisfies the following conditions:

(1) the nominal system corresponding to the system (8.26) is uniformly exponentially stable on $\mathbb{T}$, i. e. for its fundamental matrix the inequality (8.22) holds true;

(2) there exists a constant $0 < \beta < +\infty$ such that

$$\|D_{\alpha}E\| \leq \beta \quad \text{for all} \quad \alpha \in \mathcal{S}, \tag{8.27}$$

where $\mathcal{S} = \{(\alpha_1, \alpha_2, \dots, \alpha_r), \quad \alpha_i \in \mathbb{R}^{p_i \times q_i}, \quad \|\alpha_i\| \leq 1\};$

(3) the vector-function $g(\alpha, x(t)) \in C_{\mathrm{rd}}(\mathbb{T}, \mathbb{R}^n)$ for all $\alpha \in \mathcal{S}$ satisfies the estimate

$$\|g(\alpha, x(t))\| \le \|E\| \, \|x(t)\| \qquad (8.28)$$

for all $t \in \mathbb{T}$;

(4) the solution $x(t)$ of the system (8.26) is determined for all $t \ge t_0$ and rd-continuous on $\mathbb{T}$ for all $\alpha \in \mathcal{S}$.

Find the estimates of β and $\|E\|$, under which the solutions of the uncertain system (8.26) will satisfy the estimate

$$\|x(t)\| \le \gamma \|x_0\| \, e_{-\lambda^*}(t, t_0), \qquad (8.29)$$

where the constants $\gamma, \lambda^* > 0$ and $-\lambda^* \in \mathcal{R}^+$.

Apply the technique of estimates, similar to that used in the Example 8.2, and represent the solution of the system (8.26) in the form

$$x(t) = \Phi(t, t_0)x_0 + \int_{t_0}^{t} \Phi(t, \sigma(s)) \left[D_\alpha E x(s) + g(\alpha, x(s)) \right] \Delta s \qquad (8.30)$$

for all $t \ge t_0$. From the conditions of the assumption (1) it follows that there exist constants $\gamma, \lambda > 0$, $-\lambda \in \mathcal{R}^+$, such that

$$\|\Phi(t, t_0)\| \le \gamma e_{-\lambda}(t, t_0) \qquad (8.31)$$

for all $t \ge t_0$. Here $-\lambda$ is uniformly regressive on $\mathbb{T}$, i.e., there exists $\delta > 0$ such that $0 < \delta^{-1} < (1 - \mu(t)\lambda)$ for all $t \ge t_0$.

Taking into account the estimates (8.27), (8.28), and (8.31), from (8.30) we find

$$\|x(t)\| \le \|\Phi(t, t_0)\| \, \|x_0\| + \int_{t_0}^{t} \|\Phi(t, \sigma(s))\| \left[\|D_\alpha E\| \|x(s)\| + \|g(\alpha, x(s))\| \right] \Delta s$$

$$\le \gamma e_{-\lambda}(t, t_0) \, \|x_0\| + \int_{t_0}^{t} \gamma e_{-\lambda}(t, \sigma(s)) \left[\beta \|x(s)\| + \|E\| \, \|x(s)\| \right] \Delta s$$

$$\le e_{-\lambda}(t, t_0) \left[\gamma \|x_0\| + \int_{t_0}^{t} \gamma \delta (\beta + \|E\|) e_{-\lambda}(t_0, s) \, \|x(s)\| \Delta s \right]$$

$$\qquad (8.32)$$

for all $t \ge t_0$. Denote $z(t) = e_{-\lambda}(t, t_0) \|x(t)\|$ and rewrite the inequality (8.32) in the form

$$z(t) \le \gamma \|x_0\| + \int_{t_0}^{t} \gamma \delta (\beta + \|E\|) \Delta s.$$

Applying Gronwall's inequality to this estimate, we obtain

$$z(t) \le \gamma \|x_0\| e_{\gamma\delta(\beta+\|E\|)}(t, t_0) \tag{8.33}$$

for all $t \ge t_0$. From the estimate (8.33) it follows that

$$
\begin{aligned}
\|x(t)\| &\le \gamma \|x_0\| e_{\gamma\delta(\beta+\|E\|)}(t, t_0) e_{-\lambda}(t, t_0) \\
&= \gamma \|x_0\| e_{-\lambda \oplus \gamma\delta(\beta+\|E\|)}(t, t_0)
\end{aligned}
\tag{8.34}
$$

for all $t \ge t_0$. To obtain an estimate of the form (8.29) for solutions of the system (8.26), in the inequality (8.34) there should be $-\lambda \oplus \gamma\delta(\beta+\|E\|) \in \mathcal{R}^+$ and $-\lambda \oplus \gamma\delta(\beta+\|E\|) < 0$. Since $\gamma\delta(\beta+\|E\|) > 0$, then $\gamma\delta(\beta+\|E\|) \in \mathcal{R}^+$, and from the fact that $\mathcal{R}^+$ is a subgroup of $\mathcal{R}$ it follows that $-\lambda \oplus \gamma\delta(\beta+\|E\|) \in \mathcal{R}^+$. Then, from the inequality

$$-\lambda < -\lambda \oplus \gamma\delta(\beta+\|E\|) < 0$$

we find

$$0 < \beta < \frac{\lambda}{(1-\mu(t)\lambda)\gamma\delta} - \|E\|. \tag{8.35}$$

From the fact that β should be bounded for all $t \in \mathbb{T}$ and for all $\alpha \in \mathcal{S}$, we obtain

$$\|E\| < \frac{\lambda}{(1-\mu(t)\lambda)\gamma\delta}. \tag{8.36}$$

Thus, the inequalities (8.35) and (8.36) solve the problem in question, and the inequalities (8.35) and (8.36) $-\lambda^* = -\lambda \oplus \gamma\delta\,(\beta+\|E\|)$ in the estimate (8.29).

Note that equations on time scales, even if uncertainties are not taken into account, have a wider spectrum of dynamic properties as compared both with ordinary differential equations and difference equations. Here are some examples.

Example 8.4 Consider the dynamic equations

$$
\begin{aligned}
x^\Delta &= y(x+y), & x(t_0) &= x_0, \\
y^\Delta &= -x(x+y), & y(t_0) &= y_0.
\end{aligned}
\tag{8.37}
$$

For the function $V(x,y) = x^2 + y^2$ at $\mathbb{T} = \mathbb{R}$ obtain

$$\dot{V}(x(t), y(t)) = 0 \quad \text{for all} \quad t \in \mathbb{R}$$

and

$$V^\Delta(x(t), y(t))|_{(8.37)} = \mu(t)(x+y)^2(x^2+y^2). \tag{8.38}$$

From (8.38) it follows that the solution $x = y = 0$ of the system corresponding to (8.37) at $\mathbb{T} = \mathbb{R}$ is stable, whereas the solution $x = y = 0$ of the system (8.37) is unstable at any function with the granularity $0 < \mu(t) < +\infty$.

Example 8.5 Let a system of dynamic equations be specified:

$$x^\Delta = -x - y(x^2 + y^2), \quad x(t_0) = x_0,$$
$$y^\Delta = -y + x(x^2 + y^2), \quad y(t_0) = y_0. \tag{8.39}$$

For the function $V(x, y) = x^2 + y^2$ at $\mathbb{T} = \mathbb{R}$ obtain

$$\dot{V}(x(t), y(t)) = -2(x^2 + y^2) \quad \text{for all} \quad t \in \mathbb{R} \tag{8.40}$$

and on $\mathbb{T}$

$$V^\Delta(x(t), y(t))|_{(8.39)} = -2(x^2 + y^2) + \mu(t)[x^2 + y^2 + (x^2 + y^2)^3]. \tag{8.41}$$

From the analysis (8.40) and (8.41) it follows that the solution $x = y = 0$ of the system corresponding to (8.39) at $\mathbb{T} = \mathbb{R}$ is asymptotically stable.

If the scale $\mathbb{T}$ has the granularity $\mu(t) = 1$, i.e. $\mathbb{T} = \mathbb{Z}$, then at the initial values (x_0, y_0) from the domain $x_0^2 + y_0^2 < 1$ the zero solution of the system (8.39) will be asymptotically stable on $\mathbb{Z}$. If $\mu(t) = 2$, which corresponds to the time scale $\mathbb{T} = 2\mathbb{N}_0 = \{k_0, k_0 + 2, k_0 + 4, \dots\}$, then

$$V^\Delta(x(t), y(t))|_{(8.39)} = 2(x^2 + y^2)^3$$

and the equilibrium state $x = y = 0$ of (8.39) is unstable.

Chapter 9

Singularly Perturbed Systems with Uncertain Structure

It is well known that the singular perturbation methods on the basis of the two-time-scale approach are very powerful analyses of large scale systems with structural perturbations (see Grujić, Martynyuk and Ribbens-Pavella [1984]).

In this chapter we propose some development of the direct Liapunov method for the given class of systems of equations in terms of auxiliary matrix valued functions. This allows us to weaken the requirements of the dynamical properties of the individual subsystems and to extend the variation limits for the small parameters μ_i for higher derivatives of the systems of perturbed motion equations.

The chapter is arranged as follows.

Section 9.1 sets out the method of composition of large-scale systems on the basic of individual subsystems for the given model of uncertainties.

Sections 9.2 and 9.3 contain the results of development of a new method of stability and/or instability analysis of large-scale systems under structural uncertainties.

In Section 9.4. similar problems are discussed for linear singularly perturbed systems for uniform and nonuniform time scaling.

In final Section 9.5 two problems of practical importance are considered. One of the problems relates to absolute stability of large scale Lur'e–Postnukov systems under structural perturbations and the other deals with gyroscopic stabilization of orbital apparatus.

9.1 Structural Uncertainties in Singularly Perturbed Systems

The real systems in which the fast and slow variables can potentially exist are modelled by means of the systems of equations with small parameters at higher variables (singularly perturbed systems). The class of systems of equations under consideration (furtheron designated as F) is described basing on the hypotheses below (cf. Grujić, Martynyuk and Ribbens-Pavella [1984]).

H_1. System F consists of q subsystems of ordinary differential equations with structural uncertainties and r subsystems with structural uncertainties and small parameters at higher derivatives. The order of fast and slow components of the system remains unchanged during all the periods of system F functioning.

H_2. Dynamics of the i-th interconnected subsystem F_i in system F is described by the equations

$$\frac{dx_i}{dt} = f_i(t, x, y, P_i, S_i),$$

$$\mu_i \frac{dy_i}{dt} = g_i(t, x, y, M, P_{q+i}, S_{q+i}),$$

where $x_i \in R^{n_i}$, $y_i \in R^{m_i}$, f_i and g_i are continuous vector-functions of the corresponding dimensions, μ_i are small positive parameters, $\mu_i \in (0, 1]$ and $M = \text{diag}\{\mu_1, \dots, \mu_n\}$.

H_3. Dynamics of the i-th isolated subsystem $\widehat{F}_i$ in system F is described by the equations

$$\frac{dx_i}{dt} = f_i(t, x^i, y^i, P_i, S_i),$$

$$\mu_i \frac{dy_i}{dt} = g_i(t, x^i, y^i, M, P_{q+i}, S_{q+i}),$$

where $x^i = (0, 0, \dots, 0, x_i^\mathrm{T}, 0, 0, \dots, 0)^\mathrm{T} \in R^n$, $n = n_1 + n_2 + \dots + n_q$, $x_i \in R^{n_i}$, $y^i = (0, 0, \dots, 0, y_i^\mathrm{T}, 0, 0, \dots, 0)^\mathrm{T} \in R^m$, $m = m_1 + m_2 + \dots + m_r$, $y_i \in R^{m_i}$.

In the case when $q = r$, the equations

$$\frac{dx_i}{dt} = f_i(t, x^i, y^i, P_i, S_i), \tag{9.1}$$

$$0 = g_i(t, x^i, y^i, 0, P_{q+i}, S_{q+i})$$

describe the dynamics of the i-th isolated subsystem $\widehat{F}_{i0}$ of system F, and the equations

$$\frac{dy_i}{dt_i} = g_i(\alpha, b^i, y^i, 0, P_{q+i}, S_{q+i})$$

characterize the boundary layer of the fast subsystem $\widehat{F}_{t_i}$ of system F. Here $\alpha \in R$, $b^i = (0, \dots, 0, b_i^\mathrm{T}, 0, \dots, 0)^\mathrm{T} \in R^n$, $b_i \in R^{n_i}$, $t_i = \frac{t - t_0}{\mu_i}$ and $i = 1, 2, \dots, r$.

H_4. Dynamics of the whole system F is described by the equations

$$\frac{dx_i}{dt} = f_i(t, x, y, P_i, S_i), \quad i = 1, 2, \dots, q, \tag{9.2}$$

$$\mu_i \frac{dy_i}{dt} = g_i(t, x, y, M, P_{q+i}, S_{q+i}), \quad i = 1, 2, \dots, r,$$

where $x_i \in R^{n_i}$, $\sum\limits_{i=1}^{q} n_i = n$, $y_i \in R^{m_i}$, $\sum\limits_{i=1}^{r} m_i = m$, $q + r = s$, P_i, $i = 1, 2, \ldots, q$, are the parametric perturbations, and S_i, $i = 1, 2, \ldots, s$, are the structural matrices of the form $S_i = [s_{i1}I_i, s_{i2}I_i, \ldots, s_{iN}I_i]$, $I_i = \mathrm{diag}\{1, 1, \ldots, 1\} \in R^{n_i \times n_i}$ and $s_{ij} \colon R \to \{0, 1\}$ are structural parameters. Here $\mu_i \in (0, 1]$ and the set of all admissible values of M is designated as

$$\mathcal{M} = \{M \colon 0 < M \le I\}, \quad I = \mathrm{diag}\{1, 1, \ldots, 1\} \in R^{r \times r}.$$

Moreover

$$\mathcal{M}_m = \{M \colon 0 < \mu_i < \mu_{im}, \ \forall\, i \in [1, r]\},$$

where μ_{im} is an admissible upper value of μ_i.

If in the system of equations (9.2) all μ_i (formally) form a zero set, then the equations

$$\frac{dx_i}{dt} = f_i(t, x, y, P_i, S_i), \quad i = 1, 2, \ldots, q,$$
$$0 = g_i(t, x, y, 0, P_{q+i}, S_{q+i}), \quad i = 1, 2, \ldots, r, \tag{9.3}$$

describe the dynamics of the interconnected degenerated subsystem F_0 of system F, and the equations

$$\frac{dy_i}{dt_i} = \tau_i g_i(a, b, y, 0, P_{q+i}, S_{q+i}), \quad i = 1, 2, \ldots, r,$$

characterize the behavior of the interconnected fast subsystem F_t (the boundary layer of system F).

If the small parameters μ_i are not mutually connected, then the system F has r essentially independent time scales t_i:

$$t_i = \frac{t - t_0}{\mu_i}, \quad i = 1, 2, \ldots, r.$$

In this case the time scaling is nonuniform.

The time scales t_i can be interconnected through the values τ_i:

$$\frac{t_i}{t_1} = \tau_i, \quad i = 1, 2, \ldots, r, \tag{9.4}$$

which are variable within certain limits

$$\tau_i \in [\underline{\tau}_i, \overline{\tau}_u], \quad i = 1, 2, \ldots, r, \tag{9.5}$$

where $0 < \underline{\tau}_i \le \tau_i < \infty$, $\forall\, i \in [1, r]$.

In the case (9.4) and (9.5) the time scaling is uniform and

$$\tau_i = \frac{\mu_1}{\mu_i}, \quad i = 1, 2, \ldots, r.$$

Obviously, in this case $\underline{\tau}_1 = \tau_1 = \overline{\tau}_1 = 1$.

Everywhere below it is assumed that the correlations

$$0 = g_i(t, x, y, 0, P_{q+i}, S_{q+i}), \quad \forall (t, x, y) \in R \times \mathcal{N}_x \times \mathcal{N}_y,$$

are satisfied for each pair $(P, S) \in \mathcal{P} \times \mathcal{S}$ iff $y = 0$ and

$$0 = g_i(t, x^i, y^i, 0, P_{q+i}, S_{q+i}), \quad \forall (t, x^i, y^i) \in R \times \mathcal{N}_x \times \mathcal{N}_y,$$

are satisfied for each pair $(P, S) \in \mathcal{P} \times \mathcal{S}$ iff $y^i = 0$.

Therefore systems (9.3) and (9.1) are equivalent to

$$\frac{dx_i}{dt} = f_i(t, x^i, 0, P_i, S_i), \quad \frac{dx_i}{dt} = f_i(t, x, 0, P_i, S_i) \text{ and } i = 1, 2, \ldots, q,$$

respectively.

9.2   Tests for Stability Analysis

In the qualitative analysis for the given class of large scale systems, the question whether different time scales t_i are interconnected is of importance.

General purpose of our investigation is to determine conditions under which stability of zero solution of the initial system is implied by stability of some independent degenerated subsystems and stability of the independent subsystems describing boundary layer with allowance for the qualitative properties of interconnections between the subsystems.

9.2.1   Non-uniform time scaling

Assume that the correlation $q = r$ is satisfied. Then system (9.2) is represented as

$$\frac{dx_i}{dt} = f_i(t, x^i, 0, P_i, S_i) + f_i^* + f_i^{**}, \quad i = 1, 2, \ldots, q,$$

$$\mu_i \frac{dy_i}{dt} = g_i(\alpha, b^i, y^i, P_{q+i}, S_{q+i}) + g_i^* + g_i^{**}, \quad i = 1, 2, \ldots, q,$$

where

$$f_i^* = f_i(t, x^i, y^i, P_i, S_i) - f_i(t, x^i, 0, P_i, S_i),$$
$$g_i^* = g_i(t, x^i, y^i, M^i, P_{q+i}, S_{q+i}) - g_i(\alpha, b^i, y^i, 0, P_{q+i}, S_{q+i}),$$
$$f_i^{**} = f_i(t, x, y, P_i, S_i) - f_i(t, x^i, y^i, P_i, S_i),$$
$$g_i^{**} = g_i(t, x, y, M, P_{q+i}, S_{q+i}) - g_i(t, x^i, y^i, M^i, P_{q+i}, S_{q+i}).$$

Here the functions f_i^* and g_i^* describe the connections between equations of the i-th independent singularly perturbed subsystem (F_i) of the system F, and the functions f_i^{**} and g_i^{**} describe all the remaining connections in system F.

In view of results from Martynyuk and Miladzhanov [2009] we introduce some assumptions.

Assumption 9.1 There exist

(1) open connected neighborhoods $\mathcal{N}_{ix} \subseteq R^{n_i}$, $\mathcal{N}_{iy} \subseteq R^{m_i}$ of the states $x_i = 0$ and $y_i = 0$ respectively;

(2) functions φ_{ik}, ψ_{ik} of Hahn class $K(KR)$, $k = 1, 2$, $i \in [1, q]$, constants $\underline{\alpha}_{ij}$, $\overline{\alpha}_{ij}$, $\underline{\alpha}_{i, q+j}$, $\overline{\alpha}_{i, q+j}$, $\underline{\alpha}_{q+i, q+j}$, $\overline{\alpha}_{q+i, q+j}$, $i, j = 1, 2, \ldots, q$, and the matrix-function

$$U(t, x, y, M) = \begin{pmatrix} U_{11}(t, x) & U_{12}(t, x, y, M) \\ U_{12}^{\mathrm{T}}(t, x, y, M) & U_{22}(t, y, M) \end{pmatrix} \tag{9.6}$$

where

$$U_{11}(t, x) = [v_{ij}(t, \cdot)], \quad v_{ii} = v_{ii}(t, x_i),$$
$$v_{ij} = v_{ji} = v_{ij}(t, x_i, x_j), \quad i \neq j = 1, 2, \ldots, q;$$
$$U_{22}(t, y, M) = [v_{q+i, q+j}^*(t, \cdot)], \quad v_{q+i, q+i}^* = \mu_i v_{q+i, q+i}(t, y_i),$$
$$v_{q+i, q+j}^* = v_{q+j, q+i}^* = \mu_i \mu_j v_{q+i, q+j}(t, y_i, y_j), \quad i, j = 1, 2, \ldots, q;$$
$$U_{12}(t, x, y, M) = [\mu_j v_{i, q+j}(t, x_i, y_j)], \quad i, j = 1, 2, \ldots, q, \quad 2q = s,$$

such that

(a) $\underline{\alpha}_{ij} \varphi_{i1}(x_i) \varphi_{j1}(x_j) \leq v_{ij}(t, \cdot) \leq \overline{\alpha}_{ij} \varphi_{i2}(x_i) \varphi_{j2}(x_j)$, $\forall (t, x_i, x_j) \in R \times \mathcal{N}_{ix} \times \mathcal{N}_{jx}$, $i, j = 1, 2, \ldots, q$, $j \geq i$;

(b) $\underline{\alpha}_{q+i, q+j} \psi_{i1}(y_i) \psi_{j1}(y_j) \leq v_{q+i, q+j}(t, \cdot) \leq \overline{\alpha}_{q+i, q+j} \psi_{i2}(y_i) \psi_{j2}(y_j)$, $\forall (t, y_i, y_j) \in R \times \mathcal{N}_{iy} \times \mathcal{N}_{jy}$, $i, j = 1, 2, \ldots, q$, $j \geq i$;

(c) $\underline{\alpha}_{i, q+j} \varphi_{i1}(x_i) \psi_{j1}(y_j) \leq v_{i, q+j}(t, \cdot) \leq \overline{\alpha}_{i, q+j} \varphi_{i2}(x_i) \psi_{j2}(y_j)$, $\forall (t, x_i, y_j) \in R \times \mathcal{N}_{ix} \times \mathcal{N}_{jy}$, $i, j = 1, 2, \ldots, q$.

By means of the matrix-function (9.6) and the constant vector $\eta \in R_+^s$ we introduce the function

$$v(t, x, y, M) = \eta^{\mathrm{T}} U(t, x, y, M) \eta \tag{9.7}$$

and consider the expressions of the upper right Dini derivative

$$D^+ v(t, x, y, M) = \eta^{\mathrm{T}} D^+ U(t, x, y, M) \eta,$$
$$D^+ U(t, x, y, M) \overset{\mathrm{def}}{=} [D^+ v_{rk}(t, \ldots)], \quad r, k = 1, 2, \ldots, s. \tag{9.8}$$

For the function (9.7) the following assertion holds true.

Proposition 9.1 Under conditions of Assumption 9.1 for function (9.7) the bilateral estimate

$$u_1^T A(M) u_1 \leq v(t, x, y, M) \leq u_2^T B(M) u_2,$$
$$\forall\, (t, x, y, M) \in R \times \mathcal{N}_x \times \mathcal{N}_y \times \mathcal{M},$$

is satisfied, where

$$\mathcal{N}_x \subseteq \mathcal{N}_{1x} \times \mathcal{N}_{2x} \times \ldots \times \mathcal{N}_{qx}, \quad \mathcal{N}_y \subseteq \mathcal{N}_{1y} \times \mathcal{N}_{2y} \times \ldots \times \mathcal{N}_{qy},$$
$$u_k^T = (\varphi_{1k}(x_1), \ldots, \varphi_{qk}(x_q), \psi_{1k}(y_1), \ldots \psi_{qk}(y_q)), \quad k = 1, 2,$$
$$A(M) = H^T A_1(M) H, \quad B(M) = H^T A_2(M) H,$$
$$H = \mathrm{diag}\{\eta_1, \eta_2, \ldots, \eta_s\}, \quad s = 2q,$$

$$A_1(M) = \begin{pmatrix} A_{11} & A_{12}(M) \\ A_{12}^T(M) & A_{22}(M) \end{pmatrix}, \quad A_2(M) = \begin{pmatrix} \overline{A}_{11} & \overline{A}_{12}(M) \\ \overline{A}_{12}^T(M) & \overline{A}_{22}(M) \end{pmatrix},$$

$$A_{11} = [\underline{\alpha}_{ij}], \quad \underline{\alpha}_{ij} = \underline{\alpha}_{ji}, \quad \overline{A}_{11} = [\overline{\alpha}_{ij}], \quad \overline{\alpha}_{ij} = \overline{\alpha}_{ji},$$
$$A_{12}(M) = [\mu_j \underline{\alpha}_{i,q+j}], \quad \overline{A}_{12}(M) = [\mu_j \overline{\alpha}_{i,q+j}],$$
$$A_{22}(M) = [\mu_{ij}^* \underline{\alpha}_{q+i,q+j}] \quad \underline{\alpha}_{q+i,q+j} = \underline{\alpha}_{q+j,q+i},$$
$$\overline{A}_{22}(M) = [\mu_{ij}^*] \overline{\alpha}_{q+i,q+j}, \quad \overline{\alpha}_{q+i,q+j} = \overline{\alpha}_{q+j,q+i},$$

$$\mu_{ij}^* = \begin{cases} \mu_i & \text{for } i = j, \\ \mu_i \mu_j & \text{for } i \neq j, \end{cases} \quad i, j = 1, 2, \ldots, q.$$

Proof Let all conditions of Assumption 9.1 be satisfied. Then for function (9.7) we have

$$v(t, x, y, M) = \sum_{i=1}^{q} \eta_i^2 v_{ii}(t, x_i) + 2 \sum_{i=1}^{q} \sum_{j=2}^{q} \eta_i \eta_j v_{ij}(t, x_i, x_j)$$

$$+ \sum_{i=1}^{q} \eta_{q+i}^2 \mu_i v_{q+i,q+i}(t, y_i) + 2 \sum_{i=1}^{q} \sum_{\substack{j=2 \\ j>i}}^{q} \eta_{q+i} \eta_{q+j} \mu_i \mu_j v_{q+i,q+j}$$

$$+ 2 \sum_{i=1}^{q} \sum_{j=1}^{q} \eta_i \eta_{q+j} \mu_j v_{i,q+j}(t, x_i, y_j) \geq \sum_{i=1}^{q} \eta_i^2 \underline{\alpha}_{ii} \varphi_{i1}^2$$

$$+ 2 \sum_{i=1}^{q} \sum_{\substack{j=2 \\ j>i}}^{q} \eta_i \eta_j \underline{\alpha}_{ij} \varphi_{i1}(x_i) \varphi_{j1}(x_j) + \sum_{i=1}^{q} \eta_{q+i}^2 \mu_i \underline{\alpha}_{q+i,q+j} \psi_{i1}^2(y_i)$$

$$+ 2 \sum_{i=1}^{q} \sum_{\substack{j=2 \\ j>i}}^{q} \eta_{q+i} \eta_{q+j} \mu_i \mu_j \underline{\alpha}_{q+i,q+j} \psi_{i1}(y_i) \psi_{j1}(y_j)$$

$$+ 2 \sum_{i=1}^{q} \sum_{j=1}^{q} \eta_i \eta_{q+j} \mu_j \underline{\alpha}_{i,q+j} \varphi_{i1}(x_i) \psi_{j1}(y_j)$$

$$= (\varphi_{11}(x_1), \ldots, \varphi_{q1}(x_q), \psi_{11}(y_1), \ldots, \psi_{q1}(y_q))^{\mathrm{T}} \mathrm{diag}\{\eta_1, \ldots, \eta_{2q}\}$$

$$\times \begin{pmatrix} A_{11} & A_{12}(M) \\ A_{12}^{\mathrm{T}}(M) & A_{22}(M) \end{pmatrix} \mathrm{diag}\{\eta_1, \eta_2, \ldots, \eta_{2q}\}$$

$$\times (\varphi_{11}(x_1), \ldots, \varphi_{q1}(x_q), \psi_{11}(y_1), \ldots, \psi_{q1}(y_q)) = u_1^{\mathrm{T}} A(M) u_1.$$

The upper estimate is proved in the same manner.

Assumption 9.2 There exist

(1) functions φ_i, ψ_i of class K (KR), $i = 1, 2, \ldots, q$;

(2) functions v_{ij}, $v_{i,q+j}$, $v_{q+i,q+j}$, $i, j = 1, 2, \ldots, q$ satisfying the conditions mentioned in Assumption 9.1, and

    (a) functions $v_{ij}(t, x_i, x_j)$ are continuous on $(R \times \mathcal{N}_{ix_0} \times \mathcal{N}_{jx_0})$ or on $(R \times R^{n_i} \times R^{n_j})$;

    (b) functions $v_{i,q+j}(t, x_i, y_j)$ are continuous on $(R \times \mathcal{N}_{ix_0} \times \mathcal{N}_{jy_0})$ or on $(R \times R^{n_i} \times R^{m_j})$;

    (c) functions $v_{q+i,q+j}(t, y_i, y_j)$ are continuous on $(R \times \mathcal{N}_{iy_0} \times \mathcal{N}_{jy_0})$ or on $(R \times R^{m_i} \times R^{m_j})$;

(3) real numbers $\rho_{\alpha i}(P, S)$, $\rho_{\beta ij}(P, S)$, $\alpha = 1, 2, \ldots, 13$, $\beta = 1, 2, \ldots, 8$, $i, j = 1, 2 \ldots, q$, and the following conditions are satisfied

    (a) $\eta_i^2 D_t^+ v_{ii} + \eta_i^2 (D_{x_i}^+ v_{ii})^{\mathrm{T}} f_i(t, x^i, 0, P_i, S_i) \le \rho_{1i}(P, S) \varphi_i^2(x_i)$, $\forall (t, x_i, P, S) \in R \times \mathcal{N}_{ix_0} \times \mathcal{P} \times \mathcal{S}$, $i = 1, 2, \ldots, q$;

    (b) $\eta_{q+i}^2 \mu_i D_t^+ v_{q+i,q+i} + \eta_{q+i}^2 (D_{y_i}^+ v_{q+i,q+i})^{\mathrm{T}} g_i(\alpha, b^i, y^i, 0, P_{q+i}, S_{q+i}) \le \rho_{2i}(P, S) \psi_i^2(y_i)$, $\forall (t, y_i, M, P, S) \in R \times \mathcal{N}_{iy_0} \times \mathcal{M} \times \mathcal{P} \times \mathcal{S}$, $i = 1, 2, \ldots, q$;

    (c) $\eta_i^2 (D_{x_i}^+ v_{ii})^{\mathrm{T}} f_i^* + \eta_{q+i}^2 (D_{y_i}^+ v_{q+i,q+i})^{\mathrm{T}} g_i^* + 2\eta_i \eta_{q+i} \{\mu_i D_t^+ v_{i,q+i} + \mu_i (D_{x_i}^+ v_{i,q+i})^{\mathrm{T}} f_i(t, x^i, y^i, P_i, S_i) + (D_{y_i}^+ v_{i,q+i})^{\mathrm{T}} \times g_i(t, x^i, y^i, M^i, P_{q+i}, S_{q+i})\} \le (\rho_{3i}(P, S) + \mu_i \rho_{4i}(P, S)) \varphi_i^2(x_i) + (\rho_{5i}(P, S) + \mu_i \rho_{6i}(P, S)) \psi_i^2(y_i) + 2(\rho_{7i}(P, S) + \mu_i \rho_{8i}(P, S)) \times \varphi_i(x_i) \psi_i(y_i)$, $\forall (t, x_i, y_i, M, P, S) \in R \times \mathcal{N}_{ix_0} \times \mathcal{N}_{iy_0} \times \mathcal{M} \times \mathcal{P} \times \mathcal{S}$, $i = 1, 2, \ldots, q$;

    (d) $\displaystyle\sum_{i=1}^{q} \eta_i^2 (D_{x_i}^+ v_{ii})^{\mathrm{T}} f_i^{**} + \sum_{i=1}^{q} \eta_{q+i}^2 (D_{y_i}^+ v_{q+i,q+i})^{\mathrm{T}} g_i^{**} + \sum_{i=1}^{q} 2\eta_i \eta_{q+i} \times$

$$\left\{\mu_i (D_{x_i}^+ v_{i,q+i})^{\mathrm{T}} f_i^{**} + (D_{y_i}^+ v_{i,q+i})^{\mathrm{T}} g_i^{**}\right\} + \sum_{i=1}^{q} \sum_{\substack{j=2 \\ j>i}}^{q} \eta_i \eta_j \left\{D_t^+ v_{ij} + \right.$$

$$\left. (D_{x_i}^+ v_{ij})^{\mathrm{T}} f_i(t, x, y, P_i, S_i) + (D_{x_j}^+ v_{ij})^{\mathrm{T}} f_j(t, x, y, P_j, S_j)\right\} +$$

$$2 \sum_{i=1}^{q} \sum_{\substack{j=2 \\ j>i}}^{q} \eta_{q+i} \eta_{q+j} \left\{\mu_i \mu_j D_t^+ v_{q+i,q+j} + \mu_j (D_{y_i}^+ v_{q+i,q+j})^{\mathrm{T}} \times \right.$$

$$g_i(t, x, y, M, P_{q+i}, S_{q+i}) + \mu_i (D_{y_j}^+ v_{q+i,q+j})^{\mathrm{T}} \times$$

$$g_j(t, x, y, M, P_{q+j}, S_{q+j})\Big\} + 2\sum_{i=1}^{q}\sum_{\substack{j=1 \\ j\neq i}}^{q} \eta_i \eta_{q+j} \Big\{ \mu_j D_t^+ v_{i,q+j} +$$

$$\mu_j (D_{x_i}^+ v_{i,q+j})^{\mathrm{T}} f_i(t, x, y, P_i, S_i) + (D_{y_j}^+ v_{i,q+j})^{\mathrm{T}} \times$$

$$g_j(t, x, y, M, P_{q+j}, S_{q+j})\Big\} \leq \sum_{i=1}^{q} \Big\{ (\rho_{9i}(P, S) + \mu_i \rho_{10i}(P, S))\varphi_i^2(x_i) +$$

$$\Big(\rho_{11i}(P, S) + \mu_i \rho_{12i}(P, S) + \mu_i \Big(\sum_{\substack{j=2 \\ j>i}} \mu_j \Big) \rho_{13i} \Big) \psi_i^2(y_i) \Big\} +$$

$$2\sum_{i=1}^{q}\sum_{\substack{j=2 \\ j>i}}^{q} \Big\{ (\rho_{1,i,j}(P, S) + \mu_i \rho_{2,i,j}(P, S))\varphi_i(x_i)\varphi_j(x_j) + (\rho_{3,i,j}(P, S) +$$

$$\mu_i \rho_{4,i,j}(P, S) + \mu_i \mu_j \rho_{5,i,j}(P, S))\psi_i(y_i)\psi_j(y_j)\Big\} + 2\sum_{i=1}^{q}\sum_{\substack{j=1 \\ j\neq i}} (\rho_{6,i,j}(P, S)$$

$$+ \mu_i \rho_{7,i,j}(P, S) + \mu_i \mu_j \rho_{8,i,j}(P, S))\varphi_i(x_i)\psi_j(y_j),$$
$$\forall\, (t, x, y, M, P, S) \in R \times \mathcal{N}_{x_0} \times \mathcal{N}_{y_0} \times \mathcal{N} \times \mathcal{P} \times \mathcal{S}, \text{ where}$$

$$\mathcal{N}_{ix_0} = \{x_i:\ x_i \in \mathcal{N}_{ix},\ x_i \neq 0\}, \quad \mathcal{N}_{iy_0} = \{y_i:\ y_i \in \mathcal{N}_{iy},\ y_i \neq 0\},$$
$$i = 1, 2, \ldots, q, \quad 2q = s.$$

Proposition 9.2 Under conditions of Assumption 9.2 the estimate

$$D^+ v(t, x, y, M) \leq u^{\mathrm{T}} G(M, P, S) u,$$
$$\forall\, (t, x, y, M, P, S) \in R \times \mathcal{N}_{x_0} \times \mathcal{N}_{y_0} \times \mathcal{M} \times \mathcal{P} \times \mathcal{S},$$

is true, where

$$u^{\mathrm{T}} = (\varphi_1(1), \ldots, \varphi_q(x_q), \psi_1(y_1), \ldots, \psi_q(y_q)),$$
$$G(M, P, S) = [\sigma_{ij}(M, P, S)], \quad \sigma_{ij} = \sigma_{ji}, \quad i, j = 1, 2, \ldots, s,$$
$$\sigma_{ii}(M, P, S) = \rho_{1i}(P, S) + \rho_{3i}(P, S) + \rho_{9i}(P, S) + \mu_i(\rho_{4i}(P, S)$$
$$+ \rho_{10i}(P, S)), \quad i = 1, 2, \ldots, q;$$
$$\sigma_{q+i,q+i}(M, P, S) = \rho_{2i}(P, S) + \rho_{5i}(P, S) + \rho_{11i}(P, S)$$

$$+ \mu_i \Big(\rho_{6i}(P, S) + \rho_{12i}(P, S) + \Big(\sum_{\substack{j=2 \\ j>i}}^{q} \mu_j \Big) \rho_{13i}(P, S) \Big),$$

$$i = 1, 2, \ldots, q;$$
$$\sigma_{i,q+i}(M, P, S) = \rho_{7i}(P, S) + \mu_i \rho_{8i}(P, S), \quad i = 1, 2, \ldots, q;$$
$$\sigma_{ij}(M, P, S) = \rho_{1ij}(P, S) + \mu_i \rho_{2ij}, \quad i = 1, 2, \ldots, q, \quad j = 2, 3, \ldots, q, \quad j > i;$$
$$\sigma_{q+i,q+j}(M, P, S) = \rho_{3ij}(P, S) + \mu_i \rho_{4ij}(P, S) + \mu_i \mu_j \rho_{5ij}(P, S),$$
$$i = 1, 2, \ldots, q, \quad j = 2, 3, \ldots, q, \quad j > i;$$

$$\sigma_{i,q+j}(M,P,S) = \rho_{6ij}(P,S) + \mu_i\rho_{7ij}(P,S) + \mu_i\mu_j\rho_{8ij}(P,S),$$
$$i, j = 1, 2, \ldots, q, \quad i \neq j.$$

Proof Let all conditions of Assumption 9.2 be satisfied. Then for the expression (9.8) we have

$$
\begin{aligned}
D^+v(t,x,y,M) = \sum_{i=1}^{q} &\Big\{ \eta_i^2 D_t^+ v_{ii} + (D_{x_i}^+ v_{ii})^\mathrm{T} f_i(t,x^i,0,P_i,S_i) \\
&+ \eta_{q+i}^2 \mu_i D_t^+ v_{q+i,q+i} + \eta_{q+i}^2 (D_{y_i}^+ v_{q+i,q+i})^\mathrm{T} g_i(\alpha,b^i,y^i,0,P_{q+i},S_{q+i}) \\
&+ \eta_i^2 (D_{x_i}^+ v_{ii})^\mathrm{T} f_i^* + \eta_{q+i}^2 (D_{y_i}^+ v_{q+i,q+i})^\mathrm{T} g_i^* + 2\eta_i\eta_{q+i}\big(\mu_i D_t^+ v_{i,q+i} \\
&+ \mu_i (D_{x_i}^+ v_{i,q+i})^\mathrm{T} f_i(t,x^i,y^i,P_i,S_i) \\
&+ (D_{y_i}^+ v_{i,q+i})^\mathrm{T} g_i(t,x^i,y^i,M^i,P_{q+i},S_{q+i})\big) \\
&+ \eta_i^2 (D_{x_i}^+ v_{ii})^\mathrm{T} f_i^{**} + \eta_{q+i}^2 (D_{y_i}^+ v_{q+i,q+i})^\mathrm{T} g_i^{**} \\
&+ 2\eta_i\eta_{q+i}\big(\mu_i(D_{x_i}^+ v_{i,q+i})^\mathrm{T} f_i^{**} + (D_{y_i}^+ v_{i,q+i})^\mathrm{T} g_i^{**}\big) \Big\} \\
+ 2\sum_{i=1}^{q}\sum_{\substack{j=2\\j>i}}^{q} &\Big\{ \eta_i\eta_j\big(D_t^+ v_{ij} + (D_{x_i}^+ v_{ij})^\mathrm{T} f_i(t,x,y,P_i,S_i) \\
&+ (D_{x_j}^+ v_{ij})^\mathrm{T} f_j(t,x,y,P_j,S_j)\big) + \eta_{q+i}\eta_{q+j}\big(\mu_i\mu_j D_t^+ v_{q+i,q+j} \\
&+ \mu_j(D_{y_j}^+ v_{q+i,q+j})^\mathrm{T} g_i(t,x,y,M,P_{q+i},S_{q+i}) \\
&+ \mu_i(D_{y_j}^+ v_{q+i,q+j})^\mathrm{T} g_j(t,x,y,M,P_{q+j},S_{q+j})\big) \Big\} \\
+ 2\sum_{i=1}^{q}\sum_{\substack{j=1\\j\neq i}}^{q} &\eta_i\eta_{q+j}\Big\{ \mu_j D_t^+ v_{i,q+j} + \mu_j(D_{x_i}^+ v_{i,q+j})^\mathrm{T} f_i(t,x,y,P_i,S_i) \\
&+ (D_{y_j}^+ v_{i,q+j})^\mathrm{T} g_j(t,x,y,M,P_{q+j},S_{q+j}) \Big\} \\
\leq \sum_{i=1}^{q} &\Big\{ \rho_{1i}(P,S) + \rho_{3i}(P,S) + \rho_{9i}(P,S) + \mu_i(\rho_{4i}(P,S) \\
&+ \rho_{10i}(P,S)\Big\}\varphi_i^2(x_i) + \sum_{i=1}^{q}\Big\{ \rho_{2i}(P,S) + \rho_{5i}(P,S) + \rho_{11,i}(P,S) \\
&+ \mu_i\Big(\rho_{6i}(P,S) + \rho_{12}(P,S) + \Big(\sum_{\substack{j=2\\j>i}}^{q}\mu_j\Big)\rho_{13,i}(P,S)\Big)\Big\}\psi_i^2(y_i) \\
+ \sum_{i=1}^{q} &\{\rho_{7i}(P,S) + \mu_i\rho_{8i}(P,S)\}\varphi_i(x_i)\psi_i(y_i) \\
+ 2\sum_{i=1}^{q}\sum_{\substack{j=2\\j>i}}^{q} &\{\rho_{1ij}(P,S) + \mu_i\rho_{2ij}(P,S)\}\varphi_i(x_i)\varphi_j(x_j)
\end{aligned}
$$

$$+ 2 \sum_{\substack{i=1}}^{q} \sum_{\substack{j=2 \\ j>i}}^{q} \{ \rho_{3ij}(P,S) + \mu_i \rho_{4ij}(P,S) + \mu_i \mu_j \rho_{5ij}(P,S) \} \psi_i(y_i) \psi_j(y_j)$$

$$+ 2 \sum_{\substack{i=1}}^{q} \sum_{\substack{j=1 \\ j\neq i}}^{q} \{ \rho_{6ij}(P,S) + \mu_i \rho_{7ij}(P,S) + \mu_i \mu_j \rho_{8ij}(P,S) \} \varphi_i(x_i) \psi_j(y_j)$$

$$= u^{\mathrm{T}} G(M,P,S) u.$$

This estimate allows us to state the following.

Theorem 9.1 *Let the perturbed motion equations* (9.2) *be such that all conditions of Assumptions 9.1 and 9.2 are satisfied and*

(a) *matrix* $A(M)$ *is positive definite for any* $\mu_i \in (0, \widetilde{\mu}_{i1})$ *and for* $\mu_i \to 0$, $i = 1, 2, \ldots, q$;

(b) *there exists a matrix* $\overline{G}(M)$ *which is negative definite for any* $\mu_i \in (0, \widetilde{\mu}_{i2})$ *and for* $\mu_i \to 0$, $i = 1, 2, \ldots, q$, *such that for the matrix* $G(M,P,S)$ *determined in Proposition 9.2 the estimate*

$$G(M,P,S) \leq \overline{G}(M), \quad \forall (M,P,S) \in \mathcal{M} \times \mathcal{P} \times \mathcal{S}$$

is satisfied.

Then the equilibrium state $(x^{\mathrm{T}}, y^{\mathrm{T}})^{\mathrm{T}} = 0$ *of system* F *is uniformly asymptotically stable for any* $\mu_i \in (0, \widetilde{\mu}_i)$ *and for* $\mu_i \to 0$ *on* $\mathcal{P} \times \mathcal{S}$, *where* $\widetilde{\mu}_i = \min \{ 1, \widetilde{\mu}_{i1}, \widetilde{\mu}_{i2} \}$.

If, moreover, $\mathcal{N}_{i1} \times \mathcal{N}_{iy} = R^{n_i + m_i}$, *the functions* $\varphi_{ik}, \psi_{ik}, \varphi_i, \psi_i$ *are of class KR, then the equilibrium state* $(x^{\mathrm{T}}, y^{\mathrm{T}})^{\mathrm{T}} = 0$ *of system* F *is uniformly asymptotically stable in the whole for any* $\mu_i \in (0, \widetilde{\mu}_i)$ *and for* $\mu_i \to 0$ *on* $\mathcal{P} \times \mathcal{S}$.

Proof Under conditions of Assumption 9.1, Proposition 9.1 and condition (a) of Theorem 9.1 the function $v(t, x, y, M)$ is positive definite for any $\mu_i \in (0, \widetilde{\mu}_{i1})$ and for $\mu_i \to 0$ it is decreasing on $\mathcal{N}_{ix} \times \mathcal{N}_{iy}$. Conditions of Assumption 9.2, Proposition 9.2 and condition (b) of Theorem 9.1 imply that the expression $D^+ v(t, x, y, M)$ is negative definite for any $\mu_i \in (0, \widetilde{\mu}_{i2})$ and for $\mu_i \to 0$ for each $(P, S) \in \mathcal{P} \times \mathcal{S}$.

These conditions are sufficient for uniform asymptotic stability of the equilibrium state of system (9.2) for any $\mu_i \in (0, \widetilde{\mu}_i)$ and for $\mu_i \to 0$ on $\widetilde{\mathcal{M}} \times \mathcal{P} \times \mathcal{S}$ since all conditions of Theorem 7 from Chapter 1 of the monograph by Grujić, et al. [1984] are satisfied.

In the case when $\mathcal{N}_{ix} \times \mathcal{N}_{iy} = R^{n_i + m_i}$, the function $v(t, x, y, M)$ is positive definite, decreasing and radially unbounded. This fact together with the other conditions of the theorem prove the second statement.

Remark 9.1 From the condition of matrix $A(M)$ positive definiteness and matrix $G(M)$ negative definiteness the values $\tilde{\mu}_{i1}$ and $\tilde{\mu}_{i2}$ are determined respectively, since $\tilde{\mu}_i = \min\{1, \tilde{\mu}_{i1}, \tilde{\mu}_{i2}\}$ is the lower estimate of the upper bound of the admissible μ_i so that $\widetilde{\mathcal{M}} = \{M: 0 < \mu_i < \tilde{\mu}_i, \ i = 1, 2, \ldots, q\}$.

Example 9.1 Consider nonlinear and nonstationary 8-th order system consisting of two interconnected 4-th order subsystems described by the equations

$$\frac{dx_i}{dt} = (1 + \sin^2 t)(-x_i^3 + 0.1\, y_i^3) + 0.2 s_{i1}(t) y_j^3 \cos^2 t,$$

$$\mu_i \frac{dy_i}{dt} = (1 + \sin^2 t)(-y_i^3 + 0.1\, \mu_i x_i^3) + 0.2 s_{2+i,1}(t) x_j^3 \cos^2 t, \tag{9.9}$$

$$i, j = 1, 2, \quad i \neq j,$$

where $x_i = (x_{i1}, x_{i2})^{\mathrm{T}} \in R^2$, $y_i = (y_{i1}, y_{i2})^{\mathrm{T}} \in R^2$, $M = \mathrm{diag}\{\mu_1, \mu_2\}$, $\mathcal{M} = \{M: 0 < \mu_i \leq 1, \ i = 1, 2\}$, $s_{ij}(t) \in [0, 1]$, $i = 1, 2, 3, 4$, $j = 1, 2$ and

$$S_i = \begin{pmatrix} 1 & 0 & s_{i1}(t) & 0 \\ 0 & 1 & 0 & s_{i1}(t) \end{pmatrix}, \quad i = 1, 2, 3, 4.$$

For system (9.9) the elements of the matrix-function (9.6) are taken as follows

$$v_{ii}(x_i) = x_i^2, \quad v_{2+i,2+i}(y_i) = \mu_i y_i^2, \quad v_{ij} = v_{2+i,2+j} = v_{i,2+j} = 0$$
$$v_{i,2+i}(x_i, y_i) = 0.1\, \mu_i x_i y_i, \quad i, j = 1, 2, \quad i \neq j.$$

Let $\eta^{\mathrm{T}} = (1, 1, 1, 1)$. Then the matrix

$$A(M) = \begin{pmatrix} A_{11} & A_{12}(M) \\ A_{12}(M) & A_{22}(M) \end{pmatrix},$$

where

$$A_{11} = \mathrm{diag}(1, 1), \quad A_{22}(M) = \mathrm{diag}(\mu_1, \mu_2),$$
$$A_{12}(M) = \mathrm{diag}(-0.1\, \mu_1, -0.1\, \mu_2),$$

is positive definite for any $\mu_i \in (0, 1]$ and for $\mu_i \to 0$, $i = 1, 2$. The elements of the matrix $\overline{G}(M)$ are of the form

$$\overline{\sigma}_{ii}(M) = -2 + 0.26\, \mu_i, \quad i = 1, 2;$$
$$\overline{\sigma}_{2+i,2+i}(M) = -1.8 + 0.06\, \mu_i, \quad i = 1, 2;$$
$$\overline{\sigma}_{i,2+i}(M) = 0, \quad i = 1, 2; \quad \overline{\sigma}_{ij}(M) = \sigma_{2+i,2+j}(M) = 0.01\mu_i,$$
$$\overline{\sigma}_{i,2+j}(M) = 0.2(1 + \mu_i), \quad i, j = 1, 2, \quad i \neq j.$$

Moreover, the matrix $\overline{G}(M)$ is negative definite for any $\mu_i \in (0, 1]$ and for $\mu_i \to 0$, $i = 1, 2$. Therefore, by Theorem 9.1 the equilibrium state $(x^{\mathrm{T}}, y^{\mathrm{T}})^{\mathrm{T}} =$

$0 \in R^8$ of system (9.9) is uniformly asymptotically stable in the whole on $\mathcal{M} \times \mathcal{S}, ,$ where $x = (x_1^T, x_2^T)^T \in R^4$, $y = (y_1^T, y_2^T)^T \in R^4$,

$$\mathcal{S} = \left\{ S: \quad S = \mathrm{diag}(S_1, S_2, S_3, S_4), \right.$$

$$\left. \begin{pmatrix} 1 & 0 & 0 & 0 \\ 0 & 1 & 0 & 0 \end{pmatrix} \le S_i \le \begin{pmatrix} 1 & 0 & 1 & 0 \\ 0 & 1 & 0 & 1 \end{pmatrix}, \quad i = 1, 2, 3, 4 \right\}.$$

9.2.2   Uniform time scaling

In the case of uniform time scaling the system (9.2) is represented as

$$\begin{aligned}
\frac{dx_i}{dt} &= f_i(t, x, 0, P_i, S_i) + f_i^*, \quad i = 1, 2, \ldots, q, \\
\mu_1 \frac{dy_i}{dt} &= \tau_i g_i(\alpha, b, y, 0, P_{q+i}, S_{q+i}) + \tau_i g_i^*, \quad i = 1, 2, \ldots, r,
\end{aligned} \tag{9.10}$$

where

$$f_i^* = f_i(t, x, y, P_i, S_i) - f_i(t, x, 0, P_i, S_i),$$
$$g_i^* = g_i(t, x, y, M, P_{q+i}, S_{q+i}) - g_i(\alpha, b, y, 0, P_{q+i}, S_{q+i}),$$

and

$$\tau_i \in [\underline{\tau}_i, \overline{\tau}_i], \quad 0 < \underline{\tau}_i < \overline{\tau}_i < +\infty, \quad \underline{\tau}_1 = \tau_1 = \overline{\tau}_1 = 1.$$

To study system (9.10) we make some assumptions.

Assumption 9.3 There exist

(1) open connected neighborhoods $\mathcal{N}_{ix} \subseteq R^{n_i}$ and $\mathcal{N}_{jy} \subseteq R^{m_j}$ of the states $x_i = 0$ and $y_j = 0$ respectively;

(2) functions $\varphi_{ik} : \mathcal{N}_{ix} \to R_+$, $\psi_{jk} : \mathcal{N}_{jy} \to R_+$, $i = 1, 2, \ldots, q$, $j = 1, 2, \ldots, r$, $q + r = s$, $k = 1, 2$, φ_{ik}, ψ_{jk} are of class K (KR);

(3) constants $\underline{\alpha}_{ip}, \overline{\alpha}_{ip}, \underline{\alpha}_{q+j,q+l}, \overline{\alpha}_{q+j,q+l}, \underline{\alpha}_{i,q+j}, \overline{\alpha}_{i,q+j}, i, p = 1, 2, \ldots, q$, $j, l = 1, 2, \ldots, r$, $q + r = s$, and matrix-function

$$U(t, x, y, \mu_1) = \begin{pmatrix} U_{11}(t, x) & \mu_1 U_{12}(t, x, y) \\ \mu_1 U_{12}^T(t, x, y) & \mu_1 U_{22}(t, y) \end{pmatrix}, \tag{9.11}$$

where

$$U_{11(t,x)} = [v_{ip}(t, x_i, x_p)], \quad v_{ip} = v_{pi}, \quad i, p = 1, 2, \ldots, q;$$
$$U_{22}(t, y) = [v_{q+i,q+l}(t, y_j, y_l)], \quad v_{q+j,q+l} = v_{q+l,q+j}, \quad j, l = 1, 2, \ldots, r;$$
$$U_{12}(t, x, y) = [v_{i,q+j}(t, x_i, y_j)], \quad i = 1, 2, \ldots, q, \quad j = 1, 2, \ldots, r,$$

whose elements satisfy the estimates

(a) $\underline{\alpha}_{ip}\varphi_{i1}(x_i)\varphi_{p1}(x_p) \le v_{ip}(t, x_i, x_p) \le \overline{\alpha}_{ip}\varphi_{i2}(x_i)\varphi_{p2}(x_p),$
$\forall\,(t, x_i, x_p) \in R \times \mathcal{N}_{ix} \times \mathcal{N}_{px}, \ i, p = 1, 2, \ldots, q, \ i \le p;$

(b) $\underline{\alpha}_{q+j,q+l}\psi_{j1}(y_j)\psi_{l1}(y_l) \le v_{q+i,q+l}(t, y_j, y_l) \le \overline{\alpha}_{q+j,q+l}\psi_{j2}(y_j) \times$
$\psi_{l2}(y_l), \ \forall\,(t, y_j, y_l) \in R \times \mathcal{N}_{jy} \times \mathcal{N}_{ly}, \ (j \le l) \in [1, r];$

(c) $\underline{\alpha}_{i,q+j}\varphi_{i1}(x_i)\psi_{j1}(y_j) \le v_{i,q+j}(t, x_i, y_j) \le \overline{\alpha}_{i,q+j}\varphi_{i2}(x_i)\psi_{j2}(y_j)$
$\forall\,(t, x_i, y_j) \in R \times \mathcal{N}_{ix} \times \mathcal{N}_{jy}, \ i = 1, 2, \ldots, q, \ j = 1, 2, \ldots, r,$
$q + r = s.$

Matrix-function (9.11) and constant vector $\eta \in R_+^s$ allow us to construct an auxiliary function

$$v(t, x, y, \mu_1) = \eta^{\mathrm{T}} U(t, x, y, \mu_1)\eta. \tag{9.12}$$

Alongside function (9.12) we consider the expression of the upper right Dini derivative

$$D^+ v(t, x, y, \mu_1) = \eta^{\mathrm{T}} D^+ U(t, x, y, \mu_1)\eta, \tag{9.13}$$

where

$$D^+ U(t, x, y, \mu_1) \stackrel{\mathrm{def}}{=} \begin{pmatrix} D^+ U_{11}(t, x) & \mu_1 D^+ U_{12}(t, x, y) \\ \mu_1 D^+ U_{12}^{\mathrm{T}}(t, x, y) & \mu_1 D^+ U_{22}(t, y) \end{pmatrix},$$

$$D^+ U_{11} = [D^+ v_{ip}(t, \cdot)], \quad D^+ U_{12} = [D^+ v_{ij}(t, \cdot)],$$

$$D^+ U_{22} = [D^+ v_{jl}(t, \cdot)], \quad i, p = 1, 2, \ldots, q; \quad j, l = 1, 2, \ldots, r; \quad q + r = s.$$

Proposition 9.3 Under conditions of Assumption 9.3 the function (9.12) satisfies the bilateral estimate

$$u_1^{\mathrm{T}} A(\mu_1) u_1 \le v(t, x, y, \mu_1) \le u_2^{\mathrm{T}} B(\mu_1) u_2,$$
$$\forall\,(t, x, y, \mu_1) \in R \times \mathcal{N}_x \times \mathcal{N}_y \times \mathcal{M},$$

where

$$u_1^{\mathrm{T}} = (\varphi_{11}(x_1), \ldots, \varphi_{q1}(x_q), \psi_{11}(y_1), \ldots, \psi_{r1}(y_r)),$$

$$u_2^{\mathrm{T}} = (\varphi_{12}(x_1), \ldots, \varphi_{q2}(x_q), \psi_{12}(y_1), \ldots, \psi_{r2}(y_r)),$$

$$A(\mu_1) = H^{\mathrm{T}} A_1(\mu_1) H, \quad B(\mu_1) = H^{\mathrm{T}} A_2(\mu_1) H, \quad H = \mathrm{diag}\{\eta_1, \ldots, \eta_s\},$$

$$A_1(\mu_1) = \begin{pmatrix} A_{11} & \mu_1 A_{12} \\ \mu_1 A_{12}^{\mathrm{T}} & \mu_1 A_{22} \end{pmatrix}, \quad A_2(\mu_1) = \begin{pmatrix} \overline{A}_{11} & \mu_1 \overline{A}_{12} \\ \mu_1 \overline{A}_{12}^{\mathrm{T}} & \mu_1 \overline{A}_{22} \end{pmatrix},$$

$$A_{11} = [\underline{\alpha}_{ip}], \quad \underline{\alpha}_{ip} = \underline{\alpha}_{pi}, \quad \overline{A}_{11} = [\overline{\alpha}_{ip}], \quad \overline{\alpha}_{ip} = \overline{\alpha}_{pi},$$

$$A_{22} = [\underline{\alpha}_{q+j,q+l}], \quad \underline{\alpha}_{q+j,q+l} = \underline{\alpha}_{q+l,q+j},$$

$$\overline{A}_{22} = [\overline{\alpha}_{q+j,q+l}], \quad \overline{\alpha}_{q+j,q+l} = \overline{\alpha}_{q+l,q+j},$$

$$A_{12} = [\underline{\alpha}_{i,q+j}], \quad \overline{A}_{12} = [\overline{\alpha}_{i,q+j}],$$

$$i, p = 1, 2, \ldots, q, \quad j, l = 1, 2, \ldots, r, \quad q + r = s.$$

The proof of Proposition 9.3 is similar to that of Proposition 9.1.

Proposition 9.4 If in Proposition 9.3 the matrices A_{11} and A_{22} are positive definite, then the function (9.12) is positive definite for any $\mu_1 \in (0, \mu_1^*)$ and for $\mu_1 \to 0$, where

$$\mu_1^* = \min\left\{1, \ \frac{\lambda_m(A_{11}^*)\lambda_m(A_{22}^*)}{\lambda_M(A_{12}^* A_{12}^{*\mathrm{T}})}\right\},$$

$$A_{11}^* = H_1^{\mathrm{T}} A_{11} H_1, \quad A_{22}^* = H_2^{\mathrm{T}} A_{22} H_2, \quad A_{12}^* = H_1 A_{12} H_2,$$

$$H_1 = \mathrm{diag}\{\eta_1, \eta_2, \dots, \eta_q\}, \quad H_2 = \mathrm{diag}\{\eta_{q+1}, \eta_{q+2}, \dots, \eta_s\}.$$

Proposition 9.4 is proved by the immediate testing.

Assumption 9.4 There exist

(1) open connected neighborhoods $\mathcal{N}_{ix} \subseteq R^{n_i}$ and $\mathcal{N}_{jy} \subseteq R^{m_j}$ of the states $x + i = 0$ and $y_j = 0$ respectively;

(2) functions φ_i, ψ_j of class K (KR), $i = 1, 2, \dots, q$, $j = 1, 2, \dots, r$;

(3) functions $v_{ip} = v_{pi}$, $v_{q+j,q+l} = v_{q+l,q+j}$, $v_{i,q+j}$, $i, p = 1, 2, \dots, r$, $j, l = 1, 2, \dots, r$, which satisfy the conditions of Assumption 9.3, and

    (a) $v_{ip}(t, x_i, x_p) \in C$ on $(R \times \mathcal{N}_{ix0} \times \mathcal{N}_{px0})$ or on $(R \times R^{n_i} \times R^{n_p})$;

    (b) $v_{q+i,q+l}(t, y_j, y_l) \in C$ on $(R \times \mathcal{N}_{jy0} \times \mathcal{N}_{ly0})$ or on $(R \times R^{m_j} \times R^{m_l})$;

    (c) $v_{i,q+j}(t, x_i, y_j) \in C$ on $(R \times \mathcal{N}_{ix0} \times \mathcal{N}_{jy0})$ or on $(R \times R^{n_i} \times R^{m_j})$;

(4) real numbers $\rho_{\alpha i}(P, S)$, $\rho_{\alpha ip}(P, S)$, $\rho_{\alpha, q+j}(P, S)$, $\rho_{\alpha, q+j,q+l}(P, S)$, $\rho_{\beta,,i,q+j}(P, S)$, $\alpha = 1, 2, 3$, $\beta = 1, 2$, $i, p = 1, 2, \dots, q$, $j, l = 1, 2, \dots, r$, $q + r = s$, and

    (a) $\eta_i^2 D_t^+ v_{ii} + \eta_i^2 (D_{x_i}^+ v_{ii})^{\mathrm{T}} f_i(t, x, 0, P_i, S_i) \le \rho_{1i}(P, S)\varphi_i^2(x_i) +$

$$\sum_{\substack{p=1 \\ p \ne i}}^{q} \rho_{1ip}(P, S)\varphi_i(x_i)\varphi_p(x_p), \ \forall (t, x_i, P, S) \in R \times \mathcal{N}_{ix0} \times \mathcal{P} \times \mathcal{S},$$

$$i = 1, 2, \dots, q;$$

    (b) $\eta_{q+j}^2 \mu_1 D_t^+ v_{q+j,q+j} + \eta_{q+j}^2 \tau_j (D_{y_j}^+ v_{q+j,q+j})^{\mathrm{T}} g_j(\alpha, y, 0, P_{q+j}, S_{q+j}) \le$

$$\rho_{1,q+j}(P, S)\psi_j^2(y_j) + \sum_{\substack{l=1 \\ l \ne j}}^{r} \rho_{1,q+j,q+l}(P, S)\psi_j(y_j)\psi_l(y_l),$$

$$\forall (t, y_j, \mu_j, P, S) \in R \times \mathcal{N}_{jy0} \times \mathcal{M} \times \mathcal{P} \times \mathcal{S}, \ j = 1, 2, \dots, r;$$

    (c) $\sum_{i=1}^{q} \eta_i^2 (D_{x_i}^+ v_{ii})^{\mathrm{T}} f_i^* + \sum_{j=1}^{r} \eta_{q+j}^2 \tau_j (D_{y_j}^+ v_{q+j,q+j})^{\mathrm{T}} g_j^* +$

$$2 \sum_{i=1}^{q} \sum_{\substack{p=2 \\ p>i}}^{q} \eta_i \eta_p \left\{ D_t^+ v_{ip} + (D_{x_i}^+ v_{ip})^{\mathrm{T}} f_i(t, x, y, P_i, S_i) + \right.$$

$$(D_{x_p}^+ v_{ip})^{\mathrm{T}} f_p(t,x,y,P_p,S_p)\Big\} + 2 \sum_{j=1}^{r} \sum_{\substack{l=2 \\ l>j}}^{r} \eta_{q+j}\eta_{q+l}\Big\{\mu_1 D_t^+ v_{q+j,q+l} +$$

$$\tau_j (D_{y_j}^+ v_{q+j,q+l})^{\mathrm{T}} g_j(t,x,y,M,P_{q+j},S_{q+j}) + \tau_l (D_{y_l}^+ v_{q+j,q+l})^{\mathrm{T}} \times$$

$$g_l(t,x,y,M,P_{q+l},S_{q+l})\Big\} + 2 \sum_{i=1}^{q} \sum_{j=1}^{r} \eta_i \eta_{q+j}\Big\{\mu_i D_t^+ v_{v,q+j} +$$

$$\mu_1 (D_{x_i}^+ v_{i,q+j})^{\mathrm{T}} f_i(t,x,y,P_i,S_i) + \tau_j (D_{y_j}^+ v_{i,q+j})^{\mathrm{T}} \times$$

$$g_j(t,x,y,M,P_{q+j},S_{q+j})\Big\} \leq \sum_{i=1}^{q}(\rho_{2i}(P,S) + \mu_1 \rho_{3i}(P,S))\varphi_i^2(x_i) +$$

$$\sum_{j=1}^{r}(\rho_{2,q+i}(P,S) + \mu_1 \rho_{3,q+j}(P,S))\psi_j^2(y_j) + 2 \sum_{i=1}^{q} \sum_{\substack{p=2 \\ p>i}}^{q}(\rho_{2ip}(P,S) +$$

$$\mu_1 \rho_{3ip}(P,S))\varphi_i(x_i)\varphi_p(x_p) + 2 \sum_{j=1}^{r} \sum_{\substack{l=2 \\ l>j}}^{r}(\rho_{2,q+j,q+l}(P,S) +$$

$$\mu_1 \rho_{3,q+j,q+l}(P,S))\psi_j(y_j)\psi_l(y_l) + \sum_{i=1}^{q} \sum_{j=1}^{r}(\rho_{1,i,q+j}(P,S) +$$

$$\mu_1 \rho_{2,i,q+j}(P,S))\varphi_i(x_i)\psi_j(y_j),$$
$$\forall\,(t,x_i,y_j,M,P,S) \in R \times \mathcal{N}_{ix_0} \times \mathcal{N}_{jy_0} \times \mathcal{M} \times \mathcal{P} \times \mathcal{S}.$$

Proposition 9.5 Under all conditions of Assumption 9.4 for the expression (9.13) the estimate

$$D^+ v(t,x,y,\mu_1) \leq u^{\mathrm{T}} \overline{C} u + \mu_1 u^{\mathrm{T}} \overline{G} u,$$
$$\forall\,(t,x,y,\mu_1,P,S) \in R \times \mathcal{N}_{x_0} \times \mathcal{N}_{y_0} \times \mathcal{M} \times \mathcal{P} \times \mathcal{S}, \quad \forall\,\tau_j \in [\underline{\tau}_j, \overline{\tau}_j],$$

holds where

$$u^{\mathrm{T}} = (\varphi_1(x_1),\dots,\varphi_q(x_q),\psi_1(y_1),\dots,\psi_r(y_r)),$$
$$\overline{C}[\overline{c}_{ij}], \quad \overline{c}_{ij} = \overline{c}_{ji}, \quad \overline{G} = [\sigma_{ij}], \quad \overline{\sigma}_{ij}\overline{\sigma}_{ji}, \quad i,j \in [1,s],$$
$$c_{ip} = \rho_{1ip}(\overline{P},\overline{S}) + \rho_{2ip}(\overline{P},\overline{S}), \quad \overline{\sigma}_{ip} = \rho_{3ip}(\overline{P},\overline{S}), \quad i,p \in [1,q], \quad p > i,$$
$$\overline{c}_{q+j,q+j} = \rho_{1,q+j}(\overline{P},\overline{S}) + \rho_{2,q+j}(\overline{P},\overline{S}),$$
$$\overline{\sigma}_{q+j,q+j} = \rho_{3,q+j}(\overline{P},\overline{S}), \quad j = 1,2,\dots,r,$$
$$\overline{c}_{q+j,q+l} = \rho_{1,q+j,q+l}(\overline{P},\overline{S}) + \rho_{2,q+j,q+l}(\overline{P},\overline{S}),$$
$$\overline{\sigma}_{q+j,q+l} = \rho_{3,q+j,q+l}(\overline{P},\overline{S}), \quad j,l = 1,2,\dots,r, \quad j > l,$$
$$\overline{c}_{i,q+j} = \rho_{1,i,q+j}(\overline{P},\overline{S}), \quad \overline{\sigma}_{i,q+j} = \rho_{2,i,q+j}(\overline{P},\overline{S}),$$
$$i = 1,2,\dots,q, \quad j = 0,r, \quad q+r = s.$$

Here $\overline{P}, \overline{S} \in \mathcal{S}$ are constant matrices such that

$$\rho_{\alpha i}(P,S) \leq \rho_{\alpha i}(\overline{P},\overline{S}), \quad \rho_{\alpha ip}(P,S) \leq \rho_{\alpha ip}(\overline{P},\overline{S}),$$

$$\rho_{\alpha,q+j}(P,S) \leq \rho_{\alpha,q+j}(\overline{P},\overline{S}), \quad \rho_{\alpha,q+j,q+l}(P,S) \leq \rho_{\alpha,q_j,q+l}(\overline{P},\overline{S}),$$

$$\rho_{\beta,i,q+i}(P,S) \leq \rho_{\beta,i,q+i}(\overline{P},\overline{S}), \quad \alpha = 1,2,3, \quad \beta = 1,2,$$

$$i, p = 1,2,\ldots,q, \quad j, l = 1,2,\ldots,r, \quad q+r = s.$$

Proof of Proposition 9.5 is similar to that of Proposition 9.2.

Proposition 9.6 If in Proposition 9.5 the matrix $\overline{C}$ is negative-definite and $\lambda_M(\overline{G}) > 0$, then the expression $D^+v(t,x,y,\mu_1)$ defined by (9.13) is negative-definite for any $\mu_1 \in (0, \mu_1^{**})$ and for $\mu_1 \to 0$ where

$$\mu_1^{**} = \min\left\{1, -\frac{\lambda_M(\overline{C})}{\lambda_M(\overline{G})}\right\}.$$

The proof of Proposition 9.6 follows from the analysis of the inequality

$$D^+v(t,x,y,\mu_1) \leq u^{\mathrm{T}}\overline{C}u + \mu_1 u^{\mathrm{T}}\overline{G}u \leq (\lambda_M(\overline{C}) + \mu_1\lambda_M(\overline{G}))\|u\|^2.$$

Remark 9.2 If in Proposition 9.6 $\lambda_M(\overline{G}) \leq 0$, then expression (9.13) is negative definite for any $\mu_1 \in (0, 1]$ and for $\mu_1 \to 0$.

Theorem 9.2 *Let the perturbed motion equations (9.10) be such that all conditions of Assumptions 9.3 and 9.4 are satisfied and*

(1) *matrices A_{11} and A_{22} are positive definite;*

(2) *matrix $\overline{C}$ is negative definite;*

(3) $\mu_1 \in (0, \widetilde{\mu}_1)$, $\mu_i = \mu_1 \tau_i^{-1}$, $\tau_i \in [\underline{\tau}_i, \overline{\tau}_i]$, $i \in [1, r]$ *where* $\widetilde{\mu}_1 = \min\{\mu_1^*, \mu_1^{**}\}$.

Then the equilibrium state $(x^{\mathrm{T}}, y^{\mathrm{T}})^{\mathrm{T}} = 0$ of system (9.10) is uniformly asymptotically stable on $\widetilde{\mathcal{M}} \times \mathcal{P} \times \mathcal{S}$.

If all conditions of the theorem are satisfied for $\mathcal{N}_{ix} \times \mathcal{N}_{jy} = R^{n_i+m_j}$ and functions φ_i, ψ_j are of class KR, then the equilibrium state $(x^{\mathrm{T}}, y^{\mathrm{T}})^{\mathrm{T}} = 0$ of system (9.10) is uniformly asymptotically stable in the whole on $\widetilde{\mathcal{M}} \times \mathcal{P} \times \mathcal{S}$, where $\widetilde{\mathcal{M}} = \{M: 0 < \mu_1 < \widetilde{\mu}_1, \mu_i = \mu_1\tau_i^{-1}, \tau_i \in [\underline{\tau}_i, \overline{\tau}_i], i = 1,2,\ldots,r\}$.

Proof Under conditions of Assumption 9.3, Proposition 9.3 and conditions (1) and (3) of Theorem 9.2 the function $v(t,x,y,\mu_1)$ is positive definite on $\widetilde{\mathcal{M}}$ and decreasing on $\mathcal{N}_x \times \mathcal{N}_y$. Conditions of Assumption 9.4, Proposition 9.5 and conditions (2) and (3) of Theorem 9.2 imply that the expression $D^+v(t,x,y,\mu_1)$ is negative definite on $\widetilde{\mathcal{M}} \times \mathcal{P} \times \mathcal{S}$.

These conditions are sufficient for uniform asymptotic stability of the equilibrium state $(x^{\mathrm{T}}, y^{\mathrm{T}})^{\mathrm{T}} = 0$ of system (9.10) on $\widetilde{\mathcal{M}} \times \mathcal{P} \times \mathcal{S}$.

In the case when $\mathcal{N}_{ix} \times \mathcal{N}_{jy} = R^{n_i+m_j}$ the function $v(t,x,y,\mu_1)$ is positive definite, decreasing and radially unbounded. This fact together with the other conditions of Theorem 9.2 proves its second assertion.

Example 9.2 Consider a nonstationary 4-th order system consisting of two interconnected 2-nd order subsystems

$$\frac{dx_i}{dt} = \frac{1}{1+\cos^2 t}\left\{ -\frac{1-\sin 2t}{2}x_i + 0.02\,S_{i1}y_i + 0.03\,S_{i2}y_j \right\},$$

$$\mu_i\frac{dy_i}{dt} = \frac{1}{1+\cos^2 t}\left\{ -\frac{4-\mu_j\sin 2t}{2}y_i \right.$$

$$\left. + 0.01\mu_i(S_{q+i,1}x_i + S_{q+i,2}x_j) \right\},$$

$$i,j = 1,2; \quad i \neq j,$$

(9.14)

where $t, x_i, y_i \in R$, $\mathcal{M} = \{M\colon 0 < \mu_i < 1,\ i = 1,2\}$, $M = \mathrm{diag}\{\mu_1, \mu_2\}$, $\underline{\tau}_2 = \frac{1}{2}$, $\overline{\tau}_2 = 1$, so that $\tau_2 \in [\frac{1}{2}, 1]$, $S_{ij} = S_{ij}(t) \in [0,1]$, $i,j = 1,2$.

The elements of the matrix-function (9.14) are taken as follows

$$v_{ii}(t, x_i) = (1 + \cos^2 t)x_i^2, \quad i = 1,2,$$

$$v_{2+i,2+i}(t, y_i) = (1 + \cos^2 t)y_i^2, \quad i = 1,2,$$

$$v_{ip}(t, x_i, x_p)v_{2+j,2+l}(t, y_j, y_l) = 0, \quad i,j,p,l = 1,2,$$

$$v_{i,2+j}(t, x_i, y_j) = 0.1(1 + \cos^2 t)x_i y_j, \quad i,j = 1,2.$$

Let $\eta^{\mathrm{T}} = (1,1,1,1)$. Then the matrices $A_{11} = A_{22} = \mathrm{diag}\{1,1\}$ are positive definite and the matrix

$$A(\mu_1) = \begin{pmatrix} A_{11} & \mu_1 A_{12} \\ \mu_1 A_{12}^{\mathrm{T}} & \mu_1 A_{22} \end{pmatrix}, \quad \text{where} \quad A_{12} = \begin{pmatrix} -0.2 & -0.2 \\ -0.2 & -0.2 \end{pmatrix}$$

is also positive definite for any $\mu_1 \in (0,1]$ and for $\mu_1 \to 0,$, since $\mu_1^* = \min\{1, 2.5\} = 1$.

For such choice of the elements of matrix-function (9.11) we have

$$\rho_{1i} = -1, \quad i = 1,2; \quad \rho_{13} = -4; \quad \rho_{14} = -1; \quad \rho_{2j} = 0, \quad j = 1,2,3,4;$$

$$\rho_{31}(S) = 0.01(S_{31} + S_{42}); \quad \rho_{32}(S) = 0.01(S_{32} + S_{42});$$

$$\rho_{33}(S) = 0.002S_{11} + 0.003S_{22}; \quad \rho_{34}(S) = 0.003S_{12} + 0.002S_{21};$$

$$\rho_{212}(S) = 0; \quad \rho_{312}(S) = 0.01(S_{31} + S_{32} + S_{41} + S_{42});$$

$$\rho_{234}(S) = 0; \quad \rho_{334}(S) = 0.002(S_{11} + S_{21}) + 0.003(S_{12} + S_{22});$$

$$\rho_{113}(S) = 0.2 + 0.04S_{11}; \quad \rho_{213}(S) = 0.05 + 0.2S_{31};$$

$$\rho_{114}(S) = 0.1 + 0.06S_{12}; \quad \rho_{214}(S) = 0.05 + 0.1S_{42};$$

$$\rho_{123}(S) = 0.2 + 0.06S_{22}; \quad \rho_{223}(S) = 0.05 + 0.1S_{32};$$

$$\rho_{124}(S) = 0.1 + 0.04S_{21}; \quad \rho_{224}(S) = 0.05 + 0.1S_{41}.$$

The matrices $\overline{C}$ and $\overline{G}$ consist of the elements

$$\overline{c}_{11} = \overline{c}_{22} = \overline{c}_{44} = -1, \quad \overline{c}_{33} = -4, \quad \overline{c}_{12} = 0, \quad \overline{c}_{34} = 0,$$
$$\overline{c}_{13} = 0.24, \quad \overline{c}_{14} = 0.16, \quad \overline{c}_{23} = 0.26, \quad \overline{c}_{24} = 0.14;$$
$$\overline{\sigma}_{11} = 0.02, \quad \overline{\sigma}_{22} = 0.02, \quad \overline{\sigma}_{33} = 0.005, \quad \overline{\sigma}_{44} = 0.005,$$
$$\overline{\sigma}_{12} = 0.04, \quad \overline{\sigma}_{34} = 0.01, \quad \overline{\sigma}_{13} = 0.25, \quad \overline{\sigma}_{14} = 0.15,$$
$$\overline{\sigma}_{23} = 0.15, \quad \overline{\sigma}_{24} = 0.15.$$

Besides, the matrix $\overline{C}$ is negative definite and $\mu_1^{**} = \min\{1, 2, 1, \ldots\} = 1$. So, all conditions of Theorem 9.2 are satisfied, $\tilde{\mu}_1 = \min\{\mu_1^*, \mu_1^{**}\} = 1$, and therefore the equilibrium state of system (9.14) is uniformly asymptotically stable in the whole on $\mathcal{M} \times \mathcal{S}$.

9.3   Tests for Instability Analysis

9.3.1   Non-uniform time scaling

Instability of solutions is considered in two cases. First, we shall consider the case of nonuniform time scaling. To this end we need the following assumptions and estimates.

Assumption 9.5 The inequalities of Assumption 9.2 hold true when the inequality sign is reversed, i.e., "$\leq$" becomes "$\geq$".

Proposition 9.7 Under conditions of Assumption 9.5 for the expression (9.13) the estimate

$$D^+v(t, x, y, M) \geq u^{\mathrm{T}}G(M, P, S)u,$$
$$\forall\, (t, x, y, M, P, S) \in R \times \mathcal{N}_{x_0} \times \mathcal{N}_{y_0} \times \mathcal{M} \times \mathcal{P} \times \mathcal{S}$$

holds true, where u^{T} and $G(M, P, S)$ are defined in the same way as in Proposition 9.5.

The proof is similar to that of Proposition 9.5.

Theorem 9.3 *Let the perturbed motion equations (9.2) be such that all conditions of Assumptions 9.1 and 9.5 are satisfied and*

(a) *matrices $A(M)$ and $B(M)$ are positive definite for any $\mu_i \in (0, \mu_i^*)$ and for $\mu_i \to 0$, $i = 1, 2, \ldots, q$, where $\mu_i^* = \min\{\overline{\mu}_{i1}, \overline{\mu}_{i2}\}$;*

(b) *there exists a matrix $\underline{G}(M)$ which is positive definite for any $\mu_i \in$*

$(0, \overline{\mu}_{i3})$ *and for* $\mu_i \to 0$, $i = 1, 2, \ldots, q$, *such that for the matrix* $G(M, P, S)$ *defined by Proposition 9.7 the estimate*

$$G(M, P, S) \geq \underline{G}(M), \quad \forall (M, P, S) \in \mathcal{M} \times \mathcal{P} \times \mathcal{S}$$

is satisfied.

Then the equilibrium state $(x^T, y^T)^T = 0$ *of system* (9.2) *is unstable for any* $\mu_i \in (0, \overline{\mu}_i)$ *and for* $\mu_i \to 0$ *on* $\mathcal{P} \times \mathcal{S}$, *where* $\overline{\mu}_i = \min\{1, \mu_i^*, \overline{\mu}_{i3}\}$.

Proof We construct the scalar function $v(t, x, y, M)$ in the same way as in Section 9.2.1. Under the conditions of Assumption 9.1 and by condition (a) of Theorem 9.3 the function $v(t, x, y, M)$ is positive definite for any $\mu_i \in (0, \mu_i^*)$ and for $\mu_i \to 0$, $i = 1, 2, \ldots, q$, and admits infinitely small upper limits on $\mathcal{N}_x \times \mathcal{N}_y$. The conditions of Assumption 9.5, Proposition 9.7 and condition (b) of Theorem 9.3 imply that the expression $D^+ v(t, x, y, M)$ is a function that is positive definite for any $\mu_i \in (0, \overline{\mu}_{i3})$ and for $\mu_i \to 0$, for every $(P, S) \in \mathcal{P} \times \mathcal{S}$. These conditions are known to be sufficient for instability of the equilibrium state of system (9.2) for any $\mu_i \in (0, \overline{\mu}_i)$ and for $\mu_i \to 0$ on $\mathcal{M} \times \mathcal{P} \times \mathcal{S}$.

Remark 9.3 By the condition of positive definiteness of the matrices $A(M)$, $B(M)$ and $\underline{G}(M)$ the values $\overline{\mu}_{i1}$, $\overline{\mu}_{i2}$ and $\overline{\mu}_{i3}$ are determined respectively, since $\overline{\mu}_i = \min\{1, \overline{\mu}_{i1}, \overline{\mu}_{i2}, \overline{\mu}_{i3}\}$ is the lower estimate of the upper boundary of the admissible μ_i, so that $\mathcal{M} = \{M: 0 < \mu_i < \overline{\mu}_i, i = 1, 2, \ldots, q\}$.

9.3.2 Uniform time scaling

Assumption 9.6 The inequalities of Assumption 9.4 hold when the inequality sign is reversed, i.e., "$\leq$" becomes "$\geq$".

Proposition 9.8 Under all conditions of Assumption 9.6 for the expression (9.13) the estimate

$$D^+ v(t, x, y, \mu_1) \geq u^T \overline{C} u + \mu_1 u^T \overline{G} u,$$

$$\forall (t, x, y, \mu_1, P, S) \in R \times \mathcal{N}_{x_0} \times \mathcal{N}_{y_0} \times \mathcal{M} \times \mathcal{P} \times \mathcal{S}, \quad \forall \tau_i \in [\underline{\tau}_i, \overline{\tau}_i]$$

takes place, where u^T, $\overline{C}$, and $\overline{G}$ are determined as in Proposition 9.5.

The proof is similar to that of Proposition 9.2.

Proposition 9.9 If in Proposition 9.8 the matrix $\overline{C}$ is positive definite and $\lambda_m(\overline{G}) < 0$, then the expression $D^+ v(t, x, y, \mu_1)$ is positive definite for any $\mu_1 \in (0, \mu_1^{**})$ and for $\mu_1 \to 0$, where $\mu_1^{**} = \min\{1, -\lambda_m(\overline{C})\lambda_m^{-1}(\overline{G})\}$.

The proof follows from the analysis of the inequality

$$D^+ v(t, x, y, \mu_1) \geq u^T \overline{C} u + \mu_1 u^T \overline{G} u \geq (\lambda_m(\overline{C}) + \mu_1 \lambda_m(\overline{G}))\|u\|^2.$$

Remark 9.4 If in Proposition 9.9 $\lambda_m(\overline{G}) \geq 0$, then the expression $D^+ v(t, x, y, \mu_1)$ is positive definite for any $\mu_1 \in (0, 1]$ and for $\mu_1 \to 0$.

Theorem 9.4 *Let the perturbed motion equations* (9.2) *be such that all conditions of Assumption 9.6 are satisfied and*

(1) *matrices* A_{11}, A_{22}, $\bar{A}_{11}$, $\bar{A}_{22}$ *and* $\overline{C}$ *are positive definite;*

(2) $\mu_1 \in (0, \bar{\mu}_1)$, $\mu_i = \mu_1 \tau_i^{-1}$, $\tau_i \in [\underline{\tau}_i, \overline{\tau}_i]$, $i \in [1, r]$, *where*

$$\bar{\mu}_1 = \min\{\mu_1^*, \mu_1^{**}, \lambda_M(\bar{A}_{11}^*)\lambda_M(\bar{A}_{22}^*)\lambda_M^{-1}(\bar{A}_{12}^*\bar{A}_{12}^{*\mathrm{T}})\},$$
$$\bar{A}_{11}^* = H_1^{\mathrm{T}}\bar{A}_{11}H, \quad \bar{A}_{22}^* = H_2^{\mathrm{T}}\bar{A}_{22}H_2, \quad \bar{A}_{12}^* = H_1\bar{A}_{12}H_2,$$
$$H_1 = \mathrm{diag}\{\eta_1, \eta_2, \ldots, \eta_q\}, \quad H_2 = \mathrm{diag}\{\eta_{q+1}, \eta_{q+2}, \ldots, \eta_s\}.$$

Then the equilibrium state $(x^{\mathrm{T}}, y^{\mathrm{T}})^{\mathrm{T}} = 0$ *of system* (9.2) *is unstable on* $\overline{\mathcal{M}} \times \mathcal{P} \times \mathcal{S}$, *where* $\overline{\mathcal{M}} = \{M: 0 < \mu_1 < \bar{\mu}_1, \ \mu_i = \mu_1\tau_i^{-1}, \ i = 1, 2, \ldots, q\}$.

The proof is similar to that of Theorem 9.2

9.4 Linear Systems under Structural Perturbations

9.4.1 Non-uniform time scaling

Consider the linear singularly perturbed system

$$\frac{dx_i}{dt} = A_i x_i + \sum_{l=1}^{q}(S_{il}^1 A_{il} x_l + S_{il}^2 A_{il}' y_l), \quad i = 1, 2, \ldots, q,$$

$$\mu_i \frac{dy_i}{dt} = B_i y_i + \sum_{l=1}^{q}(\mu_i S_{q+i,l}^1 B_{il} x_l + S_{q+i,l}^2 B_{il}' y_l), \quad i = 1, 2, \ldots, q,$$

(9.15)

where A_i, S_i, A_{il}, A_{il}', B_{il} and B_{il}' are constant matrices, all matrices and vectors are of the corresponding order, and S_{il}^1, S_{il}^2, $S_{q+i,l}^1$ and $S_{q+i,l}^2$ are diagonal matrices, $\mu_i \in (0, 1]$, $\forall i = 1, 2, \ldots, q$. Let

$$S_i = \begin{pmatrix} S_{i1}^1 & S_{i2}^1 & \cdots & S_{i,i-1}^1 & 0 & S_{i,i+1}^1 & \cdots & S_{iq}^1 \\ S_{i1}^2 & S_{i2}^2 & \cdots & S_{i,i-1}^2 & J & S_{i,i+1}^2 & \cdots & S_{iq}^2 \\ S_{q+i,1}^1 & S_{q+i,2}^1 & \cdots & S_{q+i,i-1}^1 & J & S_{q+i,i+1}^1 & \cdots & S_{q+i,q}^1 \\ S_{q+i,1}^2 & S_{q+i,2}^2 & \cdots & S_{q+i,i-1}^2 & J & S_{q+i,i+1}^2 & \cdots & S_{q+i,q}^2 \end{pmatrix},$$

$$i = 1, 2, \ldots, q, \quad S = \mathrm{diag}\{S_1, S_2, \ldots, S_q\}.$$

The structural set is defined as

$$\mathcal{S} = \{S: \ 0 \le S_{jl}^k \le J, \ S_{ii}^1 = S_{q+i,i}^2 = 0, \ S_{ii}^2 = S_{q+i,i}^1 = J,$$
$$i, l = 1, 2, \ldots, q, \ j = 1, 2, \ldots, 2q, \ k = 1, 2\},$$

where J is an identity matrix of the corresponding dimensions.

The independent singularly perturbed subsystems corresponding to system (9.15) are obtained by substitution by x^i and y^i for x and y

$$\frac{dx_i}{dt} = A_i x_i + A'_{ii} y_i, \qquad \forall\, i = 1, 2, \ldots, q,$$

$$\mu_i \frac{dy_i}{dt} = B_i y_i + \mu_i B_{ii} x_i, \quad \forall\, i = 1, 2, \ldots, q.$$

Construct matrix $U(t, x)$ for system (9.15) with elements

$$
\begin{aligned}
v_{ij}(x_i, x_j) &= v_{ji}(x_i, x_j) = x_i^{\mathrm{T}} P_{ij} x_j, \quad i, j = 1, 2, \ldots, q; \\
v_{i,q+j}(x_i, y_j) &= x_i^{\mathrm{T}} P_{i,q+j} y_j, \quad i, j = 1, 2, \ldots, q, \quad 2q = s; \\
v_{q+i,q+j}(y_i, y_j) &= v_{q+j,q+i}(y_i, y_j) = y_i^{\mathrm{T}} P_{q+i,q+j} y_j, \quad i, j = 1, 2, \ldots, q,
\end{aligned}
\tag{9.16}
$$

where P_{ii}, $P_{q+i,q+j}$ $(i \neq j)$, $P_{i,q+j}$ are constant matrices.

For functions (9.16) the following estimates are satisfied

(a) $\lambda_m(P_{ii})\|x_i\|^2 \leq v_{ii}(x_i) \leq \lambda_M(P_{ii})\|x_i\|^2, \quad \forall\, x_i \in \mathcal{N}_{ix_0}, \quad i \in [1, q]$;

(b) $\lambda_m(P_{q+i,q+i})\|y_i\|^2 \leq v_{q+i.q+i}(y_i) \leq \lambda_M(P_{q+i,q+i})\|y_i\|^2,$
$\forall\, y_i \in \mathcal{N}_{iy_0}, \quad \forall\, i = 1, 2, \ldots, q$;

(c) $-\lambda_M^{1/2}(P_{ij}P_{ij}^{\mathrm{T}})\|x_i\|\,\|x_j\| \leq v_{ij}(x_i, x_j) \leq \lambda_M^{1/2}(P_{ij}P_{ij}^{\mathrm{T}})\|x_i\|\,\|x_j\|,$
$\forall\, (x_i, x_j) \in \mathcal{N}_{ix_0} \times \mathcal{N}_{jx_0}, \quad \forall\, i, j = 1, 2, \ldots, q, \quad i \neq j$;

(d) $-\lambda_M^{1/2}(P_{q+i,q+j})P_{q+i,q+j}^{\mathrm{T}}\|y_i\|\,\|y_j\| \leq v_{q+i,q+j}(y_i, y_j) \leq$
$\lambda_M^{1/2}(P_{q+i,q+j})P_{q+i,q+j}^{\mathrm{T}}\|y_i\|\,\|y_j\|, \quad \forall\, (y_i, y_j) \in \mathcal{N}_{iy_0} \times \mathcal{N}_{jy_0},$
$\forall\, i, j = 1, 2, \ldots, q, \quad i \neq j$;

(e) $-\lambda_M^{1/2}(P_{i,q+j}P_{i,q+j}^{\mathrm{T}})\|x_i\|\,\|y_j\| \leq v_{i,q+j}(x_i, y_j) \leq$
$\lambda_M^{1/2}(P_{i,q+j}P_{i,q+j}^{\mathrm{T}})\|x_i\|\,\|y_j\|, \quad \forall\, (x_i, y_j) \in \mathcal{N}_{ix_0} \times \mathcal{N}_{jy_0},$
$i, j = 1, 2, \ldots, q,$

$$\tag{9.17}$$

where $\lambda_m(P_{ii})$ and $\lambda_m(P_{q+i,q+i})$ are minimal eigenvalues, $\lambda_M(P_{ii})$ and $\lambda_M(P_{q+i,q+i})$ are maximal eigenvalues of matrices P_{ii} and $P_{q+i,q+i}$ respectively; $\lambda_M^{1/2}(P_{ij}P_{ij}^{\mathrm{T}})$, $\lambda_M^{1/2}(P_{q+i,q+j}P_{q+i,q+j}^{\mathrm{T}})$ and $\lambda_M^{1/2}(P_{i,q+j}P_{i,q+j}^{\mathrm{T}})$ are norms of matrices P_{ij}, $P_{q+i,q+j}$ and $P_{i,q+j}$ respectively.

When estimates (9.17) are satisfied for function (9.7) with elements (9.16) the bilateral inequality

$$u^{\mathrm{T}} A(M) u \leq v(x, y, M) \leq u^{\mathrm{T}} B(M) u$$

takes place.

Here matrices $A(M)$ and $B(M)$ are defined as in Proposition 9.1,

$$u^{\mathrm{T}} = (\|x_1\|, \|x_2\|, \dots, \|x_q\|, \|y_1\|, \|y_2\|, \dots, \|y_q\|),$$

$$\underline{\alpha}_{ii} = \lambda_m(P_{ii}), \quad \underline{\alpha}_{q+i,q+i} = \lambda_m(P_{q+i,q+i}), \quad \underline{\alpha}_{ij} = -\lambda_M^{1/2}(P_{ij}P_{ij}^{\mathrm{T}}),$$

$$\underline{\alpha}_{q+i,q+j} = -\lambda_M^{1/2}(P_{q+i,q+j}P_{q+i,q+j}^{\mathrm{T}}), \quad \underline{\alpha}_{i,q+j} = -\lambda_M^{1/2}(P_{i,q+j}P_{i,q+j}^{\mathrm{T}}),$$

$$\overline{\alpha}_{ii} = \lambda_M(P_{ii}), \quad \overline{\alpha}_{q+i,q+i} = \lambda_M(P_{q+i,q+i}), \quad \overline{\alpha}_{ij} = -\underline{\alpha}_{ij},$$

$$\overline{\alpha}_{q+i,q+j} = -\underline{\alpha}_{q+i,q+j}, \quad \overline{\alpha}_{i,q+j} = -\underline{\alpha}_{i,q+j}, \quad \forall\, i,j = 1,2,\dots,q.$$

Let $\eta^{\mathrm{T}} = (1,1,\dots,1) \in R_+^s$, then the expression of total derivative of function (9.7) with elements (9.16) is

$$DV(x,y,M) = z^{\mathrm{T}}C(S)z + z^{\mathrm{T}}G(M,S)z, \quad \forall\, (x,y) \in R^q \times R^q \qquad (9.18)$$

where

$$z = (x_1^{\mathrm{T}}, x_2^{\mathrm{T}}, \dots, x_q^{\mathrm{T}}, y_1^{\mathrm{T}}, y_2^{\mathrm{T}}, \dots, y_q^{\mathrm{T}})^{\mathrm{T}};$$

$$C(S) = [c_{ij}(S)], \quad i,j = 1,2,\dots,s;$$

$$G(M,S) = [\sigma_{ij}(M,S)], \quad i,j = 1,2,\dots,s; \quad s = 2q.$$

The elements of the matrix $C(S)$ are

$$c_{ii}(S) = P_i A_i + A_i^{\mathrm{T}} P_{ii} + \sum_{l=1}^{i-1} \left(P_{li}^{\mathrm{T}}(S_{li}^1 A_{li}) + (S_{li}^1 A_{li})^{\mathrm{T}} P_{li} \right)$$

$$+ \sum_{l=1}^{q} \left(P_{il}(S_{li}^1 A_{li}) + (S_{li}^1 A_{li})^{\mathrm{T}} P_{il}^{\mathrm{T}} \right), \quad i = 1,2,\dots,q;$$

$$c_{q+i,q+i}(S) = P_{q+i,q+i} B_i + B_i^{\mathrm{T}} P_{q+i,q+i} + \sum_{l=1}^{i-1} \left(P_{q+l,q+i}^{\mathrm{T}}(S_{q+l,i}^2 B_{li}') \right.$$

$$+ (S_{q+l,i}^2 B_{li}')^{\mathrm{T}} P_{q+l,q+i} \bigg) + \sum_{l=i}^{q} \left(P_{q+i,q+l}(S_{q+l,i}^2 B_{li}') \right.$$

$$+ (S_{q+l,i}^2 B_{li}')^{\mathrm{T}} P_{q+i,q+l}^{\mathrm{T}} \bigg), \quad i = 1,2,\dots,q;$$

$$c_{ij}(S) = c_{ji}(S) = P_{ij} A_j + A_i^{\mathrm{T}} P_{ij} + \sum_{l=1}^{i-1} \left(P_{li}^{\mathrm{T}}(S_{lj}^1 A_{lj}) \right)$$

$$+ (S_{lj}^1 A_{lj})^{\mathrm{T}} P_{lj} + \sum_{l=i}^{j-1} \left(P_{il}(S_{lj}^1 A_{lj}) + (S_{li}^1 A_{li})^{\mathrm{T}} P_{lj} \right)$$

$$+ \sum_{l=j}^{q} \left(P_{il}(S_{lj}^1 A_{lj}) + (S_{li}^1 A_{li})^{\mathrm{T}} P_{jl}^{\mathrm{T}} \right), \quad i,j = 1,2,\dots,q, \quad j > i;$$

$$c_{q+i,q+j}(S) = c_{q+j,q+i}(S) = 0, \quad i,j = 1,2,\dots,q, \quad j > i;$$

$$c_{i,q+j}(S) = P_{i,q+j} B_j + \sum_{l=1}^{i-1} P_{li}^{\mathrm{T}}(S_{lj}^2 A_{lj}') + \sum_{l=i}^{q} P_{il}(S_{lj}^2 A_{lj}')$$

$$+\sum_{l=1}^{q} P_{i,q+l}(S_{q+l,j}^{2}B_{lj}'), \quad i,j=1,2,\ldots,q.$$

The elements of the matrix $G(M,S)$ are

$$\sigma_{ii}(M,S)=\mu_i\sigma_{ii}^{*}(S), \quad i=1,2,\ldots,q;$$

$$\sigma_{ii}^{*}(S)=\sum_{l=1}^{q}\left(P_{i,q+l}(S_{q+l,i}^{1}B_{li})+(S_{q+l,i}^{1}B_{li})^{\mathrm{T}}P_{i,q+l}^{\mathrm{T}}\right), \quad i=1,2,\ldots,q;$$

$$\sigma_{q+i,q+i}(M,S)=\mu_i\sigma_{q+i,q+i}^{*}(S), \quad i=1,2,\ldots,q;$$

$$\sigma_{q+i,q+i}^{*}(S)=\sum_{l=1}^{q}\left((S_{li}^{2}A_{li}')^{\mathrm{T}}P_{l,q+i}+P_{l,q+i}^{\mathrm{T}}(S_{li}^{2}A_{li}')\right), \quad i=1,2,\ldots,q;$$

$$\sigma_{ij}(M,S)=\sigma_{ji}(M,S)=\mu_j\sigma_{ij}^{*}(S), \quad i,j=1,2,\ldots,q, \quad j>i;$$

$$\sigma_{ij}^{*}(S)=\sum_{l=1}^{q}\left(P_{i,q+l}(S_{q+l,j}^{1}B_{lj})+(S_{q+l,j}^{1}B_{lj})^{\mathrm{T}}P_{i,q+l}\right), \quad j>i=1,2,\ldots,q;$$

$$\sigma_{q+i,q+j}(M,S)=\sigma_{q+j,q+i}(M,S)=\mu_i\sigma_{q+i,q+j}^{*}(S)+\mu_j\sigma_{q+i,q+j}^{**}(S),$$
$$i,j=1,2,\ldots,q, \quad j>i;$$

$$\sigma_{q+i,q+j}^{*}(S)=P_{q+i,q+j}B_j+\sum_{l=1}^{i-1}P_{q+l,q+i}^{\mathrm{T}}(S_{q+l,j}^{2}b_{lj}')$$

$$+\sum_{l=i}^{q}P_{q+i,q+l}(S_{q+l,j}^{2}B_{lj}'), \quad i,j=1,2,\ldots,q, \quad j>i;$$

$$\sigma_{q+i,q+j}^{**}(S)=B_i^{\mathrm{T}}P_{q+i,q+j}+\sum_{l=1}^{j-1}(S_{q+l,i}^{2}B_{li}')^{\mathrm{T}}P_{q+l,q+j}$$

$$+\sum_{l=j}^{q}(S_{q+l,i}^{2}B_{li}')^{\mathrm{T}}P_{q+j,q+l}^{\mathrm{T}}+\sum_{l=1}^{q}\left((S_{li}^{2}A_{li}')^{\mathrm{T}}P_{l,q+j}\right.$$

$$\left.+P_{l,q+j}^{\mathrm{T}}(S_{li}^{2}A_{l;i}')\right), \quad j>i=1,2,\ldots,q;$$

$$\sigma_{i,q+j}(M,S)=\mu_j\sigma_{i,q+j}^{*}(S)+\mu_i\mu_j\sigma_{i,q+j}^{**}(S), \quad i,j=1,2,\ldots,q;$$

$$\sigma_{i,q+j}^{*}(S)=A_i^{\mathrm{T}}P_{i,q+j}+\sum_{l=1}^{q}(S_{li}^{1}A_{li})^{\mathrm{T}}P_{l,q+j}, \quad i,j=1,2,\ldots,q;$$

$$\sigma_{i,q+j}^{**}(S)=\sum_{l=1}^{i-1}P_{q+l,q+i}^{\mathrm{T}}(S_{q+l,j}^{1}B_{lj})+\sum_{l=i}^{q}P_{q+i,q+l}(S_{q+l,j}^{1}B_{lj}),$$
$$i,j=1,2,\ldots,q.$$

We designate the upper boundary of expression (9.17) by $DV_M(x,y,M)$ and find the estimate

$$DV_M(x,y,M)\leq u^{\mathrm{T}}\overline{G}(M)u \qquad (9.19)$$

where

$$u^{\mathrm{T}} = (\|x_1\|, \|x_2\|, \ldots, \|x_q\|, \|y_1\|, \|y_2\|, \ldots, \|y_q\|),$$
$$\overline{G}(M) = [\overline{c}_{ij} + \overline{\sigma}_{ij}(M)], \quad i, j = 1, 2, \ldots, s, \quad s = 2q.$$

The elements of the matrix $\overline{G}(M)$ are

$$\overline{c}_{ii} = \lambda_M(c_{ii}(S^*)), \quad \overline{c}_{q+i,q+i} = \lambda_M(c_{q+i,q+i}(S^*)), \quad i = 1, 2, \ldots, q;$$

$$\overline{c}_{ij} = \lambda_M^{1/2}(c_{ij}(S^*)c_{ij}^{\mathrm{T}}(S^*)) = \overline{c}_{ji}, \quad i, j = 1, 2, \ldots, q, \quad j > i;$$

$$\overline{c}_{q+i,q+j} = \overline{c}_{q+j,q+i} = 0, \quad i, j = 1, 2, \ldots, q, \quad j > i;$$

$$\overline{c}_{i,q+j} = \lambda_M^{1/2}(c_{i,q+j}(S^*)c_{i,q+j}^{\mathrm{T}}(*S^*)), \quad i, j = 1, 2, \ldots, q;$$

$$\overline{\sigma}_{ii}(M) = \mu_i \lambda_M(\sigma_{ii}^*(S^*)), \quad i = 1, 2, \ldots, q;$$

$$\overline{\sigma}_{q+i,q+i}(M) = \mu_i \lambda_M(\sigma_{q+i,q+i}^*(S^*)), \quad i = 1, 2, \ldots, q;$$

$$\overline{\sigma}_{ij}(M) = \overline{\sigma}_{ji}(M) = \mu_j \lambda_M^{1/2}(\sigma_{ij}^*(S^*)\sigma_{ij}^{*\mathrm{T}}(S^*)), \quad j > i = 1, 2, \ldots, q;$$

$$\overline{\sigma}_{q+i,q+j}(M) = \overline{\sigma}_{q+j,q+i}(M) = \mu_i \lambda_M^{1/2}(\sigma_{q+i,q+j}^*(S^*)\sigma_{q+i,q+j}^{*\mathrm{T}}(S^*))$$
$$+ \mu_j \lambda_M^{1/2}(\sigma_{q+i,q+j}^{**}(S^*)\sigma_{q+i,q+j}^{**\mathrm{T}}(S^*)), \quad i, j = 1, 2, \ldots, q, \quad j > i;$$

$$\overline{\sigma}_{i,q+j}(M) = \mu_j \lambda_M^{1/2}(\sigma_{i,q+j}^*(S^*)\sigma_{i,q+j}^{*\mathrm{T}}(S^*))$$
$$+ \mu_i \mu_j \lambda_M^{1/2}(\sigma_{i,q+j}^{**}(S^*)\sigma_{i,q_j}^{**\mathrm{T}}(S^*)), \quad i, j = 1, 2, \ldots, q.$$

Here $S^* \in \mathcal{S}$ is a constant matrix such that

$$c_{ij}(S) \leq c_{ij}(S^*), \quad \sigma_{ij}^*(S) \leq \sigma_{ij}^*(s^*), \quad i, j = 1, 2, \ldots, s,$$
$$\sigma_{i,q=j}^{**}(S) \leq \sigma_{i,q+j}^{**}(S^{**}), \quad \sigma_{q+i,q+j}^{**}(S) \leq \sigma_{q+i,q+j}^{**}(S^*), \quad i, j = 1, 2, \ldots, q.$$

Theorem 9.5 *Let the equations of linear singularly perturbed large-scale system (9.15) be such that for this system it is possible to construct matrix-function (9.6) with elements (9.16) which satisfies estimates (9.17) and for the expression (9.17) estimate (9.19) holds true and*

(1) *matrix $A(M)$ is positive definite for any $\mu_i \in (0, \tilde{\mu}_{i1})$ and for $\mu_i \to 0$, $i = 1, 2, \ldots, q$;*

(2) *matrix $\overline{G}(M)$ is negative definite for any $\mu_i \in (0, \tilde{\mu}_{i2})$ and for $\mu_i \to 0$, $i = 1, 2, \ldots, q$.*

Then the equilibrium state $(x^{\mathrm{T}}, y^{\mathrm{T}})^{\mathrm{T}} = 0$ of system (9.15) is structurally uniformly asymptotically stable in the whole for any $\mu_i \in (0, \tilde{\mu}_i)$ and for $\mu_i \to 0$ on $\mathcal{S}$, where $\tilde{\mu}_i = \min\{1, \tilde{\mu}_{i1}, \tilde{\mu}_{i2}\}$.

Here $\tilde{\mu}_{i1}$ and $\tilde{\mu}_{i2}$ are determined by conditions of matrix $A(M)$ positive definiteness and matrix $\overline{G}(M)$ negative definiteness respectively.

This theorem is proved in the same manner as Theorem 9.3.

Example 9.3 Let system (9.15) be the 12-th order system $n = m = 6$ decomposed into three $q = r = 3$ interconnected singularly perturbed subsystems determined by the matrices

$$A_1 = \begin{pmatrix} 0.1 & 0 \\ 0 & 0.1 \end{pmatrix}, \quad A_2 = \begin{pmatrix} -4 & 0 \\ 0 & -4 \end{pmatrix}, \quad A_3 = \begin{pmatrix} -3 & 0 \\ 0 & -3 \end{pmatrix},$$

$$A_{12} = \begin{pmatrix} 3 & 0 \\ 0 & 3 \end{pmatrix}, \quad A_{21} = \begin{pmatrix} -3 & 0 \\ 0 & -3 \end{pmatrix},$$

$$A_{13} = A_{31} = A_{23} = A_{32} = 10^{-1}J,$$

$$A'_{ij} = 10^{-1}J, \quad i,j = 1,2,3;$$

$$B_i = \begin{pmatrix} -2 & 0 \\ 0 & -2 \end{pmatrix}; \quad B_{ij} = 10^{-1}J, \quad B'_{ij} = 0, \quad i,j = 1,2,3;$$

$$S^k_{jl} = \mathrm{diag}\{s_{jl}k, s^k_{jl}\}, \quad k = 1,2, \quad l = 1,2,3, \quad j = 1,2,3,4,5,6;$$

$$0 \le s^k_{jl} \le 1, \quad s^1_{ii} = s_{3+i,i} = 0, \quad s^2_{ii} = s^1_{3+i,i} = 1, \quad s^1_{21} = 1, \quad i = 1,2,3.$$

In the matrix-function (9.6) the elements are taken as follows

$$v_{ii}(x_i) = x_i^T J x_i, \quad v_{ij}(x_i, x_j) = x_i^T 10^{-1} J x_j,$$

$$v_{3+i,3+i}(y_i) = y_i^T 2 J y_i, \quad v_{3+i,3+j}(y_i, y_j) = 0, \quad i,j = 1,2,3, \quad i \neq j;$$

$$v_{i,3+j}(x_i, y_j) = x_i^T 10^{-1} J y_j, \quad i,j = 1,2,3; \quad j = \mathrm{diag}\{1,1,1\}.$$

It is easy to see that for these elements

$$v_{ii}(x_i) \ge \|x_i\|^2, \quad i = 1,2,3;$$

$$v_{ij}(x_i, x_j) \ge -0.1 \|x_i\| \|x_j\|, \quad i,j = 1,2,3, \quad i \neq j;$$

$$v_{3+i,3+i}(y_i) \ge 2\|y_i\|^2, \quad i = 1,2,3;$$

$$v_{i,3+j}(x_i, y_j) \ge -0.1 \|x_i\| \|y_j\|, \quad i,j = 1,2,3.$$

Let $\eta^T = (1,1,1,1,1,1)$, then the matrix $A(M)$ becomes

$$A(M) = \begin{pmatrix} A_{11} & -A_{12}(M) \\ -A_{12}^T(M) & A_{22}(M) \end{pmatrix}$$

where

$$A_{11} \begin{pmatrix} 1 & -0.1 & -0.1 \\ -0.1 & 1 & -0.1 \\ -0.1 & -0.1 & 1 \end{pmatrix}, \quad A_{12}(M) = 0.1 \begin{pmatrix} \mu_1 & \mu_1 & \mu_3 \\ \mu_1 & \mu_1 & \mu_3 \\ \mu_1 & \mu_1 & \mu_3 \end{pmatrix},$$

$$A_{22}(M) = \mathrm{diag}\{2\mu_1, 2\mu_2, 2\mu_3\},$$

and is positive definite for any $\mu_i \in (0,1]$ and for $\mu_i \to 0$, $i = 1,2,3$.

For such choice of the elements of matrix-function (9.6) the elements of the matrix $\overline{G}(M)$ are defined as

$$\overline{c}_{11} = -0.38; \quad \overline{c}_{22} = -7.38; \quad \overline{c}_{33} = -5.96; \quad \overline{c}_{12} = \overline{c}_{21} = 0.17;$$

$$\overline{c}_{13} = \overline{c}_{31} = 0.08; \quad \overline{c}_{23} = \overline{c}_{32} = 0.19; \quad \overline{c}_{3+i,3+i} = -8, \quad i = 1,2,3;$$

$$\overline{c}_{3+i,3+j} = \overline{c}_{3+j,3+i} = 0, \quad i,j = 1,2,3, \quad i \neq j;$$

$$\overline{\sigma}_{ii}(M) = \overline{\sigma}_{3+i,3+i}(M) = 0.6 \cdot 10^{-1}\mu_i, \quad i = 1,2,3;$$

$$\overline{\sigma}_{3+i,3+j}(M) = \overline{\sigma}_{3+j,3+i}(M) = 0.1\mu_i + 0.04\mu_j, \quad i,j = 1,2,3, \quad i \neq j;$$

$$\overline{\sigma}_{ij}(M) = \overline{\sigma}_{ji}(M) = 0.6 \cdot 10^{-1}\mu_j, \quad i,j = 1,2,3, \quad i \neq j;$$

$$\overline{c}_{i,3+j} = 0.8 \cdot 10^{-1}, \quad i,j = 1,2,3;$$

$$\sigma_{i,3+j}(M) = 0.18\mu_j + 0.1\mu_i\mu_j, \quad i = 1,2, \quad j = 1,2,3;$$

$$\sigma_{2,3+j}(M) = 0.9 \cdot 10^{-1}\mu_j + 0.1\mu_2\mu_j, \quad j = 1,2,3.$$

Moreover, the matrix $\overline{G}(M)$ is negative definite for any $\mu_i \in (0,1)$ and for $\mu_i \to 0$, $i = 1,2,3$.

By Theorem 9.3 the equilibrium state $(x^{\mathrm{T}}, y^{\mathrm{T}})^{\mathrm{T}} = 0 \in R^{12}$ of the system determined in this example, is uniformly asymptotically stable in the whole on $\mathcal{M} \times \mathcal{S}$, where $\mathcal{M} = \{\mu_i \colon 0 < \mu_i \leq 1, \; i = 1,2,3\}$.

Remark 9.5   In this example the independent degenerate subsystem

$$\frac{dx_1}{dt} = \begin{pmatrix} 0.1 & 0 \\ 0 & 0.1 \end{pmatrix} x_1$$

is unstable and the independent singularly perturbed subsystem

$$\frac{dx_1}{dt} = \begin{pmatrix} 0.1 & 0 \\ 0 & 0.1 \end{pmatrix} x_1 + \begin{pmatrix} 0.1 & 0 \\ 0 & 0.1 \end{pmatrix} y_1,$$

$$\mu_1 \frac{dy_1}{dt} = \begin{pmatrix} -2 & 0 \\ 0 & -2 \end{pmatrix} y_1 + \mu_1 \begin{pmatrix} 0.1 & 0 \\ 0 & 0.1 \end{pmatrix} x_1$$

is not stable for any $\mu_1 \in (0,1]$.

9.4.2   Uniform time scaling

In the case of uniform time scaling system, (9.15) is of the form

$$\frac{dx_i}{dt} = A_i x_i + \sum_{\alpha=1}^{q} S_{i\alpha}^1 A_{i\alpha} x_\alpha + \sum_{\beta=1}^{r} S_{i\beta}^2 A'_{i\beta} y_\beta, \quad i = 1,2,\dots,q,$$

$$\mu_1 \frac{dy_j}{dt} = \tau_j B_j y_j + \mu + 1 \sum_{\alpha=1}^{q} S_{q+j,\alpha}^1 B_{j\alpha} x_\alpha \qquad (9.20)$$

$$+ \tau_j \sum_{\beta=1}^{r} S_{q+j,\beta}^2 B'_{j\beta} y_\beta, \quad j = 1,2,\dots,r,$$

where A_i, B_j, $A_{i\alpha}$, $A'_{i\beta}$, $B_{j\alpha}$ and $B'_{j\beta}$ are constant matrices. All matrices and vectors are of the corresponding order, and $S^1_{i\alpha}$, $S^2_{i\beta}$, $S^1_{q+j,\alpha}$, $S^2_{q+j,\beta} \in \mathcal{S}$ are the diagonal matrices; $\mathcal{S}$ is determined in the same way as Section 5.2, $\mu_1 \in (0, 1]$, $q + r = s$, $\tau_j \in [\underline{\tau}_j, \overline{\tau}_j]$.

Assume that $\underline{\tau}_j$ and $\overline{\tau}_j$, $j = 1, 2, \ldots, r$ are given.

We construct matrix-function (9.6) for system (9.20) with the elements

$$
\begin{aligned}
v_{ip}(x_i, x_p) &= v_{pi}(x_i, x_p) = x_i^{\mathrm{T}} P_{ip} x_p, \quad i, p = 1, 2, \ldots, q; \\
v_{q+j,q+l}(y_j, y_l) &= v_{q+l,q+j}(y_j, y_l) = y_j^{\mathrm{T}} P_{q+j,q+l} y_l, \\
v_{i,q+j}(x_i, y_j) &= x_i^{\mathrm{T}} P_{i,q+j} y_j, \quad i = 1, 2, \ldots, q, \\
& j = 1, 2, \ldots, r, \quad q + r = s,
\end{aligned}
\tag{9.21}
$$

where P_{ii}, $P_{q+j,q+j}$ are symmetric positive definite matrices; P_{ip}, $i \neq p$, $P_{q+j,q+l}$, $j \neq l$, $P_{i,q+j}$ are constant matrices.

For function (9.21) the following estimates are satisfied

(a) $\lambda_m(P_{ii})\|x_i\|^2 \leq v_{ii}(x_i) \leq \lambda_M(P_{ii})\|x_i\|^2, \quad \forall\, x_i \in \mathcal{N}_{ix_0}$,
   $i = 1, 2, \ldots, q;$

(b) $\lambda_m(P_{q+j,q+j})\|y_j\|^2 \leq v_{q+j,q+j}(y_j) \leq \lambda_M(P_{q+j,q+j})\|y_j\|^2$,
   $\forall\, y_j \in \mathcal{N}_{jy_0}, \quad j = 1, 2, \ldots, r;$

(c) $-\lambda_M^{1/2}(P_{ip}P_{ip}^{\mathrm{T}})\|x_i\|\,\|x_p\| \leq v_{ip}(x_i, x_p) \leq \lambda_M^{1/2}(P_{ip}P_{ip}^{\mathrm{T}})\|x_i\|\,\|x_p\|$,
   $\forall\, (x_i, x_p) \in \mathcal{N}_{ix_0} \times \mathcal{N}_{px_0}, \quad i, p = 1, 2, \ldots, q, \quad i \neq p;$

(d) $-\lambda_M^{1/2}(P_{q+j,q+l}P_{q+j,q+l}^{\mathrm{T}})\|y_j\|\,\|y_l\| \leq v_{q+j,q+l}(y_j, y_l) \leq$
   $\lambda_M^{1/2}(P_{q+j,q+l}P_{q+j,q+l}^{\mathrm{T}})\|y_j\|\,\|y_l\|, \quad \forall\, (y_j, y_l) \in \mathcal{N}_{jy_0} \times \mathcal{N}_{ly_0}$,
   $j, l = 1, 2, \ldots, r, \quad j \neq l;$ $\qquad\qquad$ (9.22)

(e) $-\lambda_M^{1/2}(P_{i,q+j}P_{i,q+j}^{\mathrm{T}})\|x_i\|\,\|y_j\| \leq v_{i,q+j}(x_i, y_j) \leq$
   $\lambda_M^{1/2}(P_{i,q+j}P_{i,q+j}^{\mathrm{T}})\|x_i\|\,\|y_j\|, \quad \forall\, (x_i, y_j) \in \mathcal{N}_{ix_0} \times \mathcal{N}_{jy_0}$,
   $i = 1, 2, \ldots, q, \quad j = 1, 2, \ldots, r, \quad q + r = s,$

where $\lambda_m(\cdot)$ are the minimal eigenvalues, $\lambda_M(\cdot)$ are the maximal eigenvalues, and $\lambda_M^{1/2}(\cdot, \cdot)$ is the matrix norm.

If estimates (9.22) are satisfied for function (9.6) with elements (9.21) the bilateral estimate

$$
\begin{aligned}
u^{\mathrm{T}} A(\mu_1) u &\leq v(x, y, \mu_1) \leq u^{\mathrm{T}} B(\mu_1) u, \\
&\forall\, (x_i, y_j, \mu_1) \in \mathcal{N}_{ix_0} \times \mathcal{N}_{jy_0} \times \mathcal{M}
\end{aligned}
$$

holds, where $u^{\mathrm{T}} = (\|x_1\|, \|x_2\|, \ldots, \|x_q\|, \|y_1\|, \|y_2\|, \ldots, \|y_r\|)$ and the matrices $A(\mu_1)$ and $B(\mu_1)$ are defined as in Proposition 9.3 with the elements

$$\underline{\alpha}_{ii} = \lambda_m(P_{ii}); \quad \underline{\alpha}_{ip} = \underline{\alpha}_{pi} = -\lambda_M^{1/2}(P_{ip}P_{ip}^{\mathrm{T}}), \quad i \neq p = 1, 2, \ldots, q;$$

$$\overline{\alpha}_{ii} = \lambda_M(P_{ii}); \quad \overline{\alpha}_{ip} = \overline{\alpha}_{pi} = \lambda_M^{1/2}(P_{ip}P_{ip}^{\mathrm{T}}), \quad i \neq p = 1, 2, \ldots, q;$$

$$\underline{\alpha}_{q+j,q+j} = \lambda_m(P_{q+j,q+j});$$

$$\underline{\alpha}_{q+j,q+l} = \underline{\alpha}_{q+l,q+j} = -\lambda_M^{1/2}(P_{q+j,q+l}P_{q+j,q+l}^{\mathrm{T}}), \quad j,l = 1, 2, \ldots, r, \quad j \neq l;$$

$$\overline{\alpha}_{q+j,q+j} = \lambda_M(P_{q+j,q+j});$$

$$\overline{\alpha}_{q+j,q+l} = \overline{\alpha}_{q+l,q+j} = \lambda_M^{1/2}(P_{q+j,q+l}P_{q+j,q+l}^{\mathrm{T}}), \quad j,l = 1, 2, \ldots, r, \quad j \neq l;$$

$$\underline{\alpha}_{i,q+j} = -\lambda_M^{1/2}(P_{i,q+j}P_{i,q+j}^{\mathrm{T}}), \quad \overline{\alpha}_{i,q+j} = -\underline{\alpha}_{i,q+j}, \quad i \in [1,q], \quad j \in [1,r].$$

It is easy to verify that if the matrices A_{11}^* and A_{22}^* are positive definite, then the function $V(x, y, \mu_1)$ is positive definite for any $\mu_1 \in (0, \mu_1^*)$ and for $\mu_1 \to 0$, where μ_1^* is defined as in Proposition 9.4.

Let $\eta^{\mathrm{T}} = (1, 1, \ldots, 1) \in R^s$. We designate the upper boundary of the total derivative of function (9.12) with elements (9.21) by $DV_M(x, y, \mu_1)$, and find

$$-DV_M(x, y, \mu_1) \leq u^{\mathrm{T}}\overline{C}u + \mu_1 \mathbf{1} u^{\mathrm{T}}\overline{G}u, \qquad (9.23)$$

where $\overline{C} = [\overline{c}_{ij}]$, $\overline{c}_{ij} = \overline{c}_{ji}$, $i, j = 1, 2, \ldots, s$; $\overline{G} = [\overline{\sigma}_{ij}]$, $\overline{\sigma}_{ij} = \overline{\sigma}_{ji}$, $i, j = 1, 2, \ldots, s$, the matrices with elements

$$\overline{c}_{ii} = \rho_{1i}(\overline{S}) + \rho_{2i}(\overline{S}), \quad \overline{\sigma}_{ii} = \rho_{3i}(\overline{S}), \quad i = 1, 2, \ldots, q;$$

$$\overline{c}_{ip} = \rho_{1ip}(\overline{S}) + \rho_{2ip}(\overline{S}), \quad \overline{\sigma}_{ip} = \rho_{3ip}(\overline{S}), \quad i, p = 1, 2, \ldots, q, \quad p > i;$$

$$\overline{c}_{q+j,q+j} = \rho_{1,q+j}(\overline{S}) + \rho_{2,q+j}(\overline{S}), \quad \overline{\sigma}_{q+j,q+j} = \rho_{3,q+j}(\overline{S}), \quad j = 1, 2, \ldots, r;$$

$$\overline{c}_{q+j,q+l} = \rho_{1,q+j,q+l}(\overline{S}) + \rho_{2,q+j,q+l}(\overline{S}), \quad \overline{\sigma}_{q+j,q+l} = \rho_{3,q+j,q+l}(\overline{S}),$$
$$j, l = 1, 2, \ldots, r, \quad l > j;$$

$$\overline{c}_{i,q+j} = \rho_{1,i,q+j}(\overline{S}), \quad \overline{\sigma}_{i,q+j} = \rho_{2,i,q+j}(\overline{S}),$$
$$i = 1, 2, \ldots, q, \quad j = 1, 2, \ldots, r, \quad q + r = s;$$

$$\rho_{1i}(\overline{S}) = \lambda_M(C_{ii}^1(\overline{S})), \quad \rho_{1ip}(\overline{S}) = \lambda_M^{1/2}(C_{ip}^1(\overline{S})C_{ip}^{1\mathrm{T}}(\overline{S})),$$

$$\rho_{2i}(\overline{S}) = \lambda_M(C_{ii}^2(\overline{S})), \quad \rho_{2ip}(\overline{S}) = \lambda_M^{1/2}(C_{ip}^2(\overline{S})C_{ip}^{2\mathrm{T}}(\overline{S})),$$

$$\rho_{3i}(\overline{S}) = \lambda_M(\sigma_{ii}(\overline{S})), \quad \rho_{3ip}(\overline{S}) = \lambda_M^{1/2}(\sigma_{ip}(\overline{S})\sigma_{ip}^{\mathrm{T}}(\overline{S})),$$
$$i, p = 1, 2, \ldots, q, \quad p > i;$$

$$\rho_{1,q+j}(\tau_j^*, \overline{S}) = \lambda_M(C_{q+j,q+j}^1(\tau_j^*, \overline{S})),$$

$$\rho_{1,q+j,q+l}(\tau_j^*, \overline{S}) = \lambda_M^{1/2}(C_{q+j,q+l}^1(\tau_j^*, \overline{S})C_{q+j,q+l}^{1\mathrm{T}}(\tau_j^*, \overline{S})),$$

$$\rho_{2,q+j}(\tau_j^*, \overline{S}) = \lambda_M(C_{q+j,q+j}^2(\tau_j^*, \overline{S})),$$

$$\rho_{2,q+j,q+l}(\tau_j^*, \overline{S}) = \lambda_M^{1/2}(C_{q+j,q+l}^2(\tau_j^*, \overline{S})C_{q+j,q+l}^{2\mathrm{T}}(\tau_j^*, \overline{S})),$$

$$\rho_{3,q+j}(\overline{S}) = \lambda_M(\sigma_{q+j,q+j}(\overline{S})),$$

$$\rho_{3,q+j,q+l}(\overline{S}) = \lambda_M^{1/2}(\sigma_{q+j,q+l}(\overline{S})\sigma_{q+j,q+l}^{\mathrm{T}}(\overline{S})), \quad j,l = 1,2,\ldots,r, \quad l > j;$$

$$\rho_{1ij}(\tau_j^*, \overline{S}) = \lambda_M^{1/2}(c_{i,q+j}(\tau_j^*, \overline{S})c_{i,q+j}^{\mathrm{T}}(\tau_j^*, \overline{S})),$$

$$\rho_{2ij}(\overline{S}) = \lambda_M^{1/2}(\sigma_{i,q+j}(\overline{S})\sigma_{i,q+j}^{\mathrm{T}}(\overline{S})),$$

$$i = 1,2,\ldots,q, \quad j = 1,2,\ldots,r, \quad q+r = s;$$

$$c_{ii}^1(A) = P_{ii}A_{ii} + A_{ii}^{\mathrm{T}}P_{ii},$$

$$c_{ii}^2(S) = \sum_{\alpha=1}^{i-1}\left(P_{\alpha i}^{\mathrm{T}}(S_{\alpha i}^1 A_{\alpha i}) + (S_{\alpha i}^1 A_{\alpha i})^{\mathrm{T}}P_{\alpha i}\right)$$

$$+ \sum_{\alpha=i}^{q}\left(P_{i\alpha}(S_{\alpha i}^1 A_{\alpha i}) + (S_{\alpha i}^1 A_{\alpha i})^{\mathrm{T}}P_{i\alpha}^{\mathrm{T}}\right);$$

$$\sigma_{ii}(S) = \sum_{\beta=1}^{r}\left(P_{i,q+\beta}(S_{q+\beta,i}^1 B_{\beta i}) + (S_{q+\beta,i}^1 B_{\beta i})^{\mathrm{T}}P_{i,q+\beta}^{\mathrm{T}}\right), \quad i = 1,2,\ldots,q;$$

$$c_{ip}^1(S) = \sum_{\alpha=1}^{q}\left(P_{ii}^{\mathrm{T}}(S_{\alpha p}^1 A_{\alpha p}) + (S_{\alpha p}^1 A_{\alpha p})^{\mathrm{T}}P_{ii}\right),$$

$$c_{ip}^2(S) = P_{ip}A_p + A_p^{\mathrm{T}}P_{ip} + \sum_{\alpha=1}^{i-1}\left(P_{\alpha i}^{\mathrm{T}}(S_{\alpha p}^1 A_{\alpha p}) + (S_{\alpha p}^1 A_{\alpha p})^{\mathrm{T}}P_{\alpha i}\right)$$

$$+ \sum_{\alpha=i+1}^{p-1}\left(P_{i\alpha}(S_{\alpha p}^1 A_{alp}) + (S_{\alpha p}^1 A_{alp})^{\mathrm{T}}P_{i\alpha}\right)$$

$$+ \sum_{\alpha=p+1}^{q}\left(P_{i\alpha}(S_{\alpha p}^1 A_{\alpha p}) + (S_{\alpha p}^1 A_{\alpha p})^{\mathrm{T}}P_{i\alpha}^{\mathrm{T}}\right),$$

$$\sigma_{ip}(S) = \sum_{\beta=1}^{r}\left(P_{i,q+\beta}(S_{q+\beta,p}^1 B_{\beta p}) + (S_{q+\beta,p}^1 B_{\beta p})^{\mathrm{T}}P_{i,q+\beta}\right),$$

$$i,p = 1,2,\ldots,q, \quad p > i;$$

$$c_{q+j,q+j}^1(\tau_j^*, S) = P_{q+j,q+j}\tau_j B_j + \tau_j B_j^{\mathrm{T}}P_{q+j,q+j},$$

$$c_{q+j,q+j}^2(\tau_j, S) = \sum_{\beta=1}^{j-1}\left(P_{q+\beta,q+j}^{\mathrm{T}}\tau_j(S_{q+\beta,j}^2 B_{\beta j}') + \tau_j(S_{q+\beta,j}^2 B_{\beta j}')^{\mathrm{T}}P_{q+\beta,q+j}\right)$$

$$+ \sum_{\beta=j}^{r}\left(P_{q+j,q+\beta}\tau_j(S_{q+\beta,j}^2 B_{\beta j}') + \tau_j(S_{q+\beta,j}^2 B_{\beta j}')^{\mathrm{T}}P_{q+j,q+\beta}^{\mathrm{T}}\right),$$

$$\sigma_{q+j,q+j}(S) = \sum_{\alpha=1}^{q}\left((S_{\alpha j}^2 A_{\alpha j}')^{\mathrm{T}}P_{\alpha,q+j} + P_{\alpha,q+j}^{\mathrm{T}}(S_{\alpha j}^2 A_{\alpha j}')\right), \quad j = 1,2,\ldots,r;$$

$$c_{q+j,q+l}^1(\tau_l, S) = \sum_{\beta=1}^{r}P_{q+j,q+j}^{\mathrm{T}}\tau l(S_{q+\beta,l}^2 B_{\beta l}'),$$

$$c^2_{q+j,q+l}(\tau_j, S) = P_{q+j,q+l}\tau_l B_l + \tau_j B_j^\mathrm{T} P_{q+j,q+l}$$

$$+ \sum_{\beta=1}^{j-1} \left(P^\mathrm{T}_{q+\beta,q+j}\tau_l(S^2_{q+\beta,l}B'_{\beta l}) + \tau_j(S^2 2_{q+\beta,j}B'_{\beta j})^\mathrm{T} P_{q+\beta,q+l} \right)$$

$$+ \sum_{\beta=j+1}^{l-1} \left(P_{q+j,q+\beta}\tau_l(S^2_{q+\beta,l}B'_{\beta l}) + \tau_j(S^2_{q+\beta,j}B'_{\beta j})^\mathrm{T} P_{q+\beta,q+l} \right)$$

$$+ \sum_{\beta=l+1}^{r} \left(P_{q+j,q+\beta}\tau_l(S^2_{q+\beta,l}B'_{\beta l}) + \tau_j(S^2_{q+\beta,j}B'_{\beta j})^\mathrm{T} P_{q+l,q+\beta} \right),$$

$$\sigma_{q+j,q+l}(S) = \sum_{\alpha=1}^{q} \left((S^2_{\alpha j}A'_{\alpha j})^\mathrm{T} P_{\alpha,q+l} + P^\mathrm{T}_{\alpha,q+l}(S^2_{\alpha j}A'_{\alpha j}) \right),$$

$$j,l = 1,2,\ldots,r, \quad l > j;$$

$$c_{i,q+j}(\tau_j, S) = P_{i,q+j} + \sum_{\alpha=1}^{i-1} P^\mathrm{T}_{\alpha i}(S^2_{\alpha j}A'_{\alpha j})$$

$$+ \sum_{\alpha=i}^{q} P_{i\alpha}(S^2_{\alpha j}A'_{\alpha j}) + \sum_{\beta=1}^{r} P_{i,q+\beta}(S^2_{q+\beta,j}B'_{\beta j})\tau_j,$$

$$\sigma_{i,q+j}(S) = A_i^\mathrm{T} P_{i,q+j} + \sum_{\alpha=1}^{q} (S^1_{\alpha i}A_{\alpha i})P_{\alpha,q+j}$$

$$+ \sum_{\beta=1}^{j-1} P^\mathrm{T}_{q+\beta,q+i}(S^1_{q+\beta,j}B_{\beta j}) + \sum_{\beta=j}^{q} P_{q+i,q+\beta}(S^1_{q+\beta,j}B_{\beta j}),$$

$$i = 1,2,\ldots,q, \quad j = 1,2,\ldots,2, \quad q+r = s.$$

Here $\overline{S} \in \mathcal{S}$ is a constant matrix such that

$$c^k_{ip}(S) \le c^k_{ip}(\overline{S}), \quad \forall S \in \mathcal{S}, \quad i,p = 1,2,\ldots,q, \quad p \ge i, \quad k = 1,2;$$

$$c^k_{q+j,q+l}(\tau_j, S) \le c^k_{q+j,q+l}(\tau_j^*, \overline{S}), \quad \forall S \in \mathcal{S}, \quad l \ge j = 1,2,\ldots,r, \quad k = 1,2;$$

$$\sigma_{ij}(S) \le \sigma_{ij}(\overline{S}), \quad \forall S \in \mathcal{S}, \quad i,j = 1,2,\ldots,s, \quad s = q+r;$$

$$c_{i,q+j}(\tau_j, S) \le c_{i,q+j}(\tau_j^*, \overline{S}), \quad \forall S \in \mathcal{S}, \quad i = 1,2,\ldots,q, \quad j = 1,2,\ldots,r.$$

The value τ_j^* is defined as

$$\tau_j^* = \begin{cases} \underline{\tau}_j, & \text{if the corresponding factors are negative,} \\ \overline{\tau}_j, & \text{if the corresponding factors are positive.} \end{cases}$$

Note that if the matrix $\overline{C}$ is negative definite, i.e. $\lambda_M(\overline{C}) < 0$ and $\lambda_M(G) > 0$, then the function $DV_M(x,y,\mu_1)$ is negative definite for any $\mu_1 \in (0,\mu_1^{**})$ and for $\mu_1 \to 0$, where $\mu_1^{**} = \min\{1, -\lambda_M(\overline{C})/\lambda_M(\overline{G})\}$. If $\lambda_M(\overline{C}) < 0$ and $\lambda_M(G) < 0$, then $\mu_1^{**} = 1$.

Theorem 9.6 *Let linear singularly perturbed large-scale system* (9.20) *be such that for this system it is possible to construct the matrix-function* (9.6) *with elements* (9.21) *satisfying estimates* (9.22) *and for function* $DV_M(x, y, \mu_1)$, *estimate* (9.23) *is fulfilled. Also*

(1) *matrices* A_{11}^* *and* A_{22}^* *are positive definite;*

(2) *matrix* $\overline{C}$ *is negative-definite;*

(3) $\mu_1 \in (0, \tilde{\mu}_1)$, $\mu_i = \mu_1 \tau_i^{-1}$, $i = 1, 2, \ldots, r$, *where*

$$\tau_i \in [\underline{\tau}_i, \overline{\tau}_i], \quad \tilde{\mu}_1 = \min\{1, \mu_1^*, \mu_1^{**}\}.$$

Then the equilibrium state $(x^T, y^T)^T = 0$ *of system* (9.20) *is uniformly asymptotically stable in the whole on* $\widetilde{\mathcal{M}} \times \mathcal{S}$, *where*

$$\widetilde{\mathcal{M}} = \{M: \ 0 < \mu_1 < \tilde{\mu}_1, \ \mu_i = \mu_1 \tau_i^{-1}, \ i = 1, 2, \ldots, r\}.$$

The proof of this theorem follows from Theorem 9.1.

Example 9.4 Let system (9.20) be the 8-th order system $n = m = 4$, decomposed into two interconnected singularly perturbed subsystems $q = r = 2$ defined by the matrices

$$A_i = \begin{pmatrix} -2 & 1 \\ 1 & -2 \end{pmatrix}, \quad A_{i\alpha} = A'_{i\beta} = 10^{-2} J;$$

$$B_i = \begin{pmatrix} -4 & 1 \\ 1 & -4 \end{pmatrix}, \quad B_{j\alpha} = B'_{j\beta} = 0.5 \cdot 10^{-2} J;$$

$$J = \text{diag}\{1, 1\}, \quad \underline{\tau}_2 = 0.5, \quad \overline{\tau}_2 = 1, \quad \mu_2 = \mu_1 \tau_2^{-1}.$$

In the matrix-function (9.6) the elements $v_{ij}(\cdot)$ are taken as:

$$v_{ii}(x_i) = x_i^T J x_i; \quad v_{2+i, 2+i}(y_i) = y_i^T J y_i, \quad i = 1, 2;$$

$$v_{12}(x_1, x_2) = x_1^T \cdot 10^{-1} J x_2; \quad v_{34}(y_1, y_2) = y_1^T \cdot 10^{-1} y_2,$$

$$v_{i, 2+j}(x_i, y_j) = x_i^T \cdot 10^{-1} J y_j, \quad i, j = 1, 2, \quad J = \text{diag}\{1, 1\}.$$

Obviously, for these elements the following estimates are true

$$v_{ii}(x_i) \geq \|x_i\|^2, \quad i = 1, 2; \quad v_{12}(x_1, x_2) \geq -0.1 \|x_1\| \|x_2\|;$$

$$v_{2+i, 2+i}(y_i) \geq \|y_i\|^2, \quad i = 1, 2; \quad v_{34}(y_1, y_2) \geq -0.1 \|y_1\| \|y_2\|;$$

$$v_{i, 2+j}(x_i, y_j) \geq -0.1 \|x_i\| \|y_j\|, \quad i, j = 1, 2.$$

Let $\eta^T = (1, 1, 1, 1)$, , then the matrix

$$A(\mu_1) = \begin{pmatrix} A_{11} & \mu_1 A_{12} \\ \mu_1 A_{12}^T & \mu_1 A_{22} \end{pmatrix}$$

where

$$A_{11} = A_{22} = \begin{pmatrix} 1 & -0.1 \\ -0.1 & 1 \end{pmatrix} \quad and \quad A_{12} = \begin{pmatrix} -0.1 & -0.1 \\ -0.1 & -0.1 \end{pmatrix}$$

are positive definite for any $\mu_i \in (0, 1]$ and for $\mu_1 \to 0$.

For such choice of the elements of matrix (9.6) the elements of the matrices $\overline{C}$ and $\overline{G}$ are specified as

$$\overline{c}_{ii} = -1.996, \quad i = 1, 2; \quad \overline{c}_{12} = 0.6674; \quad \overline{c}_{2+i,2+i} = -2.996, \quad i = 1, 2;$$
$$\overline{c}_{34} = 0.6474; \quad \overline{c}_{1j} = 0,2874; \quad \overline{c}_{2j} = 0.2888, \quad j = 1, 2;$$

and

$$\overline{\sigma}_{ii} = 0; \quad \overline{\sigma}_{12} = 0.002, \quad i = 1, 2; \quad \overline{\sigma}_{2+i,2+i} = \overline{\sigma}_{34} = 0.004, \quad i = 1, 2;$$
$$\overline{\sigma}_{i3} = 0.312438, \quad \overline{\sigma}_{i4} = 0.311178, \quad i = 1, 2.$$

For the elements of the matrices $\overline{C}$ and $\overline{G}$ specified in such way we have

$$\lambda_M(\overline{C}) = -1.018975; \quad \lambda_M(\overline{G}) = 0.8819733$$

and

$$\mu_1^{**} = \min \left\{ 1, \ -\frac{\lambda_M(\overline{C})}{\lambda_M(\overline{G})} \right\} = \min\{1; 1.1553354\} = 1.$$

Thus, by Theorem 9.6 the equilibrium state $(x^T, y^T)^T = 0 \in R^8$ of the system defined in Example 9.4 is uniformly asymptotically stable in the whole on $\mathcal{M} \times \mathcal{S}$.

Chapter 10

Qualitative Analysis of Solutions of Set Differential Equations

When analyzing phenomena and processes of the real world either experimentally or theoretically, one cannot represent them in a "pure" form. In other words, no matter how accurately one takes into account the forces initiating the phenomenon in question, arbitrarily small perturbations are always left unexplained. The desire to describe this situation adequately generates a need to expand the techniques applicable to the mathematical analysis of the phenomenon in question. Within the framework of the description of phenomena by using ordinary differential equations (finite-dimensional ones or equations in Banach spaces), some approaches were proposed which take into account the uncertainness of the values of the system parameters, the fuzziness of systems of differential equations, the inclusion of the derivative of the phase vector into the set of values of the right-hand part of equations of perturbed motion, etc. All those approaches are designed to take into account the fact that the real motion (the stable path) is imbedded into the set of other motions (paths) which occur under the action of unaccounted forces. N.G.Chetaev [1962] noticed that those "enveloping" motions, with an arbitrarily small difference from the stable motion, can be of oscillating nature, creating a kind of wave motion. Hence if the real motion is described by an ordinary differential equation or a system of such equations, then enveloping motions can be described both by ordinary differential equations and by equations with partial derivatives, e.g., Schredinger equations. Under the condition of connectedness of those equations, the obtained set of systems of equations is an example of a hybrid system.

One of the approaches that allow us to analyze the stability of a set of paths of nonlinear dynamics, is based on the theory of the set of system of differential equations.

In this chapter we will describe a new approach to the analysis of the stability of the set equations on the basis of the generalized direct Lyapunov method.

10.1 Some Results of the General Theory of Metric Spaces

Let $\mathbb{R}^n$ be an n-dimensional vector space with the norm $\|\cdot\|$. Let $\mathcal{K}_c(\mathbb{R}^n)$ denote a nonempty subset in $\mathbb{R}^n$, containing all nonempty compact convex subsets $\mathbb{R}^n$; let $\mathcal{K}(\mathbb{R}^n)$ contain all nonempty compact subsets in $\mathbb{R}^n$ and let $\mathcal{C}(\mathbb{R}^n)$ be the subset of all nonempty closed subsets in $\mathbb{R}^n$.

For any nonempty subset A of the space $\mathbb{R}^n$, $\mathrm{co}A$ will denote its convex hull. If A is convex, then $A \subseteq \mathrm{co}A$, and $\mathrm{co}A$ is closed if A is compact.

Let x be a point in the space $\mathbb{R}^n$ and A be a nonempty subset in $\mathbb{R}^n$. The distance from x to A is determined by the formula

$$d(x, A) = \inf\{\|x - a\|: \ a \in A\},$$

and ε-neighbourhood of the subset A is determined as follows:

$$S_\varepsilon(A) = \{x \in \mathbb{R}^n: \ d(x, A) < \varepsilon\}.$$

The closure of the ε-neighbourhood $S_\varepsilon(A)$ is the set

$$\overline{S}_\varepsilon(A) = \{x \in \mathbb{R}^n: \ d(x, A) \leq \varepsilon\}.$$

For the nonempty subsets A and B of the space $\mathbb{R}^n$ denote the Hausdorff division of those sets by the formula

$$d_H(B, A) = \sup\{d(b, A): \ b \in B\}$$

or, in an equivalent form,

$$d_H(B, A) = \inf\{\varepsilon > 0: \ B \subseteq A + \varepsilon\overline{S}_1^n\},$$

where $\overline{S}_1^n = \overline{S}_1(\theta)$, $\theta \in \mathbb{R}^n$ is the zero element in $\mathbb{R}^n$.

Note that in the general case $d_H(A, B) \neq d_H(B, A)$.

The distance between the nonempty closed subsets A and B of the space $\mathbb{R}^n$ is determined by the formula

$$D[A, B] = \max\{d_H(A, B), d_H(B, A)\}$$

and called the Hausdorff metric.

It is known that:

(a) $D[A, B] \geq 0$ and $D[A, B] = 0$, if, and only if, $\overline{A} = \overline{B}$;

(b) $D[A, B] = D[B, A]$;

(c) $D[A, B] \leq D[A, C] + D[C, B]$

for any nonempty subsets A, B, C of the space $\mathbb{R}^n$.

The pair $(\mathcal{C}(\mathbb{R}^n), D)$ is a full separable metric space in which $\mathcal{K}(\mathbb{R}^n)$ and $\mathcal{K}_c(\mathbb{R}^n)$ are closed subsets.

Let F be a mapping of the domain Q of the space $\mathbb{R}^k$ in the metric space $(\mathcal{K}_c(\mathbb{R}^n), D)$, i.e., $F \colon Q \to \mathcal{K}_c(\mathbb{R}^n)$, which is equivalent to the inclusion $F(t) \in \mathcal{K}_c(\mathbb{R}^n)$ for all $t \in Q$. Such mappings are called multivalued mappings of Q in $\mathbb{R}^n$.

If there exists a constant $L > 0$ such that

$$D[F(t^*), F(t)] \le L \|t^* - t\|$$

for all $(t^*, t) \in Q$, then the multivalued mapping F is a Lipschitz one.

Note that the distance $d(x, F(t))$ of the mapping $F(t)$ from the point $x \in \mathbb{R}^n$ satisfies the estimate

$$|d(x, F(t)) - d(y, F(t^*))| \le \|x - y\| + D[F(t), F(t^*)]$$

for all $x, y \in \mathbb{R}^n$ and $(t^*, t) \in Q$ and is continuous if the mapping $F(t)$ is continuous, or it is Lipschitz continuous if the mapping $F(t)$ is continuous and satisfies the Lipschitz condition.

The support function of the mapping $s(\cdot, F(t))$ has similar properties, which follows from the inequality

$$|s(x, F(t^*)) - s(y, F(t))| \le \|F(t)\| \, \|x - y\| + D[F(t^*), F(t)]$$

for all $(t^*, t) \in Q$ and $x, y \in \overline{S}_1^n = \{x \in \mathbb{R}^n \colon \|x\| \le 1\}$.

The selector of the multivalued mapping $F(t)$ from Q to $\mathbb{R}^n$ is the one-valued mapping $f \colon Q \to \mathbb{R}^n$ such that $f(t) \in F(t)$ for all $t \in Q$.

If the mapping $F \colon Q \to \mathcal{K}_c(\mathbb{R}^n)$ is measurable, then it has a measurable selector $f \colon Q \to \mathbb{R}^n$.

Let a multivalued function $X \colon I \to \mathcal{K}_c(\mathbb{R}^n)$ be specified on the interval $I \subset \mathbb{R}$. The function X is differentiable in the point $t_0 \in I$ in the sense of Hukuhara [1967], if there exists a value $D_H X(t_0) \in \mathcal{K}_c(\mathbb{R}^n)$ such that the limits

$$\lim\{[X(t_0 + \tau) - X(t_0)]\tau^{-1} \colon \ \tau \to 0^+\},$$
$$\lim\{[X(t_0) - X(t_0 - \tau)]\tau^{-1} \colon \ \tau \to 0^+\}$$

exist and both of them are equal to $D_H X(t_0)$.

Note that in the general case the existence of the difference $A - B$ for any $A, B \in \mathcal{K}_c(\mathbb{R}^n)$ does not imply the existence of the difference $B - A$.

According to Hukuhara [1967], for the multivalued function $F \colon [a, b] \to \mathcal{K}_c(\mathbb{R}^n)$ the integral is determined as follows:

$$D_H \int_a^t F(s)\, ds = F(t).$$

Let diam $(X(t))$ be the diameter of the set $X(t)$ for all $t \in I$. The following statement is known.

If the multivalued function $X \colon I \to \mathcal{K}_c(\mathbb{R}^n)$ is differentiable on I in the sense of Hukuhara, then the real-valued function $t \to \text{diam}\,(X(t))$, $t \in I$, is nondecreasing on I.

This result should be taken into account while setting problems of stability for a set of differential equations.

Note that the set of values of the mapping X is constant if, and only if, $D_H X = 0$ on I.

The mapping F is integrally bounded on $[0, 1]$, if there exists an integrable function $g \colon [0, 1] \to \mathbb{R}$ such that $\|F(t)\| \leq g(t)$ at almost all $t \in [0, 1]$.

Let the mapping $F \colon [, 0, 1] \to \mathcal{K}_c(\mathbb{R}^n)$ be measurable and integrally bounded. Then the mapping $A \colon [0, 1] \to \mathcal{K}_c(\mathbb{R}^n)$ determined by the expression

$$A(t) = \int_0^t F(s)\,ds,$$

for all $t \in [0, 1]$ is differentiable in the sense of Hukuhara at almost all $t_0 \in (0, 1)$ with the Hukuhara derivative $D_H A(t_0) = F(t_0)$.

10.2   Existence of Solutions of Set Differential Equations

Consider an initial value problem for the set-valued differential equation

$$D_H X = F(t, X, \alpha), \quad X(t_0) = X_0 \in \mathcal{K}_c(\mathbb{R}^n), \tag{10.1}$$

where $D_H X$ is the Hukuhara derivative of the set X, $F \in C(\mathbb{R}_+ \times \mathcal{K}_c(\mathbb{R}^n) \times S, \mathcal{K}_c(\mathbb{R}^n))$ and $\alpha \in S$ is the parameter characterizing the uncertainties of the differential equation (10.1).

The mapping $X \in C^1(I, \mathcal{K}_c(\mathbb{R}^n))$ is a solution of the set-valued equation (10.1) on I if it is differentiable in the sense of Hukuhara and satisfies the equation (10.1) on I for any $\alpha \in S$.

Since $X(t)$ is a continuously differentiable mapping in the sense of Hukuhara, then

$$X(t) = X_0 + \int_{t_0}^t D_H X(s)\,ds, \quad t \in I, \tag{10.2}$$

and according to (10.1) obtain

$$X(t) = X_0 + \int_{t_0}^t F(s, X(s), \alpha)\,ds, \quad t \in I, \tag{10.3}$$

where the integral in the equation (10.3) is understood in the sense of Hukuhara. Obviously, the mapping $X(t)$ is a solution of the initial-value problem (10.1) on I if, and only if, it satisfied the equation (10.3) on I for any $\alpha \in S$.

Let

$$F_m(t, X) = \text{co} \bigcap_{\alpha \in S} F(t, X, \alpha) \qquad (10.4)$$

and

$$F_M(t, X) = \text{co} \bigcup_{\alpha \in S} F(t, X, \alpha).$$

We will consider the set $F_\beta(t, X)$ constructed by the formula

$$F_\beta(t, X) = F_M(t, X)\beta + (1 - \beta)F_m(t, X), \quad \beta \in [0, 1]. \qquad (10.5)$$

Taking into account the notations (10.4), (10.5), put the initial-value problem for the set-valued differential equation

$$D_H Y = F_\beta(t, Y), \quad Y(t_0) = Y_0 \in \mathcal{K}_c(\mathbb{R}^n), \qquad (10.6)$$

where $F_\beta \in C(\mathbb{R}_+ \times \mathcal{K}_c(\mathbb{R}^n), \mathcal{K}_c(\mathbb{R}^n))$ for all $\beta \in [0, 1]$.

In the analysis of the initial-value problem (10.6) the comparison principle in the following form is important.

Theorem 10.1 *Assume that for the equation (10.6) the following conditions are satisfied:*

(1) $F_\beta \in C(I \times \mathcal{K}_c(\mathbb{R}^n), \mathcal{K}_c(\mathbb{R}^n))$ *and there exists a monotone function* $g(t, w)$, $g \in C(I \times \mathbb{R}_+, \mathbb{R})$ *nondecreasing with respect to* w, *such that*

$$D[F_\beta(t, Y), F_\beta(t, Z)] \leq g(t, D[Y, Z])$$

for all $Y, Z \in \mathcal{K}_c(\mathbb{R}^n)$ *and* $\beta \in [0, 1]$;

(2) *there exists a maximal solution* $r(t; t_0, w_0)$ *of the scalar equation*

$$\frac{dw}{dt} = g(t, w), \quad w(t_0) = w_0 \geq 0,$$

on I;

(3) *the initial conditions* (t_0, Y_0), (t_0, Z_0) *for the two sets of solutions* $Y(t)$, $Z(t)$ *of the equations (10.6) are such that* $D[Y_0, Z_0] \leq w_0$.

Then for all $t \in I$ *the following estimate is true:*

$$D[Y(t), Z(t)] \leq r(t; t_0, w_0).$$

Proof Let $m(t) = D[Y(t), Z(t)]$, so that

$$m(t_0) = D[Y_0, Z_0] \le \omega_0$$

according to the condition (3) of the Theorem 10.1. Taking into account the relation (10.2), for the equation (10.6) obtain

$$Y(t) = Y_0 + \int_{t_0}^{t} F_\beta(s, Y(s)) \, ds, \quad t \in I, \quad \beta \in [0, 1].$$

For the two solutions $Y(t)$ and $Z(t)$ with the initial conditions $D[Y_0, Z_0] \le \omega_0$ obtain the estimate

$$
\begin{aligned}
m(t) &= D\left[Y_0 + \int_{t_0}^{t} F_\beta(s, Y(s)) \, ds, \; Z_0 + \int_{t_0}^{t} F_\beta(s, Z(s)) \, ds \right] \\
&\le D\left[Y_0 + \int_{t_0}^{t} F_\beta(s, Y(s)) \, ds, \; Y_0 + \int_{t_0}^{t} F_\beta(s, Z(s)) \, ds \right] \\
&\quad + D\left[Y_0 + \int_{t_0}^{t} F_\beta(s, Z(s)) \, ds, \; Z_0 + \int_{t_0}^{t} F_\beta(s, Z(s)) \, ds \right] \\
&= D\left[\int_{t_0}^{t} F_\beta(s, Y(s)) \, ds, \; \int_{t_0}^{t} F_\beta(s, Z(s)) \, ds \right] + D[Y_0, Z_0].
\end{aligned}
\tag{10.7}
$$

Taking into account the condition (1) of the Theorem 10.1 and the properties of the Hukuhara integral, from (10.7) obtain

$$
\begin{aligned}
m(t) &\le m(t_0) + \int_{t_0}^{t} D[F_\beta(s, Y(s)), \; F_\beta(s, Z(s))] \, ds \\
&\le m(t_0) + \int_{t_0}^{t} g(s, D[Y(s), Z(s)]) \, ds \\
&= m(t_0) + \int_{t_0}^{t} g(s, m(s)) \, ds, \quad t \in I,
\end{aligned}
\tag{10.8}
$$

for all $\beta \in [0, 1]$. Applying the Theorem 1.6.1. from the monograph by Lakshmikantham, Leela, and Martynyuk [1988b] to the inequality (10.8) and taking into account the condition (2) of the Theorem 10.1, we find that $m(t) \le r(t; t_0, \omega_0)$ at all $t \in I$.

The Theorem 10.1 is proved.

Below we will give some results connected with the existence and uniqueness of solutions of the initial-value problem (10.6).

Theorem 10.2 *Assume that for the initial-value problem (10.6) the following conditions are satisfied:*

(1) *there exists a constant $M_0 > 0$ such that for $F_\beta \in C(I \times B(Y_0, b), \mathcal{K}_c(\mathbb{R}^n))$ at all $\beta \in [0, 1]$ the inequality $D[F_\beta(t, Y), \Theta] \leq M_0$, is true, where $B(Y_0, b) = \{Y \in \mathcal{K}_c(\mathbb{R}^n) : D[Y, Y_0] \leq b\}$;*

(2) *there exist a function $g(t, \omega)$ nondecreasing with respect to ω at all $t \in I$, and a constant $M_1 > 0$ such that $g \in C(I \times [0, 2b], \mathbb{R})$, $g(t, \omega) \leq M_1$ at all $(t, \omega) \in I \times [0, 2b]$ and $\omega(t) = 0$ is the unique solution of the initial-value problem*

$$\frac{d\omega}{dt} = g(t, \omega), \quad \omega(t_0) = 0 \quad on \quad I;$$

(3) *at all $(t, Y) \in I \times B(Y_0, b)$ the estimate*

$$D[F_\beta(t, Y), \, F_\beta(t, Z)] \leq g(t, D[Y, Z])$$

holds at all $\beta \in [0, 1]$.

Then the step-wise approximations

$$Y_{n+1}(t) = Y_0 + \int_{t_0}^{t} F_\beta(s, Y_n(s)) \, ds, \quad n = 0, 1, 2, \ldots,$$

exist on the interval $I_0 = [t_0, t_0 + \Delta]$, where $\Delta = \min(a, b/M)$, $M = \max(M_0, M_1)$, as continuous functions uniformly converging to the solution of the initial-value problem (10.6) on I_0.

The proof of this theorem is based on the application of the comparison principle and the Ascoli-Arzel theorem. As opposed to the Theorem 2.3.1 from the monograph by Lakshmikantham, Bhaskar, and Devi [2006], the condition (3) of the Theorem 10.2 should hold at all $\beta \in [0, 1]$.

Now we need the concept of partial ordering in the neighbourhood $(\mathcal{K}_c(\mathbb{R}^n), D)$.

Let $K(\mathcal{K}_c(\mathbb{R}^n))$ denote some subset in $\mathcal{K}_c(\mathbb{R}^n)$ consisting of sets $X \in \mathcal{K}_c(\mathbb{R}^n)$ such that any element $u \in X$ is a non-negative (positive) vector for which $u_i \geq 0$ ($u_i > 0$) for all $i = 1, 2, \ldots, n$.

Thus, K is a cone in $\mathcal{K}_c(\mathbb{R}^n)$, and K^0 is its nonempty interior.

If there exists a set $Z \in \mathcal{K}_c(\mathbb{R}^n)$ such that for any X and $Y \in \mathcal{K}_c(\mathbb{R}^n)$ the inclusion $Z \in K(\mathcal{K}_c(\mathbb{R}^n))$ holds and $X = Y + Z$, then $X \geq Y$ ($X > Y$).

The mapping $R(t)$ is the maximal solution of the set equation (10.6), if for any solution $Y(t)$ of the equation (10.6), which exists on I_0, the inequality $Y(t) \leq R(t)$ holds for all $t \in I_0$ and for all $\beta \in [0, 1]$.

Theorem 10.3 *Assume that for the initial-value problem* (10.6) *the following conditions are satisfied:*

(1) *the mapping $F_\beta(t, Y)$, $F_\beta \in C(\mathbb{R}_+ \times \mathcal{K}_c(\mathbb{R}^n), \mathcal{K}_c(\mathbb{R}^n))$, is not monotone decreasing with respect to Y at each $t \in \mathbb{R}_+$ and at each $\beta \in [0, 1]$, i. e., as soon as $Y \leq Z$, then $F_\beta(t, Y) \leq F_\beta(t, Z)$ at each $\beta \in [0, 1]$ and $t \in \mathbb{R}_+$;*

(2) *for any $Z, W \in C^1(\mathbb{R}_+, \mathcal{K}_c(\mathbb{R}^n))$ the inequalities*

$$D_H Z < F_\beta(t, Z) \quad and \quad D_H W \geq F_\beta(t, W)$$

hold for all $\beta \in [0, 1]$ and $t \in \mathbb{R}_+$;

(3) *at the initial point $t_0 \in \mathbb{R}_+$ the following inequality is true: $Z(t_0) < W(t_0)$.*

Then for all $t \geq t_0$ the following estimate holds:

$$Z(t) < W(t).$$

Remark 10.1 If in the conditions of the Theorem 10.3 one replaces the condition (2) by the inequalities

(2') $D_H Z \leq F_\beta(t, Z) \quad and \quad D_H W \geq F_\beta(t, W)'$

for all $t \in \mathbb{R}_+$ and for all $\beta \in [0, 1]$, then the statement of the Theorem 10.3 will not change.

Now pass over to the conditions for the global existence of a solution of the set equations (10.6).

Theorem 10.4 *Assume that for the equation* (10.6) *the following conditions are satisfied:*

(1) *there exists a scalar function $g(t, \omega)$, nondecreasing with respect to ω at each $t \in \mathbb{R}_+$, such that for the mapping $F_\beta \in C(\mathbb{R}_+ \times \mathcal{K}_c(\mathbb{R}^n), \mathcal{K}_c(\mathbb{R}^n))$ for any $\beta \in [0, 1]$ the inequality*

$$D[F_\beta(t, Y), \Theta] \leq g(t, D[Y, \Theta])$$

is true for all $(t, Y) \in \mathbb{R}_+ \times \mathcal{K}_c(\mathbb{R}^n)$, where $g \in C(\mathbb{R}_+^2, \mathbb{R}_+)$;

(2) *the maximal solution $r(t; t_0, \omega_0)$ of the equation*

$$\frac{d\omega}{dt} = g(t, \omega), \quad \omega(t_0) = \omega_0 \geq 0,$$

exists for all $t \in [t_0, \infty]$;

(3) *for any $\beta \in [0,1]$ the mapping $F_\beta(t, Y)$ satisfies the conditions for the existence of the local solution of the initial-value problem (10.6) for any $(t_0, Y_0) \in \mathbb{R}_+ \times \mathcal{K}_c(\mathbb{R}^n)$, for which $D[Y_0, \Theta] \leq w_0$.*

Then any solution $Y(t; t_0, Y_0)$ of the equation (10.6) exists on the interval $[t_0, \infty]$.

Proof Consider the solution $Y(t) = Y(t; t_0, Y_0)$ of the equation (10.6) with the initial conditions $D[Y_0, \Theta] = w_0$, which exist on the interval $[t_0, a), t_0 < a < \infty$, and assume that the value a cannot be increased. Let $m(t) = D[Y(t), \Theta]$, then, according to the Theorem 10.1, obtain the estimate

$$m(t) \leq r(t; t_0, w_0)$$

for all $t_0 \leq t < a$. For any t_1, t_2 such that $t_0 < t_1 < t_2 < a$, one can easily obtain the relations

$$D[Y(t_1), Y(t_2)] = D\left[Y_0 + \int_{t_0}^{t_1} F_\beta(s, Y(s))\, ds,\ Y_0 + \int_{t_0}^{t_2} F_\beta(s, Y(s))\, ds\right]$$

$$= D\left[\int_{t_1}^{t_2} F_\beta(s, Y(s))\, ds,\ \Theta\right] \leq \int_{t_1}^{t_2} D[F_\beta(s, Y(s)),\ \Theta]\, ds$$

$$\leq \int_{t_1}^{t_2} g(s, D[Y(s), \Theta])\, ds$$

(10.9)

for all $\beta \in [0,1]$. Taking into account the condition (1) of the Theorem 10.4 and the inequality (10.9), we obtain

$$D[Y(t_1), Y(t_2)] \leq \int_{t_1}^{t_2} g(s, r(s; t_0, w_0))\, ds = r(t_2; t_0, w_0) - r(t_1; t_0, w_0). \quad (10.10)$$

From the estimate (10.10) it follows that $\lim_{t \to a^-} Y(t; t_0, Y_0)$ exists, since $\lim_{t \to a^-} r(t; t_0, w_0)$ exists and is finite by the assumption that $t_1, t_2 \to a^-$.

Let $Y(a; t_0, Y_0) = \lim_{t \to a^-} Y(t; t_0, Y_0)$ and consider the initial-value problem

$$D_H Y = F_\beta(t, Y), \quad Y(a) = Y(a; t_0, Y_0), \quad \text{for all} \quad \beta \in [0,1].$$

According to the condition (3) of the Theorem 10.4, the solution $Y(t; t_0, Y_0)$ can be continued to an interval exceeding $[t_0, a)$, i.e., the solution $Y(t; t_0, Y_0)$ with the initial values $D[Y_0, \Theta] \leq w_0$ exists on $[t_0, \infty)$.

10.3   The Matrix-Valued Lyapunov Function and Its Application

Along with the set differential equation (10.1) we will consider the matrix-valued function

$$S(t,Y) = [u_{ij}(t,Y)], \quad i,j = 1,2, \tag{10.11}$$

with the elements $u_{ij}(t,Y)$ which can be constructed as follows:

(a) if $\beta = 0$, then the element $u_{11}(t,Y)$, $u_{11} \in C(\mathbb{R}_+ \times \mathcal{K}_c(\mathbb{R}^n), \mathbb{R}_+)$, is put into correspondence with the set-valued equation

$$D_H Y = F_m(t,Y), \quad Y(t_0) = Y_0 \in \mathcal{K}_c(\mathbb{R}^n);$$

(b) if $\beta = 1$, then the element $u_{22}(t,Y)$, $u_{22} \in C(\mathbb{R}_+ \times \mathcal{K}_c(\mathbb{R}^n), \mathbb{R}_+)$, is put into correspondence with the set-valued equation

$$D_H Y = F_M(t,Y), \quad Y(t_0) = Y_0 \in \mathcal{K}_c(\mathbb{R}^n);$$

(c) if $0 < \beta < 1$, then the element $u_{12}(t,Y) = u_{21}(t,Y)$, $u_{12} \in C(\mathbb{R}_+ \times \mathcal{K}_c(\mathbb{R}^n), \mathbb{R})$, is put into correspondence with the set-valued equation

$$D_H Y = F_\beta(t,Y), \quad Y(t_0) = Y_0 \in \mathcal{K}_c(\mathbb{R}^n).$$

Using the vector $\theta \in \mathbb{R}_+^2$, we construct the function

$$V(t,Y,\theta) = \theta^{\mathrm{T}} S(t,Y)\theta, \tag{10.12}$$

for which the full derivative, in view of the equation (10.1), can be calculated by the formula

$$D^+ V(t,A,\theta) = \lim_{h \to 0^+} \sup \frac{1}{h} \left[V(t+h,\, A+hF(t,A,\alpha),\, \theta) - V(t,A,\theta) \right] \tag{10.13}$$

for any $A \in \mathcal{K}_c(\mathbb{R}^n)$ and $\alpha \in \mathcal{S}$.

Now see below the main theorem of the principle of comparison with the function (10.12) for the set equation (10.1).

Theorem 10.5 *Assume that for the set equation (10.1) the following conditions are satisfied:*

(1) *there exist a function $S \in C(\mathbb{R}_+ \times \mathcal{K}_c(\mathbb{R}^n), \mathbb{R}^{2 \times 2})$, a vector $\theta \in \mathbb{R}_+^2$, and a constant $L > 0$ such that $V(t,Y,\theta) \in C(\mathbb{R}_+ \times \mathcal{K}_c(\mathbb{R}^n) \times \mathbb{R}_+^2, \mathbb{R}_+)$ and $|V(t,A,\theta) - V(t,B,\theta)| \le LD[A,B]$ for any $A,B \in \mathcal{K}_c(\mathbb{R}^n)$, $t \in \mathbb{R}_+$;*

(2) *there exists a function* $g(t, w)$, $g \in C(\mathbb{R}_+^2, \mathbb{R})$, *such that*

$$D^+ V(t, A, \theta)\big|_{(10.1)} \leq g(t, V(t, A, \theta))$$

for all $t \in \mathbb{R}_+$, $A \in \mathcal{K}_c(\mathbb{R}^n)$, $\theta \in \mathbb{R}_+^2$;

(3) *the maximal* $r(t; t_0, w)$ *solution*

$$\frac{dw}{dt} = g(t, w), \quad w(t_0) = w_0 \geq 0,$$

exists for all $t \in [t_0, a)$.

Then if the solution $Y(t) = Y(t; t_0, Y_0)$ *of the set equation* (10.6) *exists on the interval* $[t_0, a)$ *with the initial conditions under which* $V(t_0, Y_0, \theta) \leq w_0$, *the estimate*

$$V(t, Y(t), \theta) \leq r(t; t_0, w_0) \tag{10.14}$$

holds for all $t \in [t_0, a)$.

Proof Let the initial conditions $(t_0, Y_0) \in \mathbb{R}_+ \times \mathcal{K}_c(\mathbb{R}^n)$ be such that $V(t_0, Y_0, \theta) \leq w_0$ and let the solution $Y(t) = Y(t; t_0, Y_0)$ of the set equation (10.1) exist on the interval $[t_0, a)$. For the function $m(t) = V(t, Y(t), \theta)$ and an arbitrarily small $h > 0$ one can easily find

$$m(t + h) - m(t) \leq LD\left[Y(t + h), \ Y(t) + hF(t, Y(t), \alpha)\right] +$$
$$+ V\left(t + h, \ Y(t) + hF(t, Y(t), \alpha), \theta\right) - V(t, Y(t), \theta)$$

for all $\alpha \in \mathcal{S}$, hence the inequality

$$D^+ m(t) = \lim_{h \to 0^+} \sup \frac{1}{h}\left[m(t + h) - m(t)\right] \leq D^+ V(t, Y(t), \theta)\big|_{(10.1)} +$$
$$+ L \lim_{h \to 0^+} \sup \frac{1}{h}\left[D\left[Y(t + h), \ Y(t) + hF(t, Y(t), \alpha)\right]\right] \leq g(t, m(t)),$$

$m(t_0) \leq w_0$, since $D[D_H Y(t), F_\beta(t, Y(t))] \equiv 0$.

Hence, according to the Theorem 1.5.2 from the monograph by Lakshmikantham, Leela, and Martynyuk [1988b], it follows that for all $t \in [t_0, a)$ the estimate (10.14) holds true.

Now consider the vector function

$$L(t, Y, \theta) = AS(t, Y)\theta, \tag{10.15}$$

where A is a constant (2×2)-matrix, and prove the main theorem on the principle of comparison for the set equation (10.1).

Theorem 10.6 *Assume that for the set equation* (10.1) *the following conditions are satisfied:*

(1) *there exist a function (10.5), a vector $\theta \in \mathbb{R}_+^2$, and a constant (2×2)-matrix A such that the function (10.15) satisfies the conditions $L \in C(\mathbb{R}_+ \times \mathcal{K}_c(\mathbb{R}^n) \times \mathbb{R}_+^2, \mathbb{R}_+^2)$ and $|L(t, Y_1, \theta) - L(t, Y_2, \theta)| \le B\widetilde{D}[Y_1, Y_2]$, where B is a constant (2×2)-matrix with nonnegative elements, and the vector $\widetilde{D}[Y_1, Y_2] = (D[Y_1, Y_1'], D[Y_2, Y_2'])^{\mathrm{T}}$, $\widetilde{D}: \mathcal{K}_c(\mathbb{R}^n) \times \mathcal{K}_c(\mathbb{R}^n) \to \mathbb{R}_+^2$;*

(2) *there exists a vector function $G(t, \omega)$ nondecreasing with respect to ω at each $t \in \mathbb{R}_+$, $G \in C(\mathbb{R}_+ \times \mathbb{R}_+^2, \mathbb{R}^2)$, Such that*

$$D^+L(t, A, \theta)\big|_{(10.1)} \le G(t, L(t, A, \theta))$$

for all $(t, A, \theta) \in \mathbb{R}_+ \times \mathcal{K}_c(\mathbb{R}^n) \times \mathbb{R}_+^2$ and $\alpha \in \mathcal{S}$;

(3) *the maximal solution $R(t) = R(t; t_0, \omega_0)$ of the comparison system*

$$\frac{d\omega}{dt} = G(t, \omega), \quad \omega(t_0) = \omega_0 \ge 0,$$

exists for all $t \in [t_0, a)$.

   Then if the solution $Y(t) = Y(t; t_0, Y_0)$ of the set equation (10.1) exists on the interval $[t_0, a)$ with initial conditions for which $L(t_0, Y_0, \theta) \le \omega_0$, then the component-wise estimate

$$L(t, Y(t), \theta) \le R(t; t_0, \omega_0)$$

holds true for all $t \in [t_0, a)$.

   The proof of the Theorem 10.6 is similar to that of the Theorem 10.5; therefore it is not given here.

   Note that prototypes of the Theorems 10.5 and 10.6 are the Theorems 3.2.1. and 3.7.1 from the monograph by Lakshmikantham, Bhaskar, and Devi [2006], which were set up in terms of the existence of the Lyapunov functions.

   Now we will use the Theorems 10.5 and 10.6 for obtaining the sufficient conditions for the stability of a set of stationary solutions Θ_0 of the set equation (10.1).

10.4   Stability of a Set Stationary Solution

   For the set-valued differential equation (10.1) we introduce the following concept.

   The solution $Y(t)$ of the equation (10.1) satisfies the initial condition

$$Y(t_0) = Y_0, \quad t_0 \in \mathbb{R}_+, \quad Y_0 \in \mathcal{K}_c(\mathbb{R}^n),$$

if $a < t_0 < b$ and $Y(t)$ in the point t_0 equals Y_0.

If $F(t, \Theta_0, \alpha) = \Theta_0$, then $Y(t) = \Theta_0$ is a set of equilibrium positions or a set of stationary solutions of the equation (10.1) for any $\alpha \in S$.

Since $t \to \operatorname{diam}(Y(t))$ is nondecreasing at $t \to \infty$, the direct application of the norm $\|Y(t)\| = \operatorname{diam}(Y(t))$ at $t \geq t_0$ in the determination of the stability of solutions of the equation (10.1) is impossible. Therefore we will introduce additional assumptions on the initial-value problem (10.1).

H_1. For the set equation (10.1) there exists a set of stationary solution $\Theta_0 \in \mathcal{K}_c(\mathbb{R}^n)$, i. e. $F(t, \Theta_0, \alpha) = 0$ for all $t \in \mathbb{R}_+$ and any $\alpha \in S$.

H_2. For any initial values $Y_0, X_0 \in \mathcal{K}_c(\mathbb{R}^n)$ there exists a Hukuhara difference W_0, such that $Y_0 = X_0 + W_0$.

H_3. Solution $Y(t) = Y(t, t_0, Y_0 - X_0) = Y(t, t_0, W_0)$ of the set of equations (10.1) is unique for any $(t_0, W_0) \in \mathbb{R}_+ \times S_0(\rho_0)$, $S_0(\rho_0) \subset S(\rho) = \{Y \in \mathcal{K}_c(\mathbb{R}^n) : D[Y, W_0] < \rho\}$ for any $\alpha \in S$.

Then, while setting the problem of the stability of solutions of the equation (10.1) we will take into account the Assumptions $H_1 - H_3$.

Definition 10.1 The stationary solution Θ_0 of the equation (10.6) is:

(a) *stable*, if for any $t_0 \in \mathbb{R}_+$ and $\varepsilon > 0$ there exists $\delta = \delta(t_0, \varepsilon) > 0$ such that the inequality $D[W_0, \Theta_0] < \delta$ implies the estimate $D[Y(t; t_0, W_0), \Theta_0] < \varepsilon$ for all $t \geq t_0$;

(b) *attracting*, if for any $t_0 \in \mathbb{R}_+$ there exists $\alpha(t_0) > 0$ and for any $\xi > 0$ there exists $\tau(t_0, W_0, \xi) \in \mathbb{R}_+$ such that the inequality $D[W_0, \Theta_0] < \alpha(t_0)$ implies the estimate $D[Y(t; t_0, W_0), \Theta_0] < \xi$ for any $t \geq t_0 + \tau(t_0, W_0, \xi)$;

(c) *asymptotically stable*, if the conditions of the Definitions 10.1(a) and (b) are satisfied simultaneously.

Other types of stability, attraction, and asymptotic stability of the set of stationary solutions Θ_0 of the equation (10.1) are formulated analogously.

Example 10.1 Consider the set of differential equations

$$D_H X = (-e^\alpha)X, \quad X(0) = X_0 \in \mathcal{K}_c(\mathbb{R}), \tag{10.16}$$

where $\alpha \in [0, 1]$ is an uncertain parameter. Since the values of solutions of equations (10.16) are interval functions, equation (10.16) can be rewritten as

$$[x_1', x_2'] = (-e^\alpha)X = [(-e^\alpha)x_2, \ (-e^\alpha)x_1], \quad (') = \frac{d}{dt},$$

where $X(t) = [x_1(t), x_2(t)]$ and $X(0) = [x_{10}, x_{20}]$. From this correlation we have the system of equations

$$\begin{aligned} x_1' &= -e^\alpha x_2, & x_1(0) &= x_{10}, \\ x_2' &= -e^\alpha x_1, & x_2(0) &= x_{20}, \end{aligned}$$

whose solution are the functions

$$x_1(t) = \frac{1}{2}[x_{10} + x_{20}]e^{-t} + \frac{1}{2}[x_{10} - x_{20}]e^t,$$

$$x_2(t) = \frac{1}{2}[x_{20} + x_{10}]e^{-t} + \frac{1}{2}[x_{20} - x_{10}]e^t \qquad (10.17)$$

for all $t \geq 0$ and $\alpha = 0$.

If for given set $X_0 \in \mathcal{K}_c(\mathbb{R})$ there exist sets $V_0, W_0 \in \mathcal{K}_c(\mathbb{R})$ such that $X_0 = V_0 + W_0$, then the Hukuhara difference $X_0 - V_0 = W_0$ exists. Let $X_0 = [x_{10}, x_{20}]$, $V_0 = \frac{1}{2}([x_{10}-x_{20}], [x_{20}-x_{10}])$, then $W_0 = \frac{1}{2}([x_{10}+x_{20}], [x_{20}+x_{10}])$.
If $x_{10} \neq -x_{20}$, from (10.17) we have for $t \geq 0$:

(a) $X(t, X_0) = \frac{1}{2}(-[x_{20}-x_{10}], [x_{20}-x_{10}])e^t + \frac{1}{2}([x_{10}+x_{20}], [x_{10}+x_{20}])e^{-t}$;

(b) $X(t, V_0) = \frac{1}{2}([x_{10} - x_{20}], [x_{20} - x_{10}])e^t$;

(c) $X(t, W_0) = ([x_{10} + x_{20}], [x_{10} + x_{20}])e^{-t}$.

Hence it follows that if the Hukuhara difference exists for the initial values of problem (10.16), the zero solution $\{0\} \in \mathcal{K}_c\mathbb{R}$ is stable, while expressions (a) and (b) contain undesirable components.

10.5   Theorems on Stability

The conditions for the different types of stability of the stationary solution Θ_0 of the equation (10.1) will be determined on the basis of the function (10.12) and its full derivative along the set of solutions of the equation (10.1).

Theorem 10.7 *Assume that for the set differential equation* (10.1) *the conditions of the Assumptions* $H_1 - H_3$ *are satisfied, and, in addition,*

(1) *there exists a matrix-valued function* (10.11) *and a vector* $\theta \in \mathbb{R}_+^2$ *such that the function* (10.12) *satisfies the local Lipschitz condition*

$$|V(t, Y_1, \theta) - V(t, Y_2, \theta)| \leq LD[Y_1, Y_2],$$

where $L > 0$ *for all* $(t, Y) \in \mathbb{R}_+ \times S_0(\rho_0)$;

(2) *there exist vector comparison functions* ψ_1, $\psi_2 \in K$-*class and constant symmetric positive-definite* (2×2)-*matrices* A_1, A_2 *such that*

$$\psi_1^T(D[Y, \Theta_0]) A_1 \psi_1(D[Y, \Theta_0]) \leq V(t, Y, \theta)$$
$$\leq \psi_2^T(t, D[W_0, 0]) A_2 \psi_2(t, D[W_0, 0])$$

for all $(t, Y) \in \mathbb{R}_+ \times S_0(\rho_0)$;

(3) *for all* $(t, Y) \in \mathbb{R}_+ \times S_0(\rho_0)$ *and* $\alpha \in \mathcal{S}$ *the following inequality is true:*

$$D^+ V(t, Y, \theta)\big|_{(10.1)} \leq 0.$$

Then the stationary solution Θ_0 *of the equations* (10.1) *is stable.*

Proof Transform the condition (2) of the Theorem 10.7 to the form

$$\lambda_m(A_1)\overline{\psi}_1\left(D[Y, \Theta_0]\right) \leq V(t, Y, \theta) \leq \lambda_M(A_2)\overline{\psi}_2\left(t, D[W_0, 0]\right), \qquad (10.18)$$

where $\lambda_m(A_1)$ and $\lambda_M(A_2)$ are the minimum and the maximal eigenvalues of the matrices A_1, A_2, and $\overline{\psi}_1$, $\overline{\psi}_2 \in K$-class, such that

$$\overline{\psi}_1\left(D[Y, \Theta_0]\right) \leq \psi_1^{\mathrm{T}}\left(D[Y, \Theta_0]\right)\psi_1\left(D[Y, \Theta_0]\right),$$
$$\overline{\psi}_2\left(t, D[W_0, 0]\right) \geq \psi_2^{\mathrm{T}}\left(t, D[W_0, 0]\right)\psi_2\left(t, D[W_0, 0]\right)$$

in the range $(t, Y) \in \mathbb{R}_+ \times S_0(\rho_0)$.

Let $\varepsilon > 0$ and $t_0 \in \mathbb{R}_+$ be given. Choose the value $\delta = \delta(t_0, \varepsilon) > 0$ from the condition

$$\lambda_M(A_2)\overline{\psi}_2(t_0, \delta) < \lambda_m(A_1)\overline{\psi}_1(\varepsilon).$$

Show that with such chosen $\delta(t_0, \varepsilon) > 0$ and under the conditions of the Theorem 10.7, the stationary solution $Y(t; t_0, \Theta_0)$ of the equation (10.1) is stable. If this is not so, then there should exist a solution $Y(t) = Y(t; t_0, Y_0)$ and a value $t_1 > t_0$ such that

$$D[Y(t_1), \Theta_0] = \varepsilon \quad \text{and} \quad D[Y(t), \Theta_0] \leq \varepsilon < H_0, \quad H_0 < H$$

for all $t_0 \leq t \leq t_1$ as soon as $D[W_0, 0] < \delta$.

From the Theorem 10.5 and the condition (3) of the Theorem 10.7 it follows that

$$V(t, Y(t), \theta) \leq V(t_0, Y_0, \theta), \quad t_0 \leq t \leq t_1. \qquad (10.19)$$

Taking into account the estimate (10.18), from the inequality (10.19) we find

$$\lambda_m(A_1)\overline{\psi}_1(\varepsilon) = \lambda_m(A_1)\overline{\psi}_1(D[Y(t_1), \Theta_0]) \leq V(t_1, Y(t_1), \theta)$$
$$\leq V(t_0, W_0, \theta) \leq \lambda_M(A_2)\overline{\psi}_2(t_0, D[W_0, \theta])$$
$$\leq \lambda_M(A_2)\overline{\psi}_2(t_0, \delta) < \lambda_m(A_1)\overline{\psi}_1(\varepsilon).$$

The obtained contradiction proves the Theorem 10.7.

Theorem 10.8 *Assume that for the equation* (10.1) *the conditions of the Assumptions* $H_1 - H_3$, *the conditions* (1), (2) *of the Theorem 10.7 are satisfied, and, in addition, the condition*

(3') $D^+ V(t, Y, \theta)\big|_{(10.1)} \leq -\gamma V(t, Y, \theta)$ *for all* $(t, Y) \in \mathbb{R}_+ \times S_0(\rho_0)$ *and* $\alpha \in \mathcal{S}$

is satisfied as well.

Then the stationary solution Θ_0 of the equation (10.1) is asymptotically stable.

Proof From the conditions of the Theorem 10.8 it follows that the stationary solution Θ_0 of the equation (10.1) is stable. Let $\varepsilon = \rho_0$ and $\delta_0 = \delta(t_0, \rho_0)$. Then, according to the Theorem 10.7, the inequality $D[W_0, 0] < \delta_0$ implies the estimate $D[Y(t), \Theta_0] < \rho_0$ for all $t \geq t_0$, where $Y(t) = Y(t; t_0, W_0)$ is the solution of the equations (10.1). From the condition (3′) of the Theorem 10.8 it follows that

$$V(t, Y(t), \theta) \leq V(t_0, W_0, \theta) \exp\left[-\gamma(t - t_0)\right]$$

for all $t \geq t_0$.

For the given $\varepsilon > 0$ choose $\tau = \tau(t_0, \varepsilon)$ by the formula

$$\tau(t_0, \varepsilon) = \frac{1}{\gamma} \ln \frac{\lambda_M(A_2)\overline{\psi}_2(t_0, \delta_0)}{\lambda_m(A_1)\overline{\psi}_1(\varepsilon)} + 1.$$

It is easy to verify that the following inequality is true:

$$\lambda_m(A_1)\overline{\psi}_1\left(D[Y(t), \Theta_0]\right) \leq V(t, Y(t), \theta)$$
$$\leq \lambda_M(A_2)\overline{\psi}_2(t_0, \delta_0) \exp\left[-\gamma(t - t_0)\right] < \lambda_m(A_1)\overline{\psi}_1(\varepsilon).$$

The Theorem 10.8 is proved.

Now for the value $0 < \eta < \rho_0$ consider the set $S(\rho_0) \cap S^c(\eta)$ and assume that the function $V(t, Y, \theta)$ is defined in the range $(t, Y) \in \mathbb{R}_+ \times S(\rho_0) \cap S^c(\eta)$.

Theorem 10.9 *Assume that for the equation (10.1) the conditions of the Assumptions $H_1 - H_3$ are satisfied and, in addition,*

(1) *function (10.12) is such that*

$$|V(t, Y_1, \theta) - V(t, Y_2, \theta)| \leq L D[Y_1, Y_2],$$

where $L > 0$ for all $(t, Y) \in \mathbb{R}_+ \times S(\rho_0) \cap S^c(\eta)$;

(2) *there exist vector comparison functions ψ_1, $\psi_2 \in K$-class and constant symmetric positive-definite matrices A_1 and A_2 such that*

$$\psi_1^{\mathrm{T}}\left(D[Y, \Theta_0]\right) A_1 \psi_1\left(D[Y, \Theta_0]\right) \leq V(t, Y, \theta)$$
$$\leq \psi_2^{\mathrm{T}}\left(D[W_0, 0]\right) A_2 \psi_2\left(D[W_0, 0]\right)$$

for all $(t, Y) \in \mathbb{R}_+ \times S(\rho_0) \cap S^c(\eta)$;

(3) *for all $(t, Y) \in \mathbb{R}_+ \times S(\rho_0) \cap S^c(\eta)$ and $\alpha \in S$ the following inequality is true:*

$$D^+ V(t, Y, \theta)\big|_{(10.1)} \leq 0.$$

Then the stationary solution Θ_0 of the equation (10.1) is uniformly stable.

Proof Let $0 < \varepsilon < \rho_0$ and $t_0 \in \mathbb{R}_+$ be given. Choose $\delta = \delta(\varepsilon)$ so that

$$\lambda_M(A_2)\overline{\psi}_2(\delta) < \lambda_m(A_1)\overline{\psi}_1(\varepsilon),$$

and show that here the stationary solution Θ_0 of the equation (10.1) is uniformly stable. If this is not so, then one can find a solution $Y(t)$ of the equation (10.1), for which the relations

$$D[Y(t_1), \Theta_0] = \delta, \quad D[Y(t_2), \Theta_0] = \varepsilon \quad \text{and} \quad \delta \leq D[Y(t), \Theta_0] \leq \varepsilon < \rho_0$$

hold for all $t_1 \leq t \leq t_2$. Let $\eta = \delta$, then from the condition (3) of the Theorem 10.9 one can obtain the inequality

$$V(t_2, Y(t_2), \theta) \leq V(t_1, Y(t_1), \theta)$$

and, according to the condition (2) of the Theorem 10.9,

$$
\begin{aligned}
\lambda_m(A_1)\overline{\psi}_1(\varepsilon) &= \lambda_m(A_1)\overline{\psi}_1(D[Y(t_2), \Theta_0]) \\
&\leq V(t_2, Y(t_2), \theta) \leq V(t_1, Y(t_1), \theta) \\
&\leq \lambda_M(A_2)\overline{\psi}_2(D[Y(t_1), \Theta_0]) = \lambda_M(A_2)\overline{\psi}_2(\delta) < \lambda_m(A_1)\overline{\psi}_1(\varepsilon).
\end{aligned}
$$

The obtained contradiction proves the Theorem 10.9.

Theorem 10.10 *Assume that for the equation (10.1) all the conditions of the Assumptions $H_1 - H_3$, the conditions (1), (2) of the Theorem 10.9, and the condition (3′) are satisfied, and there exists a comparison function $\psi_3 \in K$-class such that*

$$D^+V(t, Y, \theta)\big|_{(10.1)} \leq -\psi_3(D[Y(t), \Theta_0])$$

for all $(t, Y) \in \mathbb{R}_+ \times S(\rho_0) \cap S^c(\eta)$ and $\alpha \in S$.
   Then the stationary solution Θ_0 of the equations (10.1) is uniformly asymptotically stable.

Proof Under the conditions of the Theorem 10.10 all the conditions of the Theorem 10.9 are satisfied and the stationary solution Θ_0 of the equation (10.1) is uniformly stable. Consider that $\varepsilon = \rho_0$, and choose $\delta_0 = \delta(\rho_0) > 0$. Here the condition $D[W_0, 0] < \delta_0$ implies the estimate $D[Y(t), \Theta_0] < \rho_0$ for all $t \geq t_0$ uniformly with respect to $t_0 \in \mathbb{R}_+$.
   Show that there exists $t_0 \leq t^* \leq t_0 + \tau$, where

$$\tau = 1 + \frac{\lambda_M(A_2)\overline{\psi}_2(\delta_0)}{\psi_3(\delta)},$$

such that $D[Y(t^*), \Theta_0] < \delta$. Assume that this is not so, and $D[Y(t^*), \Theta_0] \geq \delta$ at $t_0 \leq t^* \leq t_0 + \tau$. From the condition $(3')$ of the Theorem 10.10 obtain the estimate

$$V(t, Y(t), \theta) \leq V(t_0, W_0, \theta) - \int_{t_0}^{t} \psi_3(D[Y(s), \Theta_0])\, ds$$

for all $t_0 \leq t \leq t_0 + \tau$. Taking into account the chosen τ, find

$$0 \leq V(t_0 + \tau,\ Y(t_0 + \tau),\ \theta) \leq \lambda_M(A_2)\overline{\psi}_2(\delta_0) - \psi_3(\delta)\tau < 0.$$

This inequality contradicts the estimate of the function $V(t, Y, \theta)$ from the condition (2) of the Theorem 10.9. Therefore there exists t^* such that the condition $D[W_0, 0] < \delta_0$ implies the estimate $D[Y(t), \Theta_0] < \delta$ for all $t \geq t^* + \tau$.

The Theorem 10.10 is proved.

Now apply the function (10.15) and the comparison Theorem 10.6, and show the general pattern of obtaining the conditions for the stability of the stationary solution Θ_0 of the equation (10.1). An analogue of this approach in the theory of ordinary differential equation is the technique of proving theorems on the stability of the zero solution of equations of perturbed motion on the basis of the vector Lyapunov function.

Theorem 10.11 *Assume that for the set equation (10.1) all the conditions of the Assumptions H_1–H_3 are satisfied, and, in addition,*

(1) *there exist a function (10.5), a vector $\theta \in \mathbb{R}_+^2$ and a vector function $G(t, \omega)$, $G \in C(\mathbb{R}_+ \times \mathbb{R}_+^2, \mathbb{R}^2)$, $G(t, 0) = 0$, nondecreasing with respect to ω for all $t \in \mathbb{R}_+$, for which the conditions (1)–(3) of the Theorem 10.6 are satisfied for all $(t, Y) \in \mathbb{R}_+ \times S_0(\rho_0)$;*

(2) *there exist vector comparison functions ψ_1, ψ_2 belonging to the K-class and constant symmetric positive-definite (2×2)-matrices A_1 and A_2 such that for the function*

$$V_0(t, Y, \theta) = \sum_{i=1}^{2} L_i(t, Y, \theta)$$

the two-sided inequality

$$\psi_1^{\mathrm{T}}(\widetilde{D}[Y, \Theta_0]) A_1 \psi_1(\widetilde{D}[Y, \Theta_0]) \leq V_0(t, Y, \theta)$$
$$\leq \psi_2^{\mathrm{T}}(\widetilde{D}[Y, \Theta_0]) A_2 \psi_2(\widetilde{D}[Y, \Theta_0])$$

holds for all $(t, Y) \in \mathbb{R}_+ \times S_0(\rho_0)$;

(3) *the zero solution of the comparison system*

$$\frac{d\omega}{dt} = G(t,\omega), \quad \omega(t_0) = \omega_0 \geq 0$$

has a certain type of stability.

Then the stationary solution Θ_0 of the equations (10.1) has the same type of stability.

The proof of the Theorem 10.10 is similar to that of the theorem 9.10 and therefore is not given here.

10.6 The Application of the Strengthened Lyapunov Function

The condition (2) of the Theorem 10.9 may prove to be hard to satisfy, and then the lower requirements of the function (10.12) can be compensated for by the introduction of some strengthening function defined in the domain $\mathbb{R}_+ \times S(\rho_0) \cap S^c(\eta)$ at $0 < \eta < \rho_0$, where $S^c(\eta)$ denotes a complement of the set $S(\eta)$.

This idea, in combination with the comparison principle, is realized in the theorem below.

Theorem 10.12 *Assume that for the equation (10.1) the conditions of the Assumptions $H_1 - H_3$ are satisfied, and, in addition,*

(1) *there exists a function (10.12) such that $V \in C(\mathbb{R}_+ \times S(\rho_0), \mathbb{R}_+)$, $|V(t,Y_1,\theta)-V(t,Y_2,\theta)| \leq LD[Y_1,Y_2]$, where $L > 0$, and $0 \leq V(t,Y,\theta) \leq a(t, D[W_0,0])$, where $a \in C(\mathbb{R}_+ \times [0,\rho_0), \mathbb{R}_+))$, $a(t,\cdot)$ belongs to the K-class for any value of $t \in \mathbb{R}_+$;*

(2) *for all $(t,Y) \in \mathbb{R}_+ \times S(\rho_0)$ and $\alpha \in S$ the following inequality is true*

$$D^+V(t,Y,\theta)\big|_{(10.1)} \leq g_1(t, V(t,Y,\theta)),$$

where $g_1 \in C(\mathbb{R}_+^2, \mathbb{R})$ and $g_1(t,0) = 0$ for all $t \in \mathbb{R}_+$;

(3) *for any $0 < \eta < \rho_0$ there exists a function $V_\eta \in C(\mathbb{R}_+ \times S(\rho_0) \cap S^c(\eta), \mathbb{R})$ such that*

$$|V_\eta(t,Y_1) - V_\eta(t,Y_2)| \leq L_\eta D[Y_1,Y_2],$$

where $L_\eta > 0$ is constant, and for the function $V_0(t,Y,\theta) = V(t,Y,\theta) + V_\eta(t,Y)$ the two-sided inequality is true:

$$\psi_1^{\mathrm{T}}(D[Y,\Theta_0])A_1\psi_1(D[Y,\Theta_0]) \leq V_0(t,Y,\theta)$$
$$\leq \psi_2^{\mathrm{T}}(D[W_0,0])A_2\psi_2(D[W_0,0]), \tag{10.20}$$

where ψ_1, ψ_2 are vector comparison functions belonging to the K-class, A_1 and A_2 are constant symmetric positive-definite matrices;

(4) for all $(t, Y) \in \mathbb{R}_+ \times S(\rho_0) \cap S^c(\eta)$ and $\alpha \in S$ the following inequality is true

$$D^+ V_0(t, Y, \theta)\big|_{(10.1)} \leq g_2(t, V_0(t, Y, \theta)),$$

where $g_2 \in C(\mathbb{R}^2_+, \mathbb{R})$, $g_2(t, 0) = 0$ for all $t \in \mathbb{R}_+$;

(5) the zero solution $\omega_1 = 0$ of the comparison equation

$$\frac{d\omega_1}{dt} = g_1(t, \omega_1), \quad \omega_1(t_0) = \omega_{10} \geq 0 \qquad (10.21)$$

is stable;

(6) the zero solution $\omega_2 = 0$ of the comparison equation

$$\frac{d\omega_2}{dt} = g_2(t, \omega_2), \quad \omega_2(t_0) = \omega_{20} \geq 0 \qquad (10.22)$$

is stable uniformly with respect to $t_0 \in \mathbb{R}_+$.

Then the stationary solution Θ_0 of the equation (10.1) is stable.

Proof Reduce the estimates (10.20) from the condition (3) of the Theorem 10.12 to the form

$$\lambda_m(A_1)\overline{\psi}_1(D[Y, \Theta_0]) \leq V_0(t, Y, \theta) \leq \lambda_M(A_2)\overline{\psi}_2(D[W_0, 0]) \qquad (10.23)$$

where $\lambda_m(A_1) > 0$ and $\lambda_M(A_2) > 0$ are the minimum and the maximum eigenvalues of the matrices A_1 and A_2 respectively, and the functions ψ_1, ψ_2 belong to the K-class.

Let $0 < \varepsilon < \rho_0$ and $t_0 \in \mathbb{R}_+$ be given. From the uniform stability of the zero solution $\omega_2 = 0$ of the equation (10.22) it follows that for the given $\lambda_m(A_1)\overline{\psi}_1(\varepsilon) > 0$ there exists $\delta_0 = \delta_0(\varepsilon)$ such that the condition

$$0 < \omega_{20} < \delta_0$$

implies

$$\omega_2(t; t_0, \omega_{20}) < \lambda_m(A_1)\overline{\psi}_1(\varepsilon), \quad t \geq t_0$$

where $\omega_2(t; t_0, \omega_{20})$ is any solution of the equation (10.22). At the given $\varepsilon > 0$, for the function $\overline{\psi}_2 \in K$-class from the condition (10.23) one can indicate $\delta_2 = \delta_2(\varepsilon) > 0$ such that

$$\lambda_M(A_2)\overline{\psi}_2(\delta_2) < \frac{1}{2}\delta_0. \qquad (10.24)$$

Since according to the condition (5) of the Theorem 10.12, the zero solution

$\omega_1 = 0$ of the comparison equation (10.21) is stable, for the given $\frac{1}{2}\delta_0 > 0$ and $t_0 \in \mathbb{R}_+$ one can find $\delta^* = \delta^*\left(t_0, \frac{1}{2}\delta_0\right) > 0$ such that the condition

$$0 < \omega_{10} < \delta^* \tag{10.25}$$

will imply

$$\omega_1(t; t_0, \omega_{10}) < \frac{1}{2}\delta_0$$

for all $t \geq t_0$, where $\omega_1(t; t_0, \omega_{10})$ is any solution of the equation (10.21).

Let $\omega_{10} = V(t_0, W_0, \theta)$. According to the condition (1) of the Theorem 10.12, for the function $a(t, \cdot)$ one can choose $\delta_1 = \delta_1(t_0, \varepsilon) > 0$ so that the inequalities

$$D[W_0, 0] < \delta_1 \quad \text{and} \quad a(t_0, D[W_0, 0]) < \delta^* \tag{10.26}$$

will hold simultaneously.

Now choose $\delta = \min(\delta_1, \delta_2)$ and show that the condition

$$D[W_0, 0] < \delta$$

implies

$$D[Y(t), \Theta_0] < \varepsilon$$

for all $t \geq t_0$ for any solution $Y(t)$ of the equation (10.1). If this is not so, then there should exist a solution $Y(t)$ of the equation (10.1) and points of time $t_1, t_2 > t_0$ such that

$$D[Y(t_1), \Theta_0] = \delta_2, \quad D[Y(t_2), \Theta_0] = \varepsilon, \quad \delta_2 \leq D[Y(t), \Theta_0] \leq \varepsilon < \rho_0$$

for $t_1 \leq t \leq t_2$. Let $\eta = \delta_2$. Since $0 < \delta_2 < \rho_0$, the conditions (3) of the Theorem 10.12 are satisfied for the function $V_\eta(t, Y)$. For the function $m(t) = V(t, Y(t), \theta) + V_\eta(t, Y(t))$, $t \in [t_1, t_2]$, the condition (4) of the Theorem 10.12 results in the inequality

$$D^+ m(t) \leq g_2(t, m(t)), \quad t_1 \leq t \leq t_2,$$

whence follows the estimate

$$V(t_2, Y(t_2), \theta) + V_\eta(t_2, Y(t_2)) \leq r_2(t_2; t_1, \omega_{20}), \tag{10.27}$$

where $\omega_{20} = V(t_1, Y(t_1), \theta) + V_\eta(t_1, Y(t_1))$ and $r_2(t_2; t_1, \omega_{20})$ is the maximum solution of the comparison equation (10.22). Note that for the function $V(t, Y(t), \theta)$ the following estimate is true

$$V(t_1, Y(t_1), \theta) \leq r_1(t_1; t_0, \omega_{10}), \tag{10.28}$$

where $\omega_{10} = V(t_0, W_0, \theta)$ and $r_1(t_1; t_0, \omega_{10})$ is the maximum solution of the comparison equation (10.21). Taking into account the inequalities (10.26) and (10.25), from the estimate (10.28) we obtain

$$V(t_1, Y(t_1), \theta) < \frac{1}{2} \delta_0. \tag{10.29}$$

From the conditions (10.24) and (10.27) it follows that

$$V_\eta(t_1, Y(t_1)) \le \lambda_M(A_2)\overline{\psi}_2(\delta_2) < \frac{1}{2} \delta_0. \tag{10.30}$$

Taking into account (10.29) and (10.30), $\omega_{20} < \delta_0$, and therefore $\omega_2(t_2; t_1, \omega_{20}) < \lambda_m(A_1)\overline{\psi}_1(\varepsilon)$. According to the condition (1) of the Theorem 10.12, the function $V(t, Y(t), \theta) \ge 0$ for all $t \in \mathbb{R}_+$; therefore from the estimate (10.27) we obtain

$$\lambda_m(A_1)\overline{\psi}_1(\varepsilon) = \lambda_m(A_1)\overline{\psi}_1(D[Y(t_2), \Theta_0]) \le V_0(t_2, Y(t_2), \theta)$$
$$\le r_2(t_2; t_1, \omega_{20}) < \lambda_m(A_1)\overline{\psi}_1(\varepsilon).$$

The obtained contradiction proves that the stationary solution Θ_0 of the equation (10.6) is stable.

The Theorem 10.12 is proved.

An insignificant modification of the conditions of the Theorem 10.12 makes it possible to obtain sufficient conditions for the asymptotic stability of the stationary solution Θ_0 of the equation (10.1) in the following form.

Theorem 10.13 *Assume that for the set equation (10.1) the conditions of the Assumptions H_1–H_3 and the conditions (1)–(6) of the Theorem 10.12 are satisfied, and, in addition,*

(2′) *there exist a function $c \in K$-class and $\omega \in C(\mathbb{R}_+ \times S(\rho_0), \mathbb{R}_+)$ such that*

$$|\omega(t, Y_1) - \omega(t, Y_2)| \le ND[Y_1, Y_2],$$

where $N > 0$, the value $D^+\omega(t, Y)|_{(10.1)}$ is bounded above or below, and for all $(t, Y) \in \mathbb{R}_+ \times S(\rho_0)$ and $\alpha \in S$ the following estimate holds:

$$D^+V(t, Y, \theta)|_{(10.1)} \le -c(\omega(t, Y)) + g_1(t, V(t, Y, \theta)).$$

Then the stationary solution Θ_0 of the equation (10.1) is asymptotically stable, if the function $g_1(t, \omega)$ is nondecreasing with respect to ω and

$$\omega(t, Y) \ge b_0(D[Y, \Theta_0]),$$

where b_0 belongs to the K-class.

Proof With the condition (2) in the Theorem 10.12 replaced by the condition (2$'$) from the Theorem 10.13 with the given complements, the stationary solution Θ_0 of the equation (10.1) is stable. Let $\varepsilon = \rho_0$ and $\delta_0 = \delta_0(t_0, \rho_0)$. From the condition of stability of the solution Θ_0 obtain that the inequality $D[W_0, 0] < \delta_0$ implies the estimate $D[Y(t), \Theta_0] < \rho_0$ for all $t \geq t_0$.

Show that for any solution $Y(t)$ of the equation (10.1) with the initial conditions $D[W_0, 0] < \delta_0$ it follows that $\lim\limits_{t \to \infty} w(t, Y(t)) = 0$ and therefore $\lim D[Y(t), \Theta_0] = 0$ at $t \to \infty$.

Assume that $\lim\limits_{t \to \infty} \sup w(t, Y(t)) \neq 0$. Then there should exist divergent sequences $\{t_i'\}$, $\{t_i''\}$ and some quantity $\sigma > 0$ such that for all $t \in (t_i', t_i'')$

(a) $w(t_i', Y(t_i')) = \dfrac{\sigma}{2}, \quad w(t_i'', Y(t_i'')) = \sigma$ and $w(t, Y(t)) \geq \dfrac{\sigma}{2}$

or

(b) $w(t_i', Y(t_i')) = \sigma, \quad w(t_i'', Y(t_i'')) = \dfrac{\sigma}{2}$ and $w(t, Y(t)) \geq \dfrac{\sigma}{2}$.

Assume that $D^+ w(t, Y(t))\big|_{(10.1)} \leq M$, where $M > 0$ is some constant.

From the condition (a) it follows that

$$\frac{\sigma}{2} = \sigma - \frac{\sigma}{2} = w(t_i'', Y(t_i'')) - w(t_i', Y(t_i')) \leq M(t_i'' - t_i').$$

Hence, for any i, the estimate $t_i'' - t_i' \geq \dfrac{\sigma}{2M}$ is true. From the condition (2$'$) of the Theorem 10.13 it follows that

$$V(t, Y(t), \theta) \leq r_1(t; t_0, w_{10}) - \sum_{i=1}^{n} \int_{t_i'}^{t_i''} c\,(w(s, Y(s)))\, ds.$$

Since $w_{10} = V(t_0, W_0, \theta) \leq a(t_0, D[W_0, 0]) < a(t_0, \delta_0) < \delta^*(\rho_0)$ for all $t \geq t_0$, then, according to the inequality (10.25), we obtain the estimate $w_1(t; t_0, w_{10}) < \dfrac{1}{2}\delta_0(\rho_0)$ for all $t \geq t_0$ and

$$0 \leq V(t, Y(t), \theta) \leq \frac{1}{2}\delta_0(\rho_0) - c\left(\frac{\sigma}{2}\right)\frac{\sigma}{2M}\,n. \tag{10.31}$$

At $n \to \infty$ the inequality (10.31) is a contradiction. This proves that $\lim\limits_{t \to \infty} \sup w(t, Y(t)) = 0$ and therefore the stationary solution Θ_0 of the equation (10.1) is asymptotically stable.

10.7   Boundedness Theorems

Assume that the equation

$$D_H Y = F(t, Y), \quad Y(t_0) = Y_0 \in \mathcal{K}_c(\mathbb{R}^n), \tag{10.32}$$

where $Y \in \mathcal{K}_c(\mathbb{R}^n)$ and $F \in C(\mathbb{R}_+ \times \mathcal{K}_c(\mathbb{R}^n), \mathcal{K}_c(\mathbb{R}^n))$, has a solution $Y(t) = Y(t, t_0, W_0)$ defined for all $t \geq t_0$, where $W_0 = Y_0 - V_0$ for any $Y_0, V_0 \in \mathcal{K}_c(\mathbb{R}^n)$.

Definition 10.2 The solution $Y(t)$ of the equation (10.32) is:

(B_1) *equibounded*, if for any $\alpha > 0$ and $t_0 \in \mathbb{R}_+$ there exists $\beta(t_0, \alpha) > 0$ such that from the condition $\|W_0\| < \alpha$ it follows that $\|Y(t)\| < \beta(t_0, \alpha)$ for all $t \geq t_0$;

(B_2) *uniformly bounded*, if β in the definition B_1 does not depend on t_0;

(B_3) *quasi-ultimately bounded* with a boundary B, if for any $\alpha > 0$ and $t_0 \in \mathbb{R}_+$ there exist $B > 0$ and $\tau = \tau(t_0, \alpha) > 0$ such that from the condition $\|W_0\| < \alpha$ it follows that $\|Y(t)\| < B$ for all $t \geq t_0 + \tau$;

(B_4) *quasi-uniformly ultimately bounded*, if τ in the definition of B_3 does not depend on t_0;

(B_5) *equiultimately bounded*, if the conditions in the definitions B_1 and B_3 are satisfied simultaneously;

(B_6) *uniformly ultimately bounded*, if the conditions of the definitions B_2 and B_4 are satisfied simultaneously.

Now apply the function (10.12) and obtain the conditions for the boundedness of solution of the set equation (10.32).

Theorem 10.14 *Assume that for the equation (10.32) the following conditions are satisfied:*

(1) *there exist a function (10.11), $S \in C(\mathbb{R}_+ \times \mathcal{K}_c(\mathbb{R}^n), \mathbb{R}^{2 \times 2})$, a vector $\theta \in \mathbb{R}_+^2$ and a constant $L > 0$ such that $V(t, Y, \theta) \in C(\mathbb{R}_+ \times \mathcal{K}_c(\mathbb{R}^n) \times \mathbb{R}_+^2, \mathbb{R}_+)$ and $|V(t, Y_1, \theta) - V(t, Y_2, \theta)| \leq LD[Y_1, Y_2]$ for all $(t, Y) \in \mathbb{R}_+ \times \mathcal{K}_c(\mathbb{R}^n)$;*

(2) *there exist vector comparison functions $\psi_1, \psi_2 \in KR$-class and constant symmetric positive-definite (2×2)-matrices A_1 and A_2 such that*

$$\psi_1^{\mathrm{T}}(\|Y\|)A_1\psi_1(\|Y\|) \leq V(t, Y, \theta) \leq \psi_2^{\mathrm{T}}(t, \|Y\|)A_2\psi_2(t, \|Y\|)$$

for all $(t, Y) \in \mathbb{R}_+ \times \mathcal{K}_c(\mathbb{R}^n)$;

(3) *for all $(t, Y) \in \mathbb{R}_+ \times \mathcal{K}_c(\mathbb{R}^n)$ and $\alpha \in \mathcal{S}$ the following inequality holds:*

$$D^+V(t, Y, \theta)\big|_{(10.32)} \leq 0.$$

Then the solution $Y(t)$ of the equation (10.32) is equibounded.

Proof Transform the condition (2) of the Theorem 10.4 to the following form:

$$\lambda_m(A_1)\overline{\psi}_1(\|Y\|) \leq V(t,Y,\theta) \leq \lambda_M(A_2)\overline{\psi}_2(t,\|Y\|), \tag{10.33}$$

where $\overline{\psi}_1$ and $\overline{\psi}_2$ belong to the KR-class and are such that

$$\overline{\psi}_1(\|Y\|) \leq \psi_1^T(\|Y\|)\psi_1(\|Y\|) \quad \text{and} \quad \overline{\psi}_2(t,\|Y\|) \geq \psi_2^T(t,\|Y\|)\psi_2(t,\|Y\|).$$

Now choose $\beta = (t_0,\gamma) > 0$ for the given $\gamma > 0$ and $t_0 \in \mathbb{R}_+$ so that

$$\lambda_M(A_2)\overline{\psi}_2(t_0,\gamma) < \lambda_m(A_1)\overline{\psi}_1(\beta). \tag{10.34}$$

Show that with the chosen β the solution $Y(t)$ is equibounded. If this is not so, then one can find a solution $Y(t) = Y(t,t_0,W_0)$ and a value $t_1 > t_0$ such that $\|Y(t_1)\| = \beta$ and $\|Y(t)\| \leq \beta$ at $t_0 \leq t \leq t_1$.

From the condition (3) of the Theorem 10.14 it follows that

$$V(t,Y(t),\theta) \leq V(t_0,W_0,\theta), \quad t_0 \leq t \leq t_1. \tag{10.35}$$

According to the condition (10.33), from the inequalities (10.34), (10.35) we find the estimates

$$\begin{aligned} \lambda_m(A_1)\overline{\psi}_1(\beta) = \lambda_m(A_1)\overline{\psi}_1(\|Y(t)\|) \leq V(t,Y(t),\theta) \leq V(t_0,W_0,\theta)\\ \leq \lambda_M(A_2)\overline{\psi}_2(t_0,\|W_0\|) < \lambda_M(A_2)\overline{\psi}_2(t_0,\gamma) < \lambda_m(A_1)\overline{\psi}_1(\beta). \end{aligned} \tag{10.36}$$

The obtained contradiction proves the statement of the Theorem 10.14.

For the set $S(\rho) = \{Y \in \mathcal{K}_c(\mathbb{R}^n): D[Y,\Theta] < \rho\}$, where Θ is the zero element of the set $\mathcal{K}_c(\mathbb{R}^n)$, consider the complement $S^c(\rho)$ and assume that ρ can take on arbitrarily large values. Now we will prove the theorem on the uniform boundedness of the set of solutions $Y(t)$ of the equation (10.32).

Theorem 10.15 *Assume that for the equation* (10.32) *the following conditions are satisfied:*

(1) *there exist a function* (10.11), *$S \in C(\mathbb{R}_+ \times S^c(\rho), \mathbb{R}^{2\times 2})$, a vector $\theta \in \mathbb{R}_+^2$, and a constant $L > 0$ such that $V(t,Y,\theta) \in C(\mathbb{R}_+ \times S^c(\rho) \times \mathbb{R}_+^2, \mathbb{R}_+)$ and $|V(t,Y_1,\theta) - V(t,Y_2,\theta)| \leq LD[Y_1,Y_2]$ for all $(t,Y) \in \mathbb{R}_+ \times S^c(\rho)$;*

(2) *there exist vector comparison functions ψ_1, ψ_3 belonging to the KR-class, and constant symmetric positive-definite (2×2)-matrices A_1 and A_2 such that*

$$\psi_1^T(\|Y\|)A_1\psi_1(\|Y\|) \leq V(t,Y,\theta) \leq \psi_3^T(\|Y\|)A_2\psi_3(\|Y\|)$$

for all $(t,Y) \in \mathbb{R}_+ \times S^c(\rho)$;

(3) *for all $(t,Y) \in \mathbb{R}_+ \times S^c(\rho)$ the following inequality holds:*

$$D^+V(t,Y,\theta)\big|_{(10.32)} \leq 0.$$

Then the solution $Y(t)$ of the equation (10.32) is uniformly bounded.

Proof As in the proof of the Theorem 10.13, we transform the condition (2) of the Theorem 10.15 to the form (10.33), and choose $\beta = \beta(\gamma)$ from the condition

$$\lambda_M(A_2)\overline{\psi}_3(\gamma) < \lambda_m(A_1)\overline{\psi}_1(\beta).$$

If $\gamma > \rho$, then in the estimates of the form (10.35), (10.36) the set $S^c(\rho)$ is considered. If $0 < \gamma \leq \rho$, then $\beta = \beta(\rho)$ and the same reasoning is applied as in the proof of the Theorem 10.14.

Theorem 10.16 *Assume that for the equation (10.32) the condition (1) of the Theorem 10.14 is satisfied and*

(2′) there exists a constant $\eta > 0$ such that

$$D^+V(t,Y,\theta)\big|_{(10.32)} \leq -\eta V(t,Y,\theta)$$

for all $(t,Y) \in \mathbb{R}_+ \times \mathcal{K}_c(\mathbb{R}^n)$ and $\alpha \in \mathcal{S}$.

Then if the condition (3) of the Theorem 10.14 is satisfied at $\|Y\| \geq B$, where $0 < B < +\infty$, then the solution $Y(t)$ is equiultimately bounded.

Proof From the conditions of the Theorem 10.14 it follows that the solution $Y(t)$ is equibounded. Therefore from the condition $\|W_0\| < \alpha$ it follows that $\|Y(t)\| < \beta$ for all $t \geq t_0$. The condition (2′) of the Theorem 10.16 implies the estimate

$$V(t,Y(t),\theta) \leq V(t_0,W_0,\theta)\exp\left[-\eta(t-t_0)\right] \tag{10.37}$$

for all $t \geq t_0$. Let

$$\tau = \frac{1}{\eta}\ln\frac{\lambda_M(A_2)\overline{\psi}_2(t_0,\gamma)}{\lambda_m(A_1)\overline{\psi}_1(B)}$$

and for $t \geq t_0 + \tau$ the condition $\|Y(t)\| \geq B$ is satisfied. From the inequality (10.37) we obtain

$$\begin{aligned}
\lambda_m(A_1)\overline{\psi}_1(B) &\leq \lambda_m(A_1)\overline{\psi}_1(\|Y(t)\|) \leq V(t,Y(t),\theta) \\
&< \lambda_M(A_2)\overline{\psi}_2(t_0,\gamma)\exp\left[-\eta\tau\right] = \lambda_m(A_1)\overline{\psi}_1(B).
\end{aligned} \tag{10.38}$$

The contradiction (10.38) proves the Theorem 10.16.

Note that the conditions for the non-uniform boundedness of solutions of the equation (10.32) can be obtained on the basis of the strengthened Lyapunov function in the same way as analyzing the stability of the set of stationary solutions of that equation.

It is clear that the constructive application of the described conditions for the stability and boundedness of the set solutions is connected with the successive solution of the problem of construction of an appropriate Lyapunov function.

Chapter 11

Set Differential Equations with a Robust Causal Operator

A general theory of equations with a causal operator has been developed by Corduneanu [2002]. A short description of functional equation with a causal operator is as follows.

Let $E = E([0,T], \mathbb{R}^n)$ be a functional space with a norm. Let the operator $V \colon E \to E$. The V is said to be a causal operator if the following conditions are satisfied: for each pair of elements from the space E such that for $x(s) = y(s)$ when $0 \leq s \leq t$ the correlation $(Vx)(s) = (Vy)(s)$ holds true when $0 \leq s \leq t$ for arbitrary $t < T$. An example causal operator is

$$(Vx)(t) = \int_0^t K(t,s,x(s))\, ds, \quad t \in [0,T),$$

where $K(t,s,x)$ is the function with values in $\mathbb{R}^n$ determined for $0 \leq s \leq t < T$ and $x \in \mathbb{R}^n$ or

$$(Vx)(t) = f(t) + \int_0^t K(t,s,x(s))\, ds,$$

where $f(t)$ is a continuous function on $[0,T)$.

In this chapter we use the comparison principle for the set of differential equations with a robust causal operator to prove several stability theorems.

11.1 Preliminary Results

We shall consider a space of nonempty subsets of $\mathbb{R}^n$: $\mathcal{K}_c(\mathbb{R}^n)$ consisting of all nonempty compact convex subsets of $\mathbb{R}^n$. Let given constants $\beta, \gamma \in \mathbb{R}$ and $A, B \in \mathcal{K}_c(\mathbb{R}^n)$, then

$$\beta(A+B) = \beta A + \beta B, \quad \beta(\gamma A) = (\beta\gamma)A, \quad 1A = A,$$

and, if $\beta, \gamma \geq 0$, then $(\beta + \gamma)A = \beta A + \gamma A$.

Let the space $E = C([t_0, a), \mathcal{K}_c(\mathbb{R}^n))$ be equipped with the norm

$$D_0[Y, \theta] = \sup_{t_0 \leq t < a} D[Y(t), \theta],$$

where $\theta \in \mathcal{K}_c(\mathbb{R}^n)$ and

$$D[Y, \theta] = \max[\sup_{x \in \theta} d(x, Y), \sup_{y \in Y} d(y, \theta)]$$

is the Hausdorff metric.

Assume that $Q \colon E \times \mathcal{S} \to E$ is a given mapping, where $\mathcal{S}$ is a compact subset of the space $\mathbb{R}^d$ of uncertain parameters, $d \geq 1$.

Taking into account some of the results of Corduneanu [2002] and Lakshmikantham, Bhaskar and Devi [2006], we give the following definitions.

Definition 11.1 $Q \colon E \times \mathcal{S} \to E$ is a robust causal mapping if the conditions $Y(s) = V(s)$ for $t_0 \leq s < t < a$ and $Y, V \in E$ imply that $(QY)(s) = (QV)(s)$ for all $t_0 \leq s \leq t$ and all $\alpha \in \mathcal{S}$.

The initial value problem for the set of differential equations with a robust causal operator is defined as

$$D_H Y(t) = (Q(\alpha)Y)(t), \quad Y(t_0) = Y_0 \in \mathcal{K}_c(\mathbb{R}^n), \qquad (11.1)$$

where $(Q(\alpha)Y)(t) \colon \mathbb{R}_+ \times E \times \mathcal{S} \to \mathcal{K}_c(\mathbb{R}^n)$.

Along with problem (11.1), we consider the sets of equations

$$D_H W(t) = (Q_m W)(t), \quad W(t_0) = W_0 \in \mathcal{K}_c(\mathbb{R}^n), \qquad (11.2)$$

where

$$(Q_m W)(t) = \overline{co} \bigcap_{\alpha \in \mathcal{S}} (Q(\alpha)Y)(t)$$

and

$$D_H Z(t) = (Q_M Z)(t), \quad Z(t_0) = Z_0 \in \mathcal{K}_c(\mathbb{R}^n), \qquad (11.3)$$

where

$$(Q_M Z)(t) = \overline{co} \bigcup_{\alpha \in \mathcal{S}} (Q(\alpha)Y)(t).$$

We define

$$(Q_\beta X)(t) = \beta (Q_M X)(t) + (1 - \beta)(Q_m X)(t), \quad 0 \leq \beta \leq 1,$$

and consider the set equations with a causal operator

$$D_H X(t) = (Q_\beta X(t)), \quad X(t_0) = X_0 \in K(\mathbb{R}^n). \qquad (11.4)$$

11.2   Comparison Principle

Define the matrix-valued function

$$S(t, X, \beta) = [S_{ij}(t, X)], \quad i, j = 1, 2, \qquad (11.5)$$

whose elements $S_{ij}(\cdot)$ are related to equations (11.2)–(11.4). Specifically, the element $S_{11}(t, X) \in C(\mathbb{R}_+ \times \mathcal{K}_c(\mathbb{R}^n), \mathbb{R}_+)$ is associated with equation (11.2); the element $S_{22}(t, X) \in C(\mathbb{R}_+ \times \mathcal{K}_c(\mathbb{R}^n), \mathbb{R}_+)$, with equation (11.3); and the element $S_{12}(t, X, \beta) \in C(\mathbb{R}_+ \times \mathcal{K}_c(\mathbb{R}^n) \times [0, 1], \mathbb{R})$, with equation (11.4).

Given a vector $\psi \in \mathbb{R}_+^2$, $\psi > 0$, we define the function $L(t, X, \beta) = \psi^T S(t, X, \beta)\psi$ and assume that $L \in C(\mathbb{R}_+ \times \mathcal{K}_c(\mathbb{R}^n) \times [0, 1], \mathbb{R}_+)$ and $L(t, X, \beta)$ is a locally Lipschitz function in X; i.e.,

$$|L(t, X_1, \beta) - L(t, X_2, \beta)| \leq KD[X_1, X_2],$$

for all $X_1, X_2 \in \mathcal{K}_c(\mathbb{R}^n)$ and $t \in \mathbb{R}_+$.

For $L(t, X, \beta)$, define the expression

$$D_- L(t, X, \beta) = \lim_{\theta \to 0^-} \inf[L(t + \theta, X(t) + \theta((Q(\alpha)X)(t)), \beta) - L(t, X(t), \beta)]\theta^{-1}$$

for any $\beta \in [0, 1]$ and any $\alpha \in \mathcal{S}$.

Theorem 11.1 *Let equation* (11.1) *be such that the following conditions hold:*

(i) *there is a matrix-valued function $S(t, X, \beta)$ and a vector $\psi \in \mathbb{R}_+^2$ such that $L(t, X, \beta)$ is a locally Lipschitz function in X;*

(ii) *there are functions g_0 and $g_\beta \in (\mathbb{R}_+ \times \mathbb{R}_+ \times [0, 1], \mathbb{R})$ satisfying $g_0(t, w) \leq g_\beta(t, w)$ for all $t \in \mathbb{R}_+$ and $\alpha \in \mathcal{S}$;*

(iii) *the equation*

$$\frac{dv}{dt} = g_0(t, v), \quad v(\tau_0) = v_0 \geq 0, \tag{11.6}$$

has a left maximal solution $\eta(t, \tau_0, v_0)$ on the interval $[t_0, \tau_0]$, and the equation

$$\frac{dw}{dt} = g_\beta(t, w), \quad w(t_0) = w_0 \geq 0, \tag{11.7}$$

has a right maximal solution $r_\beta(t, t_0, w_0)$ on the interval $[t_0, \infty)$ for any $\beta \in [0, 1]$;

(iv) *on the set $W = \{X \in E: L(t, X(s), \beta) \leq \eta(s, t, L(t, X(t), \beta)), t_0 \leq s \leq t\}$, it is true that*

$$D_- L(t, X(t), \beta) \leq g_\beta(t, L(t, X(t), \beta)),$$

for any $\beta \in [0, 1]$.

Then, for all $t \geq t_0$, we have the estimate

$$L(t, X(t), \beta) \leq r_\beta(t; t_0, w_0)$$

whenever $L(t_0, X_0, \beta) \leq w_0$.

Proof First, we show that, under conditions (ii) and (iii) in Theorem 11.1, the left maximal solution $\eta(t, \tau_0, v_0)$ of equation (11.6) and the right maximal solution $v_\beta(t; t_0, w_0)$ of equation (11.7) satisfy

$$v_\beta(t; t_0, w_0) \leq \eta(t, \tau_0, v_0) \tag{11.8}$$

for all $t \in [t_0, \tau_0]$ and any $\beta \in [0, 1]$ whenever

$$v_\beta(\tau_0, t_0, w_0) \leq v_0. \tag{11.9}$$

For sufficiently small $\varepsilon > 0$, we have

$$\lim_{\varepsilon \to 0} w_\beta(t, \varepsilon) = r_\beta(t, t_0, u_0) \qquad \lim_{\varepsilon \to 0} v(t, \varepsilon) = \eta(t, \tau_0, v_0),$$

where $w_\beta(t, \varepsilon)$ is any solution to the equation

$$\frac{dw}{dt} = g_\beta(t, w) + \varepsilon, \quad w(t_0) = w_0 + \varepsilon \geq 0, \tag{11.10}$$

that exists to the right of t_0 for any $\beta \in [0, 1]$, and $v(t, \varepsilon)$ is any solution to the equation

$$\frac{dv}{dt} = g_0(t, v) - \varepsilon, \quad v(\tau_0) = v_0 \geq 0, \tag{11.11}$$

that exists to the left of τ_0. Inequality (11.8) will be proved if we show that $w_\beta(t, \varepsilon) < v(t, \varepsilon)$ for all $t_0 \leq t < \tau_0$ and any $\beta \in [0, 1]$.

Condition (ii) in Theorem 11.1 and estimate (11.9) imply that, for sufficiently small $\delta > 0$, it holds that $w_\beta(t, \varepsilon) < v(t, \varepsilon)$ for $\tau_0 - \delta \leq t < \tau_0$. Then $w_\beta(\tau_0 - \delta, \varepsilon) < v(\tau_0 - \delta, \varepsilon)$ for any $\beta \in [0, 1]$.

Let us show that $w_\beta(t, \varepsilon) < v(t, \varepsilon)$ for all $t_0 \leq t < \tau_0 - \delta$ and any $\beta \in [0, 1]$. If this is not true, then there exists $t^* \in [t_0, \tau_0 - \delta)$ such that $w_\beta(t, \varepsilon) < v(t, \varepsilon)$ for $t^* < t \leq \tau_0 - \delta$ and $\beta \in [0, 1]$ and $w_\beta(t^*, \varepsilon) = v(t^*, \varepsilon)$. This relation leads to the contradiction

$$g_\beta(t^*, w_\beta(t^*, \varepsilon)) + \varepsilon = \frac{dw}{dt} \leq \frac{dv}{dt} = g_0(t^*, v(t^*, \varepsilon)) - \varepsilon,$$

for any $\beta \in [0, 1]$. Therefore, $w_\beta(t, \varepsilon) < v(t, \varepsilon)$ for $t_0 \leq \tau \leq \tau_0 - \delta$ and all $\beta \in [0, 1]$.

Let $m(t) = L(t, X(t), \beta)$ for $t \geq t_0$ and $m(t_0) = L(t_0, X_0, \beta) \leq w_0$. Denote by $w_\beta(t, \varepsilon)$ any solution to equation (11.10) with an arbitrarily small $\varepsilon > 0$. Since

$$r_\beta(t, t_0, w_0) = \lim_{\varepsilon \to 0} w_\beta(t, \varepsilon)$$

for any $\beta \in [0, 1]$, it is sufficient to show that

$$m(t) < w_\beta(t, \varepsilon) \qquad t \geq t_0$$

and all $\beta \in [0,1]$. If this is not true, then there is $t_1 > t_0$ such that $m(t_1) = w_\beta(t_1, \varepsilon)$ and $m(t) < w_\beta(t, \varepsilon)$ for $t_0 \leq t < t_1$ and all $\beta \in [0,1]$. It follows that

$$D_- m(t_1) \geq \frac{dw}{dt} = g_\beta(t_1, m(t_1)) + \varepsilon. \tag{11.12}$$

Consider the left maximal solution $\eta(s, t_1, m(t_1))$ of equation (11.11) with the initial value $v(t_1) = m(t_1)$ on the interval $t_0 \leq s \leq t_1$. According to (11.8), for $s \in [t_0, t_1]$, we have

$$r_\beta(s, t_0, w_0) \leq \eta(s, t_1, m(t_1)).$$

Since

$$r_\beta(t_1; t_0, w_0) = \lim_{\varepsilon \to 0} w_\beta(t, \varepsilon) = m(t_1) = \eta(t_1; t_1, m(t_1))$$

and $m(s) \leq w_\beta(s, \varepsilon)$ for $t_0 \leq s \leq t_1$, we conclude that

$$m(s) \leq r_\beta(s; t_0, w_0) \leq \eta(s; t_1, m(t_1))$$

for $s \in [t_0, t_1]$. This inequality implies that condition (iv) in Theorem 11.1 holds for the solution $X(s; t_0, X_0)$ on $[t_0, t_1]$. Moreover, standard computations yield the estimate

$$D_- m(t_1) \leq g_\beta(t_1, m(t_1))$$

for any $\beta \in [0,1]$, which is a contradiction to (11.12). Therefore, $m(t) < r_\beta(t, t_0, w_0)$ for $t \geq t_0$, which completes the proof of the theorem.

Corollary 11.1 *If the elements $S_{ij}(t, \cdot)$ of matrix-valued function (11.5) are such that $S_{12}(t, \cdot) = S_{21}(t, \cdot) = 0$, then $L_0(t, X, \beta) = S(t, X)\psi$, $\psi \in \mathbb{R}_+^2$, is an auxiliary vector function and Theorem 11.1 remains valid with slight modifications.*

Corollary 11.2 *If the operator $Q: E \to E$ is independent of the uncertain parameter $\alpha \in S$ and function (11.5) is replaced with $V(t, X): \mathbb{R}_+ \times \mathcal{K}_c(\mathbb{R}^n) \to \mathbb{R}_+$, then Theorem 11.1 becomes the one from Lakshmikantham, Bhaskar, and Devi [2006].*

Corollary 11.3 *If $\mathcal{K}_c(\mathbb{R}^n) = \mathbb{R}^n$, $Q: E \to E$, and function (11.5) is replaced with $V: L^2_{loc}([0, \infty), \mathbb{R}^n) \to \mathbb{R}$, then Theorem 11.1 becomes the comparison principle from Corduneanu [2002] for equations with a causal operator.*

11.3 Estimates of Funnel for Solutions

The comparison principle is used to estimate the distance between the solutions to equations (11.2) and (11.3). Recall an estimate known from Lakshmikantham, Leela, and Martynyuk [1991a].

Lemma 11.1 *Let $m \in C([t_0, T), \mathbb{R}_+)$ and $g \in C([t_0, T) \times \mathbb{R}_+, \mathbb{R})$ be given functions, for all $t \in [t_0, T)$*

$$D_- m(t) \leq g(t, |m|_0(t)),$$

where $|m_0(t)| = \sup_{0 \leq s \leq t} |m(s)|$. If the scalar equation

$$\frac{dw}{dt} = g(t, w), \quad w(t_0) = w_0 \geq 0, \tag{11.13}$$

has a maximal solution $r_M(t) = r(t; t_0, w_0)$ for all $t \in [t_0, T)$ and if $m(t_0) \leq w_0$, then $m(t) \leq r_M(t)$ for all $t \in [t_0, T)$.

The proof of this result is standard for the comparison principle.

Let $D_0[W, Z](t) = \max_{t_0 \leq s \leq t} D[W(s), Z(s)]$.

Theorem 11.2 *Given causal operators $Q_m \in C(E, E)$ and $Q_M \in C(E, E)$, suppose that there exists a function $g \in C([t_0, T), \mathbb{R})$ such that $D[(Q_m W)(t), (Q_M Z)(t)] \leq g(t, D_0[W, Z](t))$ for all $t \in [t_0, T)$. If equation (11.13) has a maximal solution $r_M(t)$ on $[t_0, T)$ and the initial conditions of the solutions $W(t)$ and $Z(t)$ are such that $W(t_0) = W_0 \in \mathcal{K}_c(\mathbb{R}_n)$ and $Z(t_0) = Z_0 \in \mathcal{K}_c(\mathbb{R}_n)$, then, for all $t \in [t_0, T)$, the inequality*

$$D[W(t), Z(t)](t) \leq r_M(t; t_0, w_0),$$

holds whenever $D[W_0, Z_0] \leq w_0$.

Proof Define $m(t) = D[W(t), Z(t)]$. According to the conditions of Theorem 11.2, we have $m(t_0) = D[W_0, Z_0] \leq w_0$, where w_0 is the initial value for scalar equation (11.13). For arbitrarily small $\theta > 0$ and any $t \in [t_0, T)$, consider the expression $m(t + \theta) = D[W(t + \theta), Z(t + \theta)]$. By the property $D: D(A, B) \leq D(A, C) + D(C, B)$ of the Hausdorff metric D, it is easy to derive the estimate

$$\begin{aligned}
m(t + \theta) \leq{}& D[W(t + \theta), W(t) + \theta(Q_m W)(t)] \\
&+ D[W(t) + \theta(Q_m W)(t), Z(t + \theta)] \\
\leq{}& D[W(t + \theta), W(t) + \theta(Q_m W)(t)] \\
&+ D[W(t) + \theta(Q_m W)(t), Z(t) + \theta(Q_M Z)(t)] \\
&+ D[Z(t) + \theta(Q_M Z)(t), Z(t + \theta)] \\
\leq{}& D[W(t + \theta), W(t) + \theta(Q_m W)(t)] \\
&+ D[W(t) + \theta(Q_m W)(t), W(t) + \theta(Q_M Z)(t)] \\
&+ D[W(t) + \theta(Q_M Z)(t), Z(t) + \theta(Q_M Z)(t)] \\
&+ D[W(t) + \theta(Q_M Z)(t), Z(t + \theta)].
\end{aligned} \tag{11.14}$$

Next, applying the property $D(A + C, B + C) = D(A, B)$ for any $A, B \in \mathcal{K}_c(\mathbb{R}^n)$ and any $C \in K(\mathbb{R}^n)$ and using the fact that, for arbitrarily small

$\theta > 0$, there exist Hukuhara differences $W(t+\theta) - W(t)$ and $Z(t+\theta) - W(t)$, we rewrite inequality (11.14) in the form

$$
\begin{aligned}
m(t+\theta) \leq\ & D[W(t) + H(t,\theta), W(t) + \theta(Q_m W)(t)] \\
& + D[\theta(Q_m W)(t), \theta(Q_M Z)(t)] \\
& + D[W(t), Z(t)] + D[Z(t) + \theta(Q_M Z)(t), Z(t) + V(t,\theta)],
\end{aligned}
$$

where $W(t+\theta) = W(t) + H(t,\theta)$ and $Z(t+\theta) = Z(t) + V(t,\theta)$.

Next, we obtain

$$
\begin{aligned}
m(t+\theta) \leq\ & D[H(t,\theta), \theta(Q_m W)(t)] + D[\theta(Q_m W)(t), \theta(Q_M Z)(t)] \\
& + D[W(t), Z(t)] + D[\theta(Q_M Z)(t), V(t,\theta)].
\end{aligned}
$$

Since a Hukuhara difference exists, $H(t,\theta)$ and $V(t,\theta)$ can be replaced by $W(t+\theta) - W(t)$ and $Z(t+\theta) - Z(t)$, respectively. For $\theta > 0$, we calculate

$$
\begin{aligned}
\frac{m(t+\theta) - m(t)}{\theta} \leq\ & D\Big\{[W(t+\theta) - W(t)]\theta^{-1}, (Q_m W)(t)\Big\} + \\
& + D\big[(Q_m W)(t), (Q_M Z)(t)\big] + D\Big\{(Q_M Z)(t), [Z(t+\theta) - Z(t)]\theta^{-1}\Big\}.
\end{aligned}
$$
(11.15)

Since $W(t)$ and $Z(t)$ solve equations (11.2) and (11.3), inequality (11.15) yields, as $\theta \to 0^+$,

$$
D^+ m(t) \leq D[(Q_m W)(t), (Q_M Z)(t)] \leq g(t, D[W, Z](t)) = g(t, |m|_0(t))
$$
(11.16)

for all $t \in [t_0, T)$. Applying Lemma 11.1 to (11.16) gives the estimate from Theorem 11.2.

Let $\theta \in \mathcal{K}_c(\mathbb{R}^n)$ be a zero set. Theorem 11.2 has the following consequence.

Corollary 11.4 *Given causal operators $Q_m \in C(E, E)$ and $Q_M \in C(E, E)$, suppose that there are functions $g_i \in C([t_0, T), \mathbb{R})$, $i = 1, 2$ such that*

$$
\begin{aligned}
D[(Q_m W)(t), \theta] &\leq g_1(t, D_0[W, \theta](t)), \\
D[(Q_M Z)(t), \theta] &\geq g_2(t, D_0[Z, \theta](t)).
\end{aligned}
$$

If, for all $t \in [t_0, T)$, the equation

$$
\frac{dw}{dt} = g_1(t, w), \quad w(t_0) = w_0 \geq 0
$$

has a maximal solution and the equation

$$
\frac{dv}{dt} = g_2(t, v), \quad v(t_0) = v_0 \geq 0,
$$

has a minimal solution, then the solutions $W(t)$ and $Z(t)$ of equations (11.2) and (11.3) *with initial conditions* (t_0, W_0) *and* (t_0, Z_0), W_0, $Z_0 \in \mathcal{K}_c(\mathbb{R}^n)$, *such that* $D[W_0, \theta] \leq w_0$ *and* $D[Z_0, \theta] \geq v_0$ *satisfy the estimates*

$$D[W(t), \theta] \leq r_M(t; t_0, w_0),$$
$$D[Z(t), \theta] \geq r_m(t; t_0, v_0)$$

for all $t \in [t_0, T)$.

11.4   Test for Stability

The above comparison result in terms of matrix-valued Lyapunov like functions is a useful tool to establish some appropriate stability results of equation (11.1). In order to consider the stability properties of (11.1) we will assume that the solutions of (11.1) exist and are unique for all $\alpha \in S$ and $t \geq t_0$. In addition, in order to mach the behavior of solutions of (11.1) with those corresponding to differential equations with robust causal maps, we assume that $Y_0 = V_0 + X_0$, so the Hukuhara difference $Y_0 - V_0 = X_0$ exists. Thus we have the initial problem

$$D_H X(t) = (Q(\alpha)X)(t),$$
$$X(t_0) = X_0 \in \mathcal{K}_c(\mathbb{R}^n), \tag{11.17}$$

where $(Q(\alpha)X)(t): \mathbb{R}_+ \times E \times S \to \mathcal{K}_c(\mathbb{R}^n)$.

If $Q(\alpha)\Theta_0 = \Theta_0$ for all $\alpha \in S$, then $X(t) = \Theta_0$ is a set of equilibrium positions or a set of stationary solutions of the equation (11.17).

Since $t \to \text{diam}(X(t))$ is nondecreasing at $t \to \infty$, the direct application of the norm $\|X(t)\| = \text{diam}(X(t))$ at $t \geq t_0$ for equation (11.17) is not possible. Therefore we will introduce additional assumptions on the initial value of problem (11.17).

H_1. For all $\alpha \in S$ the equation (11.17) has a set of stationary solutions Θ_0, i.e. $Q(\alpha)\Theta_0 = \Theta_0$.

H_2. There exists a set $W_0 \in \mathcal{K}_c(\mathbb{R}^n)$ such that for any $V_0, X_0 \in \mathcal{K}_c(\mathbb{R}^n)$ there exists the Hukuhara difference $X_0 = V_0 + W_0$.

H_3. The equation (11.17) has a unique solution $X(t) = X(t, t_0, X_0 - V_0) = X(t, W_0)$ for any initial values $(t_0, W_0) \in \mathbb{R}_+ \times S_0(\rho_0)$ for all $\alpha \in S$.

Then, while setting the problem of the stability of solutions of the equation (11.17) we will take into account the Assumptions H_1–H_3.

Definition 11.2 The stationary solution Θ_0 of the equation (11.17) is robust stable, if for any $t_0 \in \mathbb{R}_+$ and $\varepsilon > 0$ there exists a positive function $\delta(t_0, \varepsilon)$ such that the inequality $D[W_0, \Theta_0] < \delta$ implies the estimate $D[X(t), \Theta_0] < \varepsilon$ at all $t \geq t_0$ and for all $\alpha \in S$.

Other definitions of Lyapunov robust stability can be formulated in a similar way following the stability definition of set differential equations.

Now we start by proving the stability results.

Theorem 11.3 *Assume that for the equation* (11.17) *the conditions of the Assumptions H_1–H_3 are satisfied, and, in addition,*

(1) *there exists a matrix-valued function* (11.5) *and a vector $\theta \in \mathbb{R}_+^2$ such that the function $V(t, X, \theta) = \theta^T S(t, X, \beta)\theta$ satisfies the local Lipschitz condition*

$$|V(t, X_1, \theta) - V(t, X_2, \theta)| \leq LD[X_1, X_2],$$

where $L > 0$ for all $(t, x) \in \mathbb{R}_+ \times B$, $B = \{X \in \mathcal{K}_c(\mathbb{R}^n): D[X, \Theta_0] \leq \rho_0\}$;

(2) *there exist vector comparison function $\overline{\psi}_1$, $\overline{\psi}_1 \in K$-class and constant symmetric positive definite (2×2)-matrix A_1 such that*

$$\overline{\psi}_1^T (D[X, \Theta_0]) A_1 \overline{\psi}_1 (D[X, \Theta_0]) \leq V(t, X, \theta)$$

for all $(t, X) \in \mathbb{R}_+ \times B$;

(3) *there exists a function $g \in C(\mathbb{R}_+ \times \mathbb{R}_+, \mathbb{R})$, $g(t, 0) = 0$ for all $t \in \mathbb{R}_+$ such that*

$$D_- V(t, X, \theta)\big|_{(11.17)} \leq g(t, V(t, X, \theta))$$

for all $t \in \mathbb{R}_+$ and $X \in E_1$, $E_1 = \{X \in \mathcal{K}_c(\mathbb{R}_+): V(s, X(s), \theta) \leq V(t, X(t), \theta), t_0 \leq s \leq t\}$ and for all $\alpha \in \mathcal{S}$.

Then the stability of the zero solution of equation

$$\frac{dw}{dt} = g(t, w), \quad w(t_0) = w_0 \geq 0 \tag{11.18}$$

implies the robust stability of the stationary set Θ_0 of (11.17).

Proof We transform the condition (2) of the Theorem 11.3 to the form

$$\lambda_m(A_1)\psi_1(D[X, \Theta_0]) \leq V(t, X, \theta), \tag{11.19}$$

where $\lambda_m(A_1)$ is the minimal eigenvalue of the matrix A_1 and $\psi_1 \in K$-class, such that

$$\psi_1^T(D[X, \Theta_0]) \leq \overline{\psi}_1^T (D[X, \Theta_0]) \overline{\psi}_1 (D[X, \Theta_0])$$

for all $(t, X) \in \mathbb{R}_+ \times B$.

Let $\varepsilon > 0$ and $t_0 \in \mathbb{R}_+$ be given. Since $V(t, X, \theta)$ is positive definite we have the estimate (11.19) for all $(t, x) \in \mathbb{R}_+ \times B$. Suppose that the zero solution of (11.18) is stable. Than given $\lambda_m(A_1)\psi_1(\varepsilon) > 0$, $t_0 \in \mathbb{R}_+$, there exists a $\delta = \delta(t_0, \varepsilon) > 0$ such that whenever $w_0 < \delta$, we have

$$w(t, t_0, w_0) < \lambda_m(A_1)\psi_1(\varepsilon) \quad \text{for all } t \geq t_0,$$

where $w(t, t_0, w_0)$ is any solution of (11.18).

We choose $w_0 = V(t_0, W_0, \theta)$. Since $V(t, X(t), \theta)$ is continuous and $V(t, \Theta_0, \theta) = 0$ there exists a positive function $\delta_1 = \delta_1(t_0, \varepsilon) > 0$ such that $D[W_0, \Theta_0] \leq \delta_1$ and $V(t, W_0, \theta) \leq \delta$ hold simultaneously. We claim that if $D[W_0, \Theta_0] \leq \delta_1$, then $D[X(t), \Theta_0] < \varepsilon$ for all $t \geq t_0$. Suppose this is not true. Then there exists a solution $X(t) = X(t, t_0, W_0)$ of (11.17) satisfying the properties $D[X(t_2), \Theta_0] = \varepsilon$ and $D[X(t), \Theta_0] < \varepsilon$ for $t_0 < t < t_2$, $t_2 \in (t_0, t_1)$. Together with (11.19), this implies that

$$V(t_2, X(t_2), \theta) \geq \lambda_m(A_1)\psi_1(\varepsilon). \tag{11.20}$$

Furthermore, $X(t) \in B$ for $t \in [t_0, t_2]$. Hence, the choice of $w_0 = V(t_0, W_0, \theta)$ and condition(3) of the Theorem 11.3 give the estimate

$$V(t, X(t), \theta) \leq r(t), \quad t \in [t_0, t_2], \tag{11.21}$$

where $r(t) = r(t, t_0, w_0)$ is the maximal solution of the comparison problem (11.18). Now from estimates (11.19)–(11.21) we have

$$\lambda_m(A_1)\psi_1(\varepsilon) \leq V(t_2, X(t_2), \theta) \leq r(t_2) < \lambda_m(A_1)\psi_1(\varepsilon)$$

which is contradiction.

The proof of the Theorem 11.3 is complete.

Further we define the following set:

$$E_2 = \{X \in \mathcal{K}_c(\mathbb{R}^n): \; V(s, X(s), \theta)\beta(s) \leq V(t, X(t), \theta)\beta(t), \quad t_0 \leq s \leq t\},$$

where $\beta(t) > 0$ is a continuous function on $\mathbb{R}_+$.

Theorem 11.4 *Assume that:*

(1) *there exist functions $V(t, X, \theta)$ and $g(t, w)$ satisfying the conditions of Theorem 11.3;*

(2) *there exists a function $\beta(t)$ such that $\beta(t) > 0$ is continuous for $t \in \mathbb{R}_+$ and $\beta(t) \to \infty$ as $t \to \infty$;*

(3) *the condition (3) of the Theorem 11.3 holds for $t > t_0$ and $X(t) \in E_2$.*

Then, if the zero solution of (11.18) is stable, then the set Θ_0 of stationary solution of (11.17) is robust asymptotic stable.

Proof Let $0 < \varepsilon < \rho$ and $t_0 \in \mathbb{R}_+$. Set $\beta_0 = \min_{t \in \mathbb{R}_+} \beta(t)$, then $\beta_0 > 0$ follows from condition (2) of the Theorem 11.4. Since $V(t, X, \theta)$ is positive definite we can define

$$\varepsilon_1 = \beta_0 \lambda_m(A_1)\psi_1(\varepsilon).$$

Then, the stability of zero solution of (11.18) implies that, given $\varepsilon_1 > 0$ and $t_0 \in \mathbb{R}_+$, there exists a $\delta = \delta(t_0, \varepsilon_1) > 0$ such that $w_0 < \delta$ implies that

$$w(t, t_0, w_0) < \varepsilon_1 \quad \text{for all} \quad t \geq t_0, \tag{11.22}$$

where $w(t, t_0, w_0)$ is any solution of (11.18). We chose $w_0 = V(t_0, W_0, \theta)$. Then proceeding as in the proof of Theorem 11.3 with ε_1 instead of $\lambda_m(A_1)\psi_1(\varepsilon)$, we can prove that the set Θ_0 of stationary solution of (11.17) is robust stable.

Let $X(t, t_0, W_0)$ be any solution of (11.17) such that $D[W_0, \Theta_0] \leq \delta_0$, where $\delta_0 = \delta(t_0, \frac{1}{2}\rho)$. Since the zero solution of (11.18) is stable, it follows that $D[X(t), \Theta_0] < \dfrac{1}{2}\rho$, for $t \geq t_0$. Since $\beta(t) \to \infty$ as $t \to \infty$, there exists a number $T = T(t_0, \varepsilon) > 0$ such that

$$\beta(t)\lambda_m(A_1)\psi_1(\varepsilon) > \varepsilon_1 \quad \text{for} \quad t \geq t_0 + T. \tag{11.23}$$

Now from the relation (11.19), we get

$$\beta(t)\lambda_m(A_1)\psi(D[X(t), \Theta_0]) \leq \beta(t)V(t, X(t), \theta) \leq r(t), \quad t \geq t_0, \tag{11.24}$$

where $X(t) = X(t, t_0, X_0)$ is any solution of (11.17) such that $D[W_0, \Theta_0] \leq \delta_0$.

If the set Θ_0 of stationary solution of (11.17) is not robust asymptotically stable, then there exists a sequence $\{t_k\}$, $t_k \geq t_0 + T$, $t_k \to \infty$ as $k \to \infty$, such that $D[X(t_k), \Theta_0] \geq \varepsilon$ for some solution $X(t)$ satisfying $D[W_0, \Theta_0] \leq \delta_0$.

The inequalities (11.22), (11.24) yield $\beta(t_k)\lambda_m(A_1)\psi_1(\varepsilon) \leq \varepsilon_1$, a contradiction to (11.23). Thus, the set Θ_0 of (11.17) is robust asymptotically stable.

Now we define the set

$$E_3 = \{X \in \mathcal{K}_c(\mathbb{R}^n): V(s, X(S), \theta) \leq W(V(t, X(t), \theta)), \; t_1 \leq s \leq t, \; t_0 \geq t_0\},$$

where $W(r)$ is continuous on $\mathbb{R}_+$ and nondecreasing in r, $W(r) > r$ for all $r > 0$.

The next theorem gives sufficient conditions for the uniform robust asymptotic stability of (11.17).

Theorem 11.5 *Assume that for the equation (11.17) all conditions of Assumptions H_1–H_3 are fulfilled and a function $V(t, X, \theta)$ satisfies the following properties:*

(1) *there exist vector comparison functions $\overline{\psi}_1, \overline{\psi}_2, \overline{\psi}_3 \in K$-class and constant symmetric positive definite (2×2)-matrices A_1, A_2 such that the condition (2) of the Theorem 11.3 has the form*

$$\overline{\psi}_1^{\mathrm{T}}(D[X, \Theta_0])A_1\overline{\psi}_1(D[X, \Theta_0]) \leq V(t, X, \theta)$$
$$\leq \overline{\psi}_2^{\mathrm{T}}(D[X, \Theta_0])A_2\overline{\psi}_2(D[X, \Theta_0])$$

for all $(t, x) \in \mathbb{R}_+ \times B$;

(2) *there exist a* (2×2)-*matrix* $A_3(\alpha)$ *and a function* $c \in K$-*class such that*

$$D_-V(t, X(t), \theta)\big|_{(11.17)} \leq c^{\mathrm{T}}(D[X(t), \Theta_0])A_3(\alpha)c(D[X(t), \Theta_0])$$

for all $t \geq t_0$, $X(t) \in E_3$ *and any* $\alpha \in S$;

(3) *there exists a* (2×2)-*matrix* $\overline{A}_3$ *such that* $\dfrac{1}{2}(A_3^{\mathrm{T}}(\alpha) + A_3(\alpha)) \leq \overline{A}_3$ *for all* $\alpha \in S$.

Then, if matrix $\overline{A}_3$ *is negative definite, the set* Θ_0 *of stationary solutions of* (11.17) *is uniformly robust asymptotically stable.*

Proof Since $V(t, x, \theta)$ is positive definite and decreasing, the condition (1) of the Theorem 11.5, we transform to the form

$$\lambda_m(A_1)\psi_1(D[X, \Theta_0]) \leq V(t, X, \theta) \leq \lambda_M(A_2)\psi_2(D[X, \Theta_0]), \qquad (11.25)$$

where $\psi_2 \in K$-class and is such that

$$\psi_2(D[X, \Theta_0]) \geq \overline{\psi}_2^{\mathrm{T}}(D[X, \Theta_0])\overline{\psi}_2(D[X, \Theta_0])$$

and $\lambda_M(A_2)$ is a maximal eigenvalue of the matrix A_2. Let $0 < \varepsilon < \rho$ and $t_0 \in \mathbb{R}_+$ be given. We choose $\delta = \delta(\varepsilon) > 0$ such that

$$\lambda_M(A_2)\psi_2(\delta) < \lambda_m(A_1)\psi_1(\varepsilon). \qquad (11.26)$$

We claim that if $D[W_0, \Theta_0] \leq \delta$, then $D[X(t), \Theta_0] < \varepsilon$ for all $t \geq t_0$, where $X(t) = X(t, t_0, W_0)$ is any solution of (11.17).

Suppose this is not true. Then there exists a solution $X(t)$ of (11.17) with $D[W_0, \Theta_0] \leq \delta$ and $t_2 > t_0$, such that $D[X(t_2, t_0, W_0), \Theta_0] = \varepsilon$ and $D[X(t, t_0, W_0), \Theta_0] \leq \varepsilon$ for $t \in [t_0, t_2]$. Thus, in view of (11.25), we have

$$V(t_2, X(t_2), \theta) \geq \lambda_m(A_1)\psi_1(\varepsilon). \qquad (11.27)$$

It is clear, since $0 < \varepsilon < \rho$, then $X(t) \in B$ for all $t \geq t_0$. By our choice $w_0 = V(t_0, W_0, \theta)$ and by the conditions (2) and (3) of the Theorem 11.5 for $t > t_0$ and $X(t) \in E_3$ we have the estimate

$$V(t, X(t), \theta) \leq V(t_0, W_0, \theta), \quad t \in [t_0, t_2]. \qquad (11.28)$$

We see that the relations (11.26)–(11.28) lead to the contradiction

$$\lambda_m(A_1)\psi_1(\varepsilon) \leq V(t_2, X(t_2), \theta) \leq \lambda_M(A_2)\psi_2(D[W_0, \Theta_0])$$
$$\leq \lambda_M(A_2)\psi_2(\delta) < \lambda_m(A_1)\psi_1(\varepsilon).$$

This proves the uniform robust stability of (11.17).

Further let $X(t) = X(t, t_0, W_0)$ be any solution of (11.17) such that $D[W_0, \Theta_0] \leq \delta_0$, where $\delta_0 = \delta(\frac{1}{2}\rho)$ and δ are the same as before. It then follows from uniform stability that $D[X(t), \Theta_0] \leq \frac{1}{2}\rho$ for all $t \geq t_0$ and hence $X(t) \in B$ for all $t \geq t_0$.

Let $0 < \eta < \delta_0$ be given. Clearly, we have

$$\lambda_m(A_1)\psi_1(\eta) \leq \lambda_M(A_2)\psi_2(\delta_0).$$

In view of the assumption on $w(r)$, there exists a $\gamma = \gamma(\eta) > 0$ such that $w(r) > r + \gamma$ if $\lambda_m(A_1)\psi_1(\eta) \leq r \leq \lambda_M(A_2)\psi_2(\delta_0)$. Furthermore, there exists a positive integer $N = N(\eta)$ such that

$$\lambda_m(A_1)\psi_1(\eta) + N(\eta)\gamma > \lambda_M(A_2)\psi_2(\delta_0).$$

If we have, for some $t \geq t_0$, $V(t, X(t), \theta) \geq \lambda_m(A_1)\psi_1(\eta)$, it follows from (11.25) that there exists a $\delta_2 = \delta_2(\eta) > 0$ such that $D[X(t), \Theta_0] \geq \delta_2$. From conditions (2) and (3) of the Theorem 11.5 it follows that

$$c^{\mathrm{T}}(D[X(t), \Theta_0])A_3(\alpha)c(D[X(t), \Theta_0]) \leq \lambda_M(\overline{A}_3)\bar{c}(D[X(t), \Theta_0]),$$

where $\lambda_M(\overline{A}_3) < 0$ is a maximal eigenvalue of the matrix $\overline{A}_3$, and $\bar{c} \in K$-class is such that

$$\bar{c}(D[X(t), \Theta_0]) \geq c^{\mathrm{T}}(D[X(t), \Theta_0])c(D[X(t), \Theta_0]).$$

In a result we have

$$\lambda_M(\overline{A}_3)\bar{c}(D[X(t), \Theta_0]) \geq \lambda_M(\overline{A}_3)\bar{c}(\delta_2) = \delta_3,$$

where $\delta_3 = \delta_3(\eta)$.

For $N + 1$ we construct numbers $t_k = t_k(t_0, \eta)$ such that $t_0(t_0, \eta) = t_0$ and $t_{k+1}(t_0, \eta) = t_k(t_0, \eta) + \gamma/\delta_3$, $k = 0, 1, 2, \ldots, N$. By letting $T(\eta) = N\gamma/\delta_3$, we have $t_k(t_0, \eta) = t_0 + T(\eta)$.

Now to prove uniform asymptotic stability we still have to prove that $D[X(t), \Theta_0] < \eta$ for all $t \geq t_0 + T(\eta)$. It is therefore sufficient to show that

$$V(t, X(t), \theta) < \lambda_M(A_2)\psi_2(\eta) + (N - k)\gamma \qquad (11.29)$$

for all $t \geq t_k$, $k = 0, 1, 2, \ldots, N$.

We will prove (11.29) by induction. For $k = 0$, $t \geq t_0$, using (11.25) and (11.26),

$$V(t, X(t), \theta) \leq V(t_0, W_0, \theta) \leq \lambda_M(A_2)\psi_2(\delta_0) < \lambda_m(A_1)\psi_1(\eta) + N\gamma.$$

Suppose we have, for some k

$$V(s, X(s), \theta) < \lambda_M(A_2)\psi_2(\eta) + (N - k)\gamma, \quad s \geq t_k,$$

and we assume that for $t \in [t_k, t_{k+1}]$

$$V(t, X(t), \theta) \geq \lambda_M(A_2)\psi_2(\eta) + (N - k - 1)\gamma.$$

It then follows that

$$\lambda_M(A_2)\psi_2(\delta_0) \geq \lambda_M(A_2)\psi_2(D[X(s),\Theta_0]) \geq V(s,X(s),\theta)$$
$$\geq \lambda_m(A_1)\psi_1(\eta) + N(\eta)\gamma - (k+1)\gamma \geq \lambda_m(A_1)\psi_1(\eta). \tag{11.30}$$

From this we have

$$w(V(s,X(s),\theta)) \geq V(s,X(s),\theta) + \gamma$$
$$> \lambda_m(A_1)\psi_1(\eta) + (N-k)\gamma > V(s,X(s),\theta)$$

for $t_k < s < t, t \in [t_k, t_{k+1}]$. In turn, this implies that $X(t) \in E_3$ for $t_k < s < t$, $t \in [t_k, t_{k+1}]$. Hence, we obtain from assumptions (2)–(3) and (11.30) that

$$V(t_{k+1}, X(t_{k+1}), \theta) \leq V(t_k, X(t_k), \theta) - \int_{t_k}^{t_{k+1}} \lambda_M(\overline{A}_3)\bar{c}(D[X(s),\Theta_0])\,ds$$

$$\leq \lambda_M(A_2)\psi_2(\eta) + (N-k)\gamma - \delta_3(\eta)(t_{k+1} - t_k)$$
$$< \lambda_m(A_1)\psi_1(\eta) + (N-k)\gamma.$$

This contradiction shows that $t^* \in [t_k, t_{k+1}]$ such that

$$V(t^*, X(t^*), \theta) < \lambda_m(A_1)\psi_1(\eta) + (N-k-1)\gamma. \tag{11.31}$$

Now we show that (11.31) implies that

$$V(t, X(t), \theta) < \lambda_m(A_1)\psi_1(\eta) + (N-k-1)\gamma, \quad t \geq t^*.$$

If it is not true, then there exists $t_1 > t^*$ such that $V(t_1, X(t_1), \theta) = \lambda_m(A_1)\psi_1(\eta) + (N-k-1)\gamma$ or for a small $h < 0$, $V(t_1 + h, X(t_1 + h), \theta) < \lambda_m(A_1)\psi_1(\eta) + (N-k-1)\gamma$, which implies that

$$D_- V(t_1, X(t_1), \theta)\big|_{(11.17)} \geq 0. \tag{11.32}$$

As we have done before, we can show that $X(t) \in B$ for $t^* \leq s \leq t_1$, and $D_- V(t_1, X(t_1), \theta) \leq -\delta_3 < 0$. This contradicts (11.32) and hence

$$V(t, X(t), \theta) < \lambda_m(A_1)\psi_1(\eta) + (N-k-1)\gamma$$

for all $t \geq t_{k+1}$.

This completes the proof of the Theorem 11.5.

Example 11.1 We consider a simple example in $\mathcal{K}(\mathbb{R})$:

$$D_H X(t) = (-e^\alpha) \int_0^t X(s)\,ds,$$

$$X(0) = X_0 \in \mathcal{K}_c(\mathbb{R}).$$

Then using interval methods, we get

$$\frac{dx_1}{dt} = -e^\alpha \int_0^t x_2(s)\, ds$$

$$\frac{dx_2}{dt} = -e^\alpha \int_0^t x_1(s)\, ds$$

where $X(t) = [x_1(t), x_2(t)]$, $X_0 = [x_{10}, x_{20}]$. Therefore if $\alpha = 0$ this yields

$$\frac{d^4 x_1}{dt^4} = x_1, \quad x_1(0) = x_{10},$$

$$\frac{d^4 x_2}{dt^4} = x_2, \quad x_2(0) = x_{20},$$

whose solutions are given by

$$x_1(t) = \left[\frac{1}{2}(x_{10} - x_{20})\right]\left[\frac{1}{2}(e^t + e^{-t})\right] + \left[\frac{1}{2}(x_{10} + x_{20})\right]\cos t,$$

$$x_2(t) = \left[\frac{1}{2}(x_{20} - x_{10})\right]\left[\frac{1}{2}(e^t + e^{-t})\right] + \left[\frac{1}{2}(x_{10} + x_{20})\right]\cos t.$$

Further, for $t \geq 0$ we have

$$X(t, 0, X_0) = \left[-\frac{1}{2}(x_{20} - x_{10}), \frac{1}{2}(x_{20} - x_{10})\right]\left[\frac{1}{2}(e^t + e^{-t})\right]$$

$$+ \left[\frac{1}{2}(x_{10} + x_{20}), \frac{1}{2}(x_{10} + x_{20})\right]\cos t.$$

If we choose

$$V_0 = \left[-\frac{1}{2}(x_{20} - x_{10}), \frac{1}{2}(x_{20} - x_{10})\right]$$

then

$$X(t, 0, W_0) = \left[\frac{1}{2}(x_{10} + x_{20}), \frac{1}{2}(x_{10} + x_{20})\right]\cos t$$

for all $t \geq 0$.

We see that the set of stationary solutions $\Theta_0 \in \mathcal{K}_c(\mathbb{R})$ is stable.

Example 11.2 We consider equation (11.17) with

$$(Q(\alpha)X)(t) = -e^\alpha X - b\int_0^t X(s)\, ds,$$

$$X(0) = X_0 \in \mathcal{K}_c(\mathbb{R}^2),$$

where $b = \text{const} > 0$, $\alpha \in [0,1]$ is uncertain parameter.

Let $X(t) = [x_1(t), x_2(t)]$ and $X_0 = [x_{10}, x_{20}]$. Then equation

$$D_H X(t) = -e^\alpha X - b \int_0^t X(s)\,ds \qquad (11.33)$$

is reduced to

$$\frac{dx_1}{dt} = -e^\alpha x_2 - b \int_0^t x_2(s)\,ds,$$

$$\frac{dx_2}{dt} = -e^\alpha x_1 - b \int_0^t x_1(s)\,ds,$$

and

$$\frac{d^4 x_1}{dt^4} = e^{2\alpha} x_1'' + 2e^\alpha b x_1' + b^2 x_1, \quad x_1(0) = x_{10},$$
$$\frac{d^4 x_2}{dt^4} = e^{2\alpha} x_2' + 2e^\alpha b x_2' + b^2 x_2, \quad x_2(0) = x_{20}. \qquad (11.34)$$

From (11.34) for $\alpha = 0$ and $b = 2$ we obtain

$$x_1(t) = \frac{1}{6}(x_{10} - x_{20})e^{-t} + \frac{1}{3}(x_{10} - x_{20})e^{2t}$$
$$+ e^{-\frac{1}{2}t}\left[\frac{1}{2}(x_{10} + x_{20})\cos\left(\frac{\sqrt{7}}{2}t\right) - \frac{1}{2\sqrt{7}}(x_{10} + x_{20})\sin\left(\frac{\sqrt{7}}{2}t\right)\right];$$

$$x_2(t) = \frac{1}{6}(x_{20} - x_{10})e^{-t} + \frac{1}{3}(x_{20} - x_{10})e^{2t}$$
$$+ e^{-\frac{1}{2}t}\left[\frac{1}{2}(x_{10} + x_{20})\cos\left(\frac{\sqrt{7}}{2}t\right) - \frac{1}{2\sqrt{7}}(x_{10} + x_{20})\sin\left(\frac{\sqrt{7}}{2}t\right)\right].$$

Therefore it follows that

$$X(t, 0, X_0) = (x_{20} - x_{10})\left[-\frac{1}{6}, \frac{1}{6}\right]e^t + (x_{20} - x_{10})\left[-\frac{1}{3}, \frac{1}{3}\right]e^{2t}$$
$$+ (x_{20} + x_{10})\left[\frac{1}{2}, \frac{1}{2}\right]e^{-\frac{1}{2}t}\cos\left(\frac{\sqrt{7}}{2}t\right)$$
$$- (x_{20} + x_{10})\left[-\frac{1}{2\sqrt{7}}, \frac{1}{2\sqrt{7}}\right]e^{-\frac{1}{2}t}\sin\left(\frac{\sqrt{7}}{2}t\right), \quad t \geq 0.$$

Now, choosing $x_{10} = x_{20}$, we eliminate the undesirable terms and, therefore, we get asymptotic stability of stationary solution $\Theta_0 \in \mathcal{K}_c(\mathbb{R}^2)$ of equation (11.33).

Chapter 12

Stability of a Set of
Impulsive Equations

In Chapter 7, development of the direct Lyapunov method is proposed for impulse systems with uncertain parameter values.

The aim of this chapter is to establish stability conditions for a set of impulsive systems of equations in terms of the heterogeneous matrix-valued Lyapunov function.

12.1 Auxiliary Results

Let $\mathbb{R}^n$ be a Euclidean space and $\mathbb{R}_+ = [0, \infty)$ as usually. Assume that on $\mathbb{R}_+$ a sequence $\{t_k\}$ is given such that $0 < t_1 < t_2 < \ldots < t_k < \ldots$ and $\lim\limits_{k \to \infty} t_k = \infty$. Then we use the following notations:

(1) If $F \in PC(\mathbb{R}_+ \times \mathcal{K}_c, \mathcal{K}_c)$, then the function $F \colon \mathbb{R}_+ \times \mathcal{K}_c \to \mathcal{K}_c$ is continuous on $(t_{k-1}, t_k] \times \mathcal{K}_c$ for any value of $k = 1, 2, \ldots$ for any $X \in \mathcal{K}_c$ and the correlation

$$\lim_{(t, Y) \to (y_k^+, X)} F(t, Y) = F(y_k^+, X)$$

exists for any $k = 1, 2, \ldots$.

(2) If $g \in PC(\mathbb{R}_+ \times \mathbb{R}_+, \mathbb{R})$, then the function $g \colon (t_{k-1}, t_k] \times \mathbb{R}_+ \to \mathbb{R}$ is continuous for any $w \in \mathbb{R}_+$ and there exists

$$\lim_{(t, z) \to (t_k^+, w)} g(t, z) = g(t_k^+, w).$$

(3) If $\Phi \in PC^1(\mathbb{R}_+ \times \mathcal{K}_c)$, then the function Φ is differentiable over each interval (t_{k-1}, t_k), $k = 1, 2, \ldots$.

Consider the set of impulsive differential equations

$$D_H X = F(t, X), \quad t \neq t_k, \tag{12.1}$$

$$X(t_k^+) = I_k(X(t_k)), \quad t = t_k, \tag{12.2}$$

$$X(t_0) = X_0 \in \mathcal{K}_c,$$

where $X \in \mathcal{K}_c$, $F \in PC(\mathbb{R}_+ \times \mathcal{K}_c, \mathcal{K}_c)$, $I_k \colon \mathcal{K}_c \to \mathcal{K}_c$ for any $k = 1, 2, \ldots$ and the sequence $\{t_k\}$ is determined above. Solution of equations (12.1)–(12.2) is the function $X(t; t_0, X_0)$ which is piece-wise continuous on $[t_0, \infty)$ and continuous from the left on any subinterval $(t_k, t_{k+1}]$, $k = 1, 2, \ldots$, determined by the correlation

$$X(t; t_0, X_0) = \begin{cases} X_0(t; t_0, X_0), t_0 \le t \le t_1; \\ X_1(t; t_1, X_1^+), & t_1 < t \le t_2; \\ \cdots\cdots\cdots\cdots\cdots\cdots\cdots\cdots \\ X_k(t; t_k, X_k^+), & t_k < t \le t_{k+1}, \end{cases}$$

where $X_k(t; t_k, X_k^+)$ is the solution of equations (12.1)–(12.2) on the subinterval $[t_{k-1}, t_k)$.

Further, together with the set of impulse equations (12.1)–(12.2) we shall consider the impulsive scalar equation

$$\frac{dw}{dt} = g(t, w), \quad t \ne t_k, \tag{12.3}$$

$$w(t_k^+) = \Phi_k(w(t_k)), \quad t = t_k, \tag{12.4}$$

$$w(t_0) = w_0,$$

where the function $g \in PC(\mathbb{R}_+^2, \mathbb{R})$, $g(t, w)$ is nondecreasing in w for all $t \in \mathbb{R}_+$, $\psi_k \colon \mathbb{R}_+ \to \mathbb{R}$ for any $k = 1, 2, \ldots$ and $\psi_k(w)$ in nondecreasing in w.

Maximal solution of equations (12.3)–(12.4) is the function $r(t; t_0, w_0)$ determined by the correlations

$$r(t; t_0, w_0) = \begin{cases} r_0(t; t_0, r_0), & t_0 \le t \le t_1, \\ r_1(t; t_1, r_1^+), & t_1 < t \le t_2, \\ \cdots\cdots\cdots\cdots\cdots\cdots\cdots \\ r_k(t; t_k, r_k^+), & t_k < t \le t_{k+1}, \end{cases}$$

and satisfying the inequality

$$w(t; t_0, w_0) \le r(t; t_0, w_0), \quad t = t_k,$$

for all $t \in \mathbb{R}_+$ for any solution $w(t; t_0, w_0)$ of equations (12.3)–(12.4) with the initial values $r_k^+ = \Phi_k(r_{k-1}(t_k))$, $k = 1, 2, \ldots$.

12.2   Heterogeneous Lyapunov Function

The set of impulsive differential equations (12.1)–(12.2) is added with the additional set equation

$$D_H Y = F_\beta(t, Y), \quad Y(t_0) = Y_0 \in \mathcal{K}_c, \tag{12.5}$$

where $F_\beta(t, Y) = \beta F(t, Y) + (1 - \beta)I_k(Y(t_k))$, $k = 1, 2, \ldots$, $0 < \beta < 1$. We shall consider the matrix-valued function

$$U(t, X) = [V_{ij}(t, \cdot)], \quad i, j = 1, 2, \tag{12.6}$$

whose elements are chosen as follows:

$V_{11}\colon \mathbb{R}_+ \times \mathcal{K}_c \to \mathbb{R}_+$ is correlated with equation (12.1),

$V_{22}\colon \mathbb{R}_+ \times \mathcal{K}_c \to \mathbb{R}_+$ is correlated with equation (12.2),

$V_{12} = V_{21}\colon \mathbb{R}_+ \times \mathcal{K}_c \to \mathbb{R}$ is correlated with the set of equations (12.5).

The scalar function

$$V(t, X, \psi) = \psi^{\mathrm{T}} U(t, X)\psi, \quad \psi \in \mathbb{R}_+^2, \tag{12.7}$$

is called a heterogeneous Lyapunov function if along with the derivative $D^+V(t, X, \psi)$ it allows us to solve the stability (instability) of stationary solution Θ_0 of the set of impulsive equations (12.1)–(12.2).

We say that the function $V(t, X, \psi)$ belongs to the class F_0 if the following conditions are satisfied:

(a) $V(t, X, \psi)$ is continuous on $(t_{k-1}, t_k] \times \mathcal{K}_c$ for any $k = 1, 2, \ldots$ and for every $X \in \mathcal{K}_c$ there exists the limit

$$\lim_{(t,X) \to (t_k^+, W)} V(t, X, \psi) = V(t_k^+, W, \psi), \quad k = 1, 2, \ldots;$$

(b) for any $A, B \in \mathcal{K}_c$ and all $t \in \mathbb{R}_+$ the estimate $|V(t, A, \psi) - V(t, B, \psi)| \leq LD[A, B]$ takes place, where L is the Lipschitz constant.

Remark 12.1 If the elements $V_{12} = V_{21} = 0$ in matrix-valued function (12.6), then the function

$$L(t, X, \psi) = U(t, X)\psi, \quad \psi \in \mathbb{R}_+^2 \tag{12.8}$$

is the heterogeneous vector Lyapunov function, i.e., the vector function with heterogeneous components.

Remark 12.2 If $V_{12} = V_{21} = V_{22} = 0$ and $\mathcal{K}_c = \mathbb{R}^n$ in matrix function (12.6), then the function $V(t, X, \psi) = V(t, X)$ is a Lyapunov-like function for equations (12.1)–(12.2), the application of which is shown by Lakshmikantham, Bhaskar and Devi [2006].

Then we present a theorem of the principle of comparison with heterogeneous function (12.7) for system (12.1)–(12.2).

Theorem 12.1 *Assume that for the set of impulsive differential equations (12.1)–(12.2) the following conditions hold:*

(1) *there exists heterogeneous function (12.7) of class F_0;*

(2) *there exists a function* $g(t, w)$ *nondecreasing in* w, $g \in PC(\mathbb{R}_+^n, \mathbb{R})$, *such that*

$$D^+V(t, X, \psi)|_{(12.1)} \leq g(t, V(t, X, \psi)), \quad t \neq t_k,$$

for all $X \in \mathcal{K}_c$, *where*

$$D^+V(t, X, \psi)|_{(12.1)}$$
$$= \lim \sup\{[V(t + \theta, X + \theta F(t, X), \psi) - V(t, X, \psi)]\theta^{-1} : \ \theta \to 0^+\};$$

(3) *there exist functions* $\Phi_k(w)$, $k = 1, 2, \ldots$, *nondecreasing in* w, *such that*

$$V(t_k^+, X(t_k^+), \psi) \leq \Phi_k(V(t_k^+, X(t_k^+), \psi)), \quad t = t_k,$$

for all $k = 1, 2, \ldots$;

(4) *there exists maximal solution of impulsive scalar equation* (12.3)–(12.4) *for all* $t \in \mathbb{R}_+$.

Then, if the solution $X(t) = X(t; t_0, X_0)$ *of the set of impulse equations* (12.1)–(12.2) *exists on* $[t_0, a)$ *and* $V(t_0, X_0, \psi) \leq w_0$, *the estimate*

$$V(t, X(t), \psi) \leq r(t; t_0, w_0) \tag{12.9}$$

holds true for all $t \in [t_0, a)$.

This theorem is proved by the application of Theorem 4.2.1 from the monograph by Lakshmikantham, Leela and Martynyuk [1991a] on every segment $[t_{k-1}, t_k]$, $k = 1, 2, \ldots$. In contrast to Theorem 5.2.4 from Lakshmikantham, Bhaskar and Devi [2006] function (12.7) is applied.

Theorem 12.1 has several corollaries which can be useful in the investigation of special systems of the (12.1)–(12.2) form.

Corollary 12.1 *Assume that in conditions (2) and (3) of Theorem 12.1 the functions* $g(t, V(t, X, \psi)$ *and* $I_k(V(t_k, X(t_k), \psi))$ *are such that:*

(a) $g(t, V(t, X, \psi) = 0$ *and* $\Phi_k(V(t_k, X(t_k), \psi)) = 0$ *for all* $k = 1, 2, \ldots$, *then estimate* (12.9) *becomes*

$$V(t, X(t), \psi) \leq V(t_0^+; w_0, \psi), \quad t \geq t_0;$$

(b) $g(t, V(t, X, \psi) = 0$ *and* $\Phi_k(V(t_k, X(t_k), \psi)) = D_k V(t_k, X(t_k), \psi)$, $D_k \geq 0$ *for all* $k = 1, 2, \ldots$, *then estimate* (12.9) *becomes*

$$V(t, X(t), \psi) \leq V(t_0^+, w_0, \psi) \prod_{t_0 \leq t_k \leq t} D_k, \quad t \geq t_0;$$

(c) $g(t, V(t, X, \psi)) = -\beta V(t, X, \psi)$, $\beta > 0$ and $\Phi_k(V(t_k, X(t_k), \psi)) = D_k V(t_k, X(t_k), \psi)$, $D_k \geq 0$ for all $k = 1, 2, \ldots$, then estimate (12.9) becomes

$$V(t, X(t), \psi) \leq V(t_0^+, w_0, \psi) \prod_{t_0 \leq t_k \leq t} D_k \, e^{-\beta(t-t_0)}, \quad t \geq t_0;$$

(d) $g(t, V(t, X, \psi)) = \dfrac{d\gamma(t)}{dt} V(t, X, \psi)$, the function $\gamma(t)$ is decreasing on $[0, \infty)$, and $\Phi_k(V(t_k, X(t_k), \psi)) = D_k V(t_k, X(t_k), \psi)$, $D_k \geq 0$ for all $k = 1, 2, \ldots$, then estimate (12.9) becomes

$$V(t, X(t), \psi) \leq V(t_0^+, w_0, \psi) \prod_{t_0 \leq t_k \leq t} D_k \, e^{\gamma(t)-\gamma(t_0)}, \quad t \geq t_0.$$

12.3   Sufficient Stability Conditions

For the set of systems of impulsive equations (12.1)–(12.2) we introduce the following assumptions.

H_1. For the set of systems (12.1)–(12.2) the correlations $F(t, \Theta_0) = \Theta_0$ and $I_k(\Theta_0) = \Theta_0$ are true where $\Theta_0 \in \mathcal{K}_c(\mathbb{R}^n)$ is a set of stationary solutions of system (12.1)–(12.2) for $k = 1, 2, \ldots$.

H_2. For any values $X_0, V_0 \in \mathcal{K}_c(\mathbb{R}^n)$ of the set of impulsive equations (12.1)–(12.2) there exists a Hukuhara difference W_0 such that $X_0 = V_0 + W_0$.

H_3. Solution of equations (12.1)–(12.2) $X(t) = X(t, t_0, X_0 - V_0) = X(t, t_0, W_0)$ is unique for any $t \neq t_k$ and $W_0 \in S_0(\rho_0)$, $k = 1, 2, \ldots$.

Under conditions H_1–H_3 we shall consider a problem on the stability of stationary solution $X(t, t_0, \Theta_0) = \Theta_0$ for the set of systems of impulsive equations (12.1)–(12.2) for all $t \geq t_0$.

Definition 12.1 Stationary solution $X(t, t_0, \Theta_0)$ of the set of systems of impulsive equations (12.1)–(12.2) is:

(a) *stable*, if and only if for any $t_0 \in \mathbb{R}$ and any $\varepsilon > 0$ there exists $\delta = \delta(t_0, \varepsilon) > 0$ such that condition $D[W_0, \Theta_0] < \delta$ implies $D[X(t, t_0, Y_0), \Theta_0] < \varepsilon$ for all $t \geq t_0$;

(b) *attractive*, iff for any $t_0 \in \mathbb{R}$ there exists $\alpha(t_0) > 0$ and for any $\xi > 0$ there exists $\tau(t_0; W_0, \xi) \in R_+$ such that the condition $D[W_0, \Theta_0] < \alpha(t_0)$ implies $D[X(t; t_0, Y_0), \Theta_0] < \xi$ for any $t \geq t_0 + \tau(t_0, W_0, \xi)$;

(c) *asymptotically stable*, iff conditions 1(a) and 1(b) are satisfied.

In terms of the principle of comparison for the set of systems of impulsive equations, the stability conditions for stationary solution Θ_0 are formulated as follows.

Theorem 12.2 *Assume that for impulse differential equations (12.1)–(12.2) the following conditions are fulfilled:*

(1) *there exist heterogeneous function (12.7), comparison functions φ_i of class K, $i = 1, 2$, and constant 2×2-matrices A_i, $i = 1, 2$, such that*

(a) $|V(t, X_1, \psi) - V(t, X_2, \psi)| \leq LD[X_1, X_2]$, $L > 0$, *for all* $(t, X) \in R_+ \times \mathcal{K}_c$;

(b) $\varphi_1^T(\|X\|)A_1\varphi_1(\|X\|) \leq V(t, X, \psi) \leq \varphi_2^T(\|X\|)A_2\varphi_2(\|X\|)$ *for all* $(t, X) \in R_+ \times \mathcal{K}_c$;

(2) *there exist functions $g(t, u)$, $g \in C(R_+^2, R)$ and $\psi_k \colon R_+ \to R_+$, $k = 1, 2, \ldots$, such that*

(a) $D^+V(t, X, \psi) \leq g(t, V(t, X, \psi))$, $t \neq t_k$, *where $g(t, 0) = 0$ for all* $(t, X) \in \mathbb{R}_+ \times B(\Theta_0, \beta)$;

(b) *for any $X \in B(\Theta_0, \beta_0)$, $\beta > \beta_0 > 0$, for all k the inclusion $X + I_k(X) \in B(\Theta_0, \beta_0)$ takes place and $V(t_k, I_k(X(t_k)), \psi) \leq \Phi_k(V(t_k, X(t_k), \psi))$, where $\Phi_k(u)$ are functions nondecreasing in u, $\Phi_k(0) = 0$ for all $k = 1, 2, \ldots$.*

Then, if the matrices A_1 and A_2 in condition 1(b) are positive definite and zero solution of scalar comparison equations (12.3)–(12.4) possesses a certain type of stability, then stationary solution Θ_0 of the set of systems of impulsive equations (12.1)–(12.2) possesses the same type of stability.

Proof Under conditions of Theorem 12.2 we transform condition 1(b) to the form

$$\lambda_m(A_1)\bar{\varphi}_1(\|X\|) \leq V(t, X, \psi) \leq \lambda_M(A_2)\bar{\varphi}_2(\|X\|),$$

where $\lambda_m(A_1)$ and $\lambda_M(A_2)$ are the minimal and maximal eigenvalues of the matrices A_1 and A_2 respectively, and $\bar{\varphi}_1(\|X\|) \leq \varphi_1^T(\|X\|)\bar{\varphi}_1(\|X\|)$ and $\bar{\varphi}_2(\|X\|) \geq \varphi_2^T(\|X\|)\bar{\varphi}_2(\|X\|)$ for all $X \in B(\Theta_0, \beta)$, $\bar{\varphi}_1, \bar{\varphi}$ are of class K.

Let $0 < \varepsilon < \beta^* = \min(\beta, \beta_0)$ and $t_0 \in \mathbb{R}_+$ be given. Assume that Θ_0, which is the zero solution of equations (12.3)–(12.4) is stable. Besides, given $\lambda_m(A_1)\bar{\varphi}_1(\varepsilon) > 0$, there exists $\delta_1(t_0, \varepsilon) > 0$ such that the condition $0 \leq \omega_0 < \delta_1$ implies $\omega(t, t_0, \omega_0) < \lambda_m(A_1)\bar{\varphi}_1(\varepsilon)$ for all $t \geq t_0$, where $\omega(t, t_0, \omega_0)$ is any solution of scalar impulsive equations (12.3)–(12.4).

Let $\omega_0 = \lambda_M(A_2)\bar{\varphi}_2(D[X_0, \Theta_0])$ and take $\delta_2 = \delta_2(\varepsilon)$ so that inequality $\lambda_M(A_2)\bar{\varphi}_2(\delta_2) < \delta_1$ is satisfied. Then we take $\delta = \min(\delta_1, \delta_2)$ and show that if $D[X_0, \Theta_0] < \delta$, then for all $t \geq t_0$ the estimate $D[X(t), \Theta_0] < \varepsilon$ is true, where $X(t) = X(t, t_0, x_0)$ is any solution of the set of systems (12.1)–(12.2). Assume this is not true. Then for solution $X(t, t_0, X_0)$ with the initial conditions satisfying the inequality $D[X_0, \Theta_0] < \delta$ a $t^* > t_0$, $t_k < t^* < t_{k+1}$ for some k for which $D[X(t^*), \Theta_0] \geq \varepsilon$ and $D[X(t), \Theta_0] < \varepsilon$ for $t_0 \leq t \leq t_k$. Since $0 < \varepsilon < \beta^*$, according to condition 2(b) of Theorem 12.2, estimates

$$D[X(t_k^+), \Theta_0] = D[I_k(X(t_k)), \Theta_0] < \beta_0$$

and $D[X(t_k), \Theta_0] < \varepsilon$ are valid. Hence, there exists $\hat{t}$ such that $t_k < \hat{t} < t^*$ and

$$\varepsilon \leq D[X(\hat{t}), \Theta_0] < \beta_0.$$

Designate $m(t) = V(t, X(t), \psi)$ and consider the behavior of function $m(t)$ on the interval $[t_0, \hat{t}]$. From Theorem 12.1 it is easy to see that

$$V(t, X(t), \psi) \leq r(t, t_0, \omega_0) \quad \text{for all} \quad t_0 \leq t \leq \hat{t},$$

where $\omega_0 = \lambda_M(A_2)\bar{\varphi}_2(D[X_0, \Theta_0])$ and $r(t, t_0, \omega_0)$ are the maximal solutions of scalar comparison equations (12.3)–(12.4).

According to estimate (12.9) we have the inequality

$$\lambda_m(A_1)\bar{\varphi}_1(D[X(\hat{t}), \Theta_0]) \leq V(\hat{t}, X(\hat{t}), \psi) \leq r(\hat{t}, t_0, \omega_0) < \lambda_m(A_1)\bar{\varphi}_1(\varepsilon),$$

which contradicts the existence of $\hat{t} \in [t_k, t_{k+1}]$ for any k. Therefore, the stationary solution Θ_0 of the set of impulse systems (12.1)–(12.2) is stable.

Theorem 12.2 has several corollaries presented below.

Corollary 12.2 *Assume that in conditions 2(a) and (b) of Theorem 12.2 the functions $g(t, V(t, X, \psi))$ and $\Phi_k(V(t_k, X(t_k), \psi))$ are of the form indicated in conditions (a) and (b) of Corollary 12.1. Then, under the rest of the conditions of Theorem 12.2 the stationary solution Θ_0 of the set of impulse equations (12.1)–(12.2) is stable.*

Corollary 12.3 *Assume that in conditions 2(a) and (b) of Theorem 12.2 the functions $g(t, V(t, X, \psi))$ and $\Phi_k(V(t_k, X(t_k), \psi))$ are of the form indicated in conditions (c) and (d) of Corollary 12.1. Then, under the rest of the conditions of Theorem 12.2 the stationary solution Θ_0 of the set of impulse equations (12.1)–(12.2) is exponentially stable (limiting exponentially stable).*

As for vector Lyapunov functions with homogeneous components (see Matrosov, et al. [1981]), it is possible to construct for function (12.8) the algorithms of establishing stability conditions in terms of the dynamical properties of a comparison system.

12.4 Impulsive Equations with Delay under Small Perturbations

The stabilities of impulsive systems with delay are well studied, and impulsive systems with delay have been investigated in several papers. In the latter case a principle of comparison with scalar or matrix Lyapunov function was developed and general method of stability analysis of a set of stationary solutions was elaborated.

It is of interest to study the stability of a set of stationary solutions to this class of equations under small persistent perturbations. In this section this problem is analyzed by using perturbed Lyapunov functions defined on space products.

In n-dimensional Euclidean space $\mathbb{R}^n$ we prescribe a family of all nonempty convex subsets $\mathcal{K}_c(\mathbb{R}^n)$. The set $\mathcal{K}_c(\mathbb{R}^n)$ with the Hausdorff metric

$$D[A, B] = \max\Big\{ \sup_{y \in B} d(y, A), \sup_{x \in A} d(x, B) \Big\},$$

where $d(y, A) = \inf [d(y, x) \colon x \in A]$, A and B are bounded subsets in $\mathbb{R}^n$, is a metric space.

For a given $\tau > 0$ define the set $C_0 = C([-\tau, 0], \mathcal{K}_c(\mathbb{R}^n))$ and the interval $J_0 = [t_0 - \tau, t_0 + a]$ for some $a \geq \tau$. For the set $X \colon J_0 \to \mathcal{K}_c(\mathbb{R}^n)$ we designate by X_t the contraction of the set X to the interval $[t - \tau, t]$ and denote the element $X_t \in C_0$ for which $X_t(s) = X(t + s)$, $-\tau \leq s \leq 0$. For any sets A and B from C_0 we introduce a metric by the formula

$$D_0[A, B] = \max_{-\tau \leq s \leq 0} D[A, B],$$

where $D[A, B]$ is the Hausdorff metric.

We consider a set of impulsive equations with delay in the form

$$\begin{aligned}
D_H X(t) &= F(t, X_t) + \mu R(t, X_t), \quad t \neq t_k, \\
X_{t_k^+} &= I_k(X_{t_k}), \quad t = t_k, \\
X_{t_0} &= \Phi_0 \in C_0,
\end{aligned} \tag{12.10}$$

where $0 \leq t_1 < t_2 < \ldots < t_k < \ldots$ and $\lim_{k \to \infty} t_k = \infty$, $F \in \mathrm{PC}\,(\mathbb{R}_+ \times C_0, \mathcal{K}_c(\mathbb{R}^n))$ and $R \in \mathrm{PC}\,(\mathbb{R}_+ \times C_0, \mathcal{K}_c(\mathbb{R}^n))$ and μ is a small positive parameter. Assume that $F(t, X_t) = \Theta_0$ iff $X_t = \Theta_0$, where Θ_0 is a set of stationary solutions of the set of equations (12.10) for $\mu = 0$, the set of perturbations $R(t, X_t) \neq 0$ for $X_t = \Theta_0$ but there exists a summable function $m(t)$ on any finite interval such that

$$D[R(t, X_t), \Theta_0] \leq m(t) \quad \text{for all} \quad t \geq t_0.$$

The function $m(t)$ characterizes the order of possible growth of the perturbing forces at receding of the "representative point" of the state of system (12.10) for $\mu = 0$ in the matrix space $(\mathcal{K}_c(\mathbb{R}^n), D[A, B])$ from the stationary solution Θ_0.

The mapping $X \colon J_0 \to \mathcal{K}_c(\mathbb{R}^n)$ is a solution of the set of equations (12.10) if the function $X(t) = X(t_0, \Phi_0)(t)$ which is piecewise continuous on $[t_0, \infty)$

is continuous on any interval $(t_k, t_{k-1}]$, $k = 0, 1, 2, \ldots$, and is determined by

$$X(t_0, \Phi_0)(t) = \begin{cases} \Phi_0, & t_0 - \tau \leq t \leq t_0, \\ X_0(t_0, \Phi_0)(t), & t_0 \leq t \leq t_1, \\ X_1(t_1, \Phi_1)(t), & t_1 < t \leq t_2, \\ \cdots\cdots\cdots\cdots\cdots\cdots \\ X_k(t_k, \Phi_k)(t), & t_k < t \leq t_{k+1}, \\ \cdots\cdots\cdots\cdots\cdots\cdots \end{cases}$$

Here the sequence $X_k(t_k, \Phi_k)(t)$ is a solution of the set of differential equations with delay

$$\begin{aligned} D_H X(t) &= F(t, X_t) + \mu R(t, X_t), \\ X_{t_k^+} &= \Phi_k, \quad k = 0, 1, 2, \ldots, \end{aligned} \qquad (12.11)$$

where Φ_k is the value of function $X(t)$ for $t = t_k^+$.

First, we shall establish existence conditions of the set of equations (12.10). Further, $F \in \mathrm{PC}\,(\mathbb{R}_+ \times C_0, \mathcal{K}_c(\mathbb{R}^n))$ means that the mapping $F \colon \mathbb{R}_+ \times C_0 \to \mathcal{K}_c(\mathbb{R}^n)$ is continuous on $(t_{k-1}, t_k] \times \mathcal{K}_c(\mathbb{R}^n)$ for each $k = 1, 2, \ldots$ and each $X_t \in C_0$ and there exists a $\lim\limits_{(t, Y_t) \to (t_k^+, X_t)} F(t, Y_t) = F(t_k^+, X_t)$. Moreover, $g \in \mathrm{PC}\,(\mathbb{R}_+ \times \mathbb{R}_+, \mathbb{R}^n)$ means that the function $g \colon (t_{k-1}, t_k] \times \mathbb{R}_+ \to \mathbb{R}$ is continuous and for each $w \in \mathbb{R}_+$ there exists a $\lim\limits_{(t, z) \to (t_k^+, w)} g(t, z) = g(t_k^+, w)$.

The following assertion is valid.

Theorem 12.3 *Assume that*

(1) *the mapping* $F \in \mathrm{PC}\,(\mathbb{R}_+ \times C_0, \mathcal{K}_c(\mathbb{R}^n))$ *and there exists a summable function* $m(t)$ *on any finite interval such that for any* $X_t \in C_0$

$$D[R(t, X_t), \Theta_0] \leq m(t) \quad \text{for all} \quad t \geq t_0;$$

(2) *there exist a comparison function* $g \in \mathrm{PC}\,(\mathbb{R}_+^2 \times [0, 1], \mathbb{R}^n)$, $g(t, w, \mu)$ *is nondecreasing in* w *for any* t, *and a value* $\mu_1 > 0$ *such that*

$$\begin{aligned} D[(F(t, \Phi) + \mu R(t, \Phi)), \Theta_0] &\leq g(t, D_0[\Phi, \Theta_0], \mu) \\ \text{for} \quad t \neq t_k \quad &\text{and} \quad 0 < \mu < \mu_1, \end{aligned}$$

where $\Phi = (\Phi_0, \Phi_1, \Phi_2, \ldots)$, $k = 0, 1, 2, \ldots$;

(3) *there exists a sequence of functions* $\Phi_k(w)$ *nondecreasing in* w *for all* $k = 0, 1, 2, \ldots$ *and such that*

$$D_0[I_k(X_{t_k}), \Theta_0] \leq Q_k(D_0[X_{t_k}, \Theta_0]) \quad \text{for} \quad t = t_k, \quad k = 0, 1, 2, \ldots;$$

(4) *on the interval* $[t_0, \infty)$ *for* $0 < \mu < \mu_2$ *there exists a maximal solution* $r(t, t_0, w_0)$ *of the scalar impulsive differential equation*

$$\frac{dw}{dt} = g(t, w, \mu), \quad t \neq t_k,$$
$$w(t_k^+) = Q_k(w(t_k^+)), \quad t = t_k, \qquad (12.12)$$
$$w(t_0) = w_0.$$

Then, on the interval $[t_0, \infty)$ *for* $0 < \mu < \min(\mu_1, \mu_2)$ *there exists at least one solution* $X(t)$ *of the set of equations* (12.10).

Proof Let $J_0 = [t_0, t_1]$ and the mappings F and R are considered on the product $J_0 \times C_0$. By hypothesis (1) of Theorem 12.3 the mappings F and R are continuous $J_0 \times C_0$. Consider problem (12.11) on J_0 for $0 < \mu < \min(\mu_1, \mu_2)$. Under hypotheses (2) and (4) of Theorem 12.3 all conditions of Theorem 5.4.3 from monograph by Lakshmikantham, Bainov and Simeonov [1989] are satisfied and solution $X(t_0, \Phi_0)(t)$ of the set of equations (12.11) exists on J_0. Then, setting $J_1 = (t_1, t_2]$ we consider mappings F and R on $J_1 \times C_0$. Similarly to the above, we get the solution $X_1(t_1, \Phi_0)(t)$ on J_1. Proceeding in this way we arrive at solution $X_k(t_k, \Phi_0)(t)$ on J_k for any $k = 0, 1, 2, \ldots$. According to the accepted definition of solution for the set of equations (12.10) we conclude that the set $X(t_0, \Theta_0)(t)$ is a solution on $[t_0, \infty)$.

Theorem 12.4 *Assume that the following conditions are satisfied:*

(1) *the mappings* $F \in PC(\mathbb{R}_+ \times C_0, \mathcal{K}_c(\mathbb{R}^n))$ *and* $R \in PC(\mathbb{R}_+ \times C_0, \mathcal{K}_c(\mathbb{R}^n))$;

(2) *there exist a function* $g \in PC(\mathbb{R}_+^2 \times [0, 1], \mathbb{R})$ *and a value* $\mu_1 > 0$ *such that*

$$D[F(t, \Phi), (F(t, \Phi) + \mu R(t, \Phi))] \leq g(t, D_0[\Phi, \Theta_0], \mu)$$

for all $t \neq t_k$ *and* $0 < \mu < \mu_1$;

(3) *there exists a sequence of functions* $Q_k(w)$ *which are nondecreasing in* w *for each* $k = 0, 1, 2, \ldots$ *such that*

$$D_0[I_k(X_{t_k}), I_k(Y_{t_k})] \leq Q_k(D_0[X_{t_k}, Y_{t_k}])$$

for all $t = t_k$, $k = 0, 1, 2, \ldots$;

(4) *for* $0 < \mu < \mu_2$ *there exists a maximal solution of comparison equation* (12.12) *on the interval* $[t_0, \infty)$.

Then the distance $D[X(t), Y(t)]$ *between solutions* $X(t) = X(t_0, \Phi_0)(t)$ *of the set of equations* (12.10) *and* $Y(t) = Y(t_0, \Phi_0)(t)$ *of the set of equations* (12.10) *for* $\mu = 0$ *is estimated by the inequality*

$$D[X(t), Y(t)] \leq r(t, t_0, w_0) \qquad (12.13)$$

for all $t \geq t_0$ *and* $0 < \mu < \min(\mu_1, \mu_2)$ *whenever* $D_0[\Phi_0, \Theta_0] \leq w_0$.

Proof Let $J_0 = [t_0, t_1]$. The mappings F and R on the product $J_0 \times C_0$ are continuous. Then, by Theorem 5.4.2 from Lakshmikantham, Bhaskar and Devi [2006] we get the estimate

$$D[X(t), Y(t)] \leq r(t, t_0, w_0), \quad t \in J_0,$$

for $0 < \mu < \min(\mu_1, \mu_2)$. This estimate yields

$$D[X(t_1), Y(t_1)] \leq r(t_1, t_0, w_0)$$

for $0 < \mu < \min(\mu_1, \mu_2)$. By hypothesis (3) of Theorem 12.4 for $t = t_1^+$ we get

$$D_0[X(t_1^+), Y(t_1^+)] = D_0[I_1(X_{t_1}), I_1(Y_{t_1})] \leq Q_1(D_0[X_{t_1}, Y_{t_1}])$$
$$\leq Q_1(r(t_1, t_0, w_0)) \equiv r(t_1^+, t_0, w_0).$$

Therefore, for $t = t_1^+$ we have

$$D_0[X_{t_1^+}, Y_{t_1^+}] \leq r(t_1^+, t_0, w_0).$$

Further, on the interval $J_1 = (t_1, t_2]$ we again consider the mappings F and R. Under hypotheses (1)–(4) on the product $J_1 \times C_0$ we obtain

$$D[X(t), Y(t)] \leq r(t, t_0, w_0)$$

for all $t \in J_1$ and $0 < \mu < \min(\mu_1, \mu_2)$. Having reapeated this procedure we come to the conclusion that estimate (12.13) is valid for all $t \geq t_0$ and $0 < \mu < \min(\mu_1, \mu_2)$.

For the set of equations (12.10) for $\mu = 0$ we consider the function $V_0(t, X, \Phi)$ defined on the product of spaces $\mathcal{K}_c(\mathbb{R}^n)$ and C_0. The function $V_0 \colon \mathbb{R}_+ \times \mathcal{K}_c(\mathbb{R}^n) \times C_0 \to \mathbb{R}_+$ is of the class SL provided that:

(a) $V_0(t, X, \Theta)$ is continuous on $(t_{k-1}, t_k] \times \mathcal{K}_c(\mathbb{R}^n) \times C_0$ and for any $X \in \mathcal{K}_c(\mathbb{R}^n)$ and $\Theta \in C_0$ there exists

$$\lim_{(t, Y, \Theta) \to (t_k^+, X, \Theta)} V_0(t, Y, \Theta) = V_0(t_k^+, X, \Theta);$$

(b) for the function $V_0(t, X, \Theta)$ there exists a constant $L > 0$ such that

$$|V_0(t, X, \Theta) - V_0(t, Y, \Theta)| \leq LD[X, Y]$$

on the product $(t_{k-1}, t_k] \times \mathcal{K}_c(\mathbb{R}^n) \times C_0$.

The function $V_0(t, X, \Theta)$ which, together with the derivative

$$D^+V_0(t, X, \Theta) = \limsup\{[V_0(t + h, X + hF(t, X_t), X_{t+h})$$
$$- V_0(t, X, X_t)]h^{-1} \colon h \to 0^+\},$$

solves the problem on stability of stationary solutions Θ_0 of the set of equations

$$D_H X = F(t, X_t), \quad t \neq t_k,$$
$$X(t_k^+) = I_k(X_{t_k}), \quad t = t_k, \quad (12.14)$$
$$X_{t_0} = \Phi_0 \in C_0,$$

is called the Lyapunov function for the set of impulsive equations with delay. Further we shall consider the sets

$$S(H) = \{X \in \mathcal{K}_c(\mathbb{R}^n): D[X, \Theta_0] < H\},$$
$$Q(H_1) = \{\Phi \in C_0: D[\Phi, \Theta_0] < H_1\},$$

for $0 < H_1 \leq H < +\infty$.

For the set of equations (12.11) we consider the functions $V_0(t, X, \Phi)$ and $V_1(t, X, \Theta, \mu)$ satisfying the following conditions:

A_1. The function $V_0(t, X, \Phi)$, $V_0 \in PC(\mathbb{R}_+ \times S(H) \times Q_1(H_1), \mathbb{R}_+)$, $V_0(t, 0, 0) = 0$, is constructed for the set of equations (12.14) and for it there exist comparison functions a and b of Hahn class K such that for $t \neq t_k$

  (a) $b(D[X(t), \Theta_0]) \leq V_0(t, X, \Phi) \leq a(D[X(t), \Theta_0])$;

  (b) $D^+ V_0(t, X, \Phi)|_{(12.14)} \leq g_0(t, V_0(t, X, \Theta))$, where $g_0 \in PC(\mathbb{R}_+^2, \mathbb{R})$, $g(t, 0) = 0$;

  (c) there exists $H_0 < H$ such that for $t = t_k$ condition $X_{t_k} \in S(H_0)$ implies $I_k(X_{t_k}) \in S(H_0)$ for all $k = 0, 1, 2, \ldots$ and

$$V_0(t_k^+, X(t_0, \Phi_0)(t_k^+), X_{t_k^+}(t_0, \Phi_0))$$
$$\leq Q_k(t_k, V_0(t_k, X(t_0, \Phi_0)(t_k), X_{t_k}(t_0, \Phi_0))), \quad k = 0, 1, 2, \ldots,$$

  where $Q_k \in C(\mathbb{R}_+^2, \mathbb{R})$.

A_2. For some $H_2 > 0$ there exists a perturbation $V_1(t, X, \Phi, \mu)$ of the function $V_0(t, X, \Phi)$ such that $V_1 \in PC(\mathbb{R}_+ \times S(H) \cap S^c(H_2) \times Q(H_1) \times [0, 1], \mathbb{R}_+)$, $V_1(t, 0, \Phi, \mu) = 0$, and moreover,

  (a) $V_1(t, X, \Phi, \mu) \leq c(\mu)$ for $t \in \mathbb{R}_+$, $X \in S(H) \cap S^c(H_2)$, $\Phi \in Q(H_1)$, $c(\mu)$ is a nondecreasing function of μ, $\lim_{\mu \to 0} c(\mu) = 0$;

  (b) $D^+ (V_0(t, X, \Phi) + V_1(t, X, \Phi, \mu))|_{(12.10)} \leq g_1(t, V_0(t, X, \Phi) + V_1(t, X, \Phi, \mu), \mu)$ for $t \neq t_k$, $k = 0, 1, 2, \ldots$;

  (c) $V_0(t_k^+, X(t_0, \Phi_0)(t_k^+), X_{t_k^+}(t_0, \Phi_0)) + V_1(t_k^+, X(t_0, \Phi_0)(t_k^+), X_{t_k^+}(t_0, \Phi_0)) \leq \overline{Q}_k(t_k, V_0(t_k, X(t_0, \Phi_0)(t_k), X_{t_k}(t_0, \Phi_0)) + V_1(t_k, X(t_0, \Phi_0)(t_k), X_{t_k}(t_0, \Phi_0)))$ for $t = t_k$, where $g_1 \in PC(\mathbb{R}_+^2, \mathbb{R})$, $g_1(t, 0, \mu)$, and $\overline{Q}_k \in C(\mathbb{R}_+^2, \mathbb{R})$.

We prove the following assertion.

Theorem 12.5 *For the set of equations (12.10) assume that the functions V_0 and V_1 are constructed which satisfy conditions of assumptions A_1 and A_2. If zero solution of the scalar impulsive equation*

$$\frac{dw}{dt} = g_0(t, w), \quad t \neq t_k,$$

$$w(t_k^+) = Q_k(t_k, w(t_k)), \quad t = t_k, \qquad (12.15)$$

$$w(t_0) = w_0 \geq 0, \quad k = 1, 2, \ldots,$$

is uniformly stable and zero solution of the equation

$$\frac{dv}{dt} = g_1(t, v, \mu), \quad t \neq t_k,$$

$$v(t_k^+) = \overline{Q}_k(t_k, v(t_k)), \quad t = t_k, \qquad (12.16)$$

$$v(t_0) = v_0 \geq 0, \quad k = 1, 2, \ldots,$$

is μ-stable, then the set of stationary solutions Θ_0 of equations (12.14) is stable under small perturbations.

Proof Let the value $0 < \varepsilon < \min(H, H_1, H_2)$ and the value $t_0 \in \mathbb{R}_+$ be given. The uniform stability of zero solution to equation (12.15) implies that for any comparison function $b(\varepsilon) > 0$, b is of Hahn class K and there exists a $\delta_0 = \delta_0(b(\varepsilon)) > 0$ such that for any solution to equation (12.15) the estimate

$$w(t, t_0, w_0) < b(\varepsilon) \quad \text{for all} \quad t \geq t_0,$$

holds true whenever $w_0 < \delta_0$.

Then, the μ-stability of zero solution to equation (12.16) implies that for any solution to this equation, given $b(\varepsilon) - c(\mu) > 0$, $\lim\limits_{\mu \to 0} c(\mu) = 0$, there are $\delta^* = \delta^*(b(\varepsilon) - c(\mu), t_0) > 0$ and $\mu_1 > 0$ such that

$$v(t, t_0, v_0, \mu) < b(\varepsilon) - c(\mu)$$

for all $t \geq t_0$ and $0 < \mu < \mu_1$ whenever $v_0 < \delta^*$.

Let $w_0 = V_0(t_0, \Theta_0, \Phi_0)$. We take $\delta_1 > 0$ so that the inequalities

$$D_0[\Phi_0, \Theta_0] < \delta_1 \quad \text{and} \quad V_0(t_0, \Phi_0, \Theta_0) < \delta^*$$

are satisfied simultaneously.

Let $\delta = \min(\delta_0, \delta^*)$. We show that for $D_0[\Phi_0, \Theta_0] < \delta$ the estimate $D[X(t_0, \Theta_0)(t), \Theta_0] < \varepsilon$ holds true for all $t \geq t_0$, where $X(t_0, \Theta_0)(t)$ is any solution of the set of equations (12.10). If this is not true, then for some solution $X(t_0, \Theta_0)(t)$ of equations (12.10) for the initial values $D_0[\Phi_0, \Theta_0] < \delta$ there must exist t_1 and t_2 such that for some k, $t_k < t_1 < t_2 < t_{k+1}$, the estimate

$$D[X(t_0, \Theta_0)(t), \Theta_0] < \varepsilon \quad \text{for} \quad t_k \geq t > t_0$$

is valid and for $t = t_2$ and $t = t_1$

$$D[X(t_0, \Theta_0)(t_2), \Theta_0] \geq \varepsilon,$$
$$D[X(t_0, \Theta_0)(t_1), \Theta_0] = \delta^*,$$

respectively. Here $X(t_0, \Theta_0)(t) \in \overline{S}(\varepsilon)$ on the interval $[t_1, t_2]$.

Let $H_2 = \delta^*$ and the function $m(t) = V_0(t, X(t_0, \Theta_0)(t), X_t(t_0, \Theta_0)) + V_1(t, X(t_0, \Theta_0)(t), X_t(t_0, \Theta_0), \mu)$ be defined on the interval $[t_1, t_2]$. By conditions A_1 and A_2 for the function $m(t)$ we have

$$m(t_1) \leq v(t_1, t_0, v_0, \mu), \tag{12.17}$$

where $v(t_1, t_0, v_0, \mu)$ is a maximal solution of comparison equation (12.16). Conditions (b) and (c) of A_2 imply

$$m(t) \leq v(t_1, t_0, v_0, \mu) \quad \text{for all} \quad t \in [t_1, t_2]. \tag{12.18}$$

Conditions $A_1(a)$ and $A_2(a)$ provide the existence of $\mu_2 > 0$ such that

$$b(\varepsilon) - c(\mu) \leq V_0(t, X(t_0, \Theta_0)(t), X_t(t_0, \Theta_0)) + V_1(t, X(t_0, \Theta_0)(t), X_t(t_0, \Theta_0), \mu) \tag{12.19}$$

for all $t \geq t_0$ and $0 < \mu < \mu_2$. In view of inequalities (12.17) and (12.18) we present estimate (12.19) in the form

$$b(\varepsilon) - c(\mu) \leq v(t_2, t_1, v_0, \mu).$$

For the value $t = t_2$ we get

$$\begin{aligned}
b(\varepsilon) - c(\mu) &\leq b(D[X(t_0, \Theta_0)(t_2), \Theta_0]) - c(\mu) \\
&\leq V_0(t_2, X(t_0, \Theta_0)(t_2), X_{t_2}(t_0, \Theta_0)) \\
&\quad + V_1(t_2, X(t_0, \Theta_0)(t_2), X_{t_2}(t_0, \Theta_0), \mu) \\
&\leq v(t_2, t_1, v_0, \mu) < b(\varepsilon) - c(\mu)
\end{aligned}$$

for $0 < \mu < \min(\mu_1, \mu_2)$. The contradiction obtained shows that the inequality $D[X(t_0, \Theta_0)(t_2), \Theta_0] \geq \varepsilon$ does not hold at the instant $t_2 < t_{k+1}$ for some k. This proves theorem 12.5.

It is known that for systems of equations with delay and/or impulsive perturbations both the Lyapunov–Razumikhin functions and the Lyapunov–Krasovskiy functionals are applied (see Burton and Hatvani [1989] and bibliography therein). However, both approaches have some disadvantages which are not characteristic of the direct Lyapunov method. The application of Lyapunov functions given on the product of spaces $\mathcal{K}_c(\mathbb{R}^n)$ and C_0 enables us to study the set of systems with delay and impulsive perturbations (12.10) by using the Dini derivatives $D^+V_0(t, X, \Phi)$ and $D^+V_1(t, X, \Phi, \mu)$, the calculation of which neither involves estimation of minimal classes of the initial functions for system (12.10) nor requires knowledge of its solutions. Also, note that the application of perturbed Lyapunov function in the analysis of system (12.10) allows us to establish stability conditions for the set of stationary solutions Θ_0 under small perturbations in the case when Θ_0 for the set of systems (12.14) is uniformly stable only.

Chapter 13

Comments and References

Chapter 1 The fundamental problems of the dynamics and the stability of systems with uncertain parameter values attract the attention of many specialists. During recent years the progress in the study of the stability of nonlinear and controlled uncertain systems is connected with the constructive application of the ideas of the direct Lyapunov method. Actual results obtained in this direction are given in the review by Corless [1993b], which contains a large bibliography (also see Garofalo and Glielmo [1996]). The concept of the parametric stability (see Ikeda et al. [1991]) and the original idea of the article by Lakshmikantham and Vatsala [1997] were incentives for conducting a series of studies of the dynamic behavior of nonlinear systems with uncertain values of their parameters.

Chapter 2 The bases of this chapter are the articles by Martynyuk-Chernienko [1998a, 1999b, 1999c, 2000e, 2002f, 2001g, 2003j, 2005k, 2007l] and Martynyuk-Chernienko and Slyn'ko [2002a]. Some results of the monographs by Demidovich [1967], Grujić et al. [1984], Krasovskiy [1959], Martynyuk [2000a] and also the articles by Lakshmikantham and Vatsala [1997], Martynyuk [1998c], Zubov [1994a], Corless and Leitmann [1996a] were used in the setting of the problem of the stability of uncertain systems. The Theorem 2.5 was proved in the article by Lakshmikantham and Shahzad [1996], using the Lemma 2.1 from the article by Pucci and Serrin [1995], where it was given without a proof. The version of the Theorem 2.11 takes into account the results of the original article by Leela [1997]. The Corollary 2.1 is given according to the article by Corless and Leitmann [1996a], where it is stated in the form of a separate theorem for a controlled system.

Chapter 3 The results of this chapter were published in the articles by Martynyk-Chernienko and Slynko [2003]. Some assumptions on the nominal system (3.2) correspond to those introduced in the article by Chen [1996]. The example of the uncertain system (3.22) is given according to the article by Lakshmikantham and Leela [1995b]. The uncertain system with a neuron regulator (3.40) is analyzed, following the approach of the article by Khlaeoom and Kuntanapreeda [2005]. The results of paragraph 3.6 given in the article by Khoroshun [2008] are from studies by Otha and Šiljak [1994]. In the estimation of the a priori domains of the parametric stability the approach described in Appendix is used, according to the article by Martynyuk and Khoroshun [2007a].

Chapter 4 In this chapter, in accordance with the articles by Martynyuk-

Chernienko [1999c, 1999d], Martynyuk-Chernienko [2003j], and Martynyuk and Martynyuk-Chernienko [2007], the approach to the analysis of uncertain quasilinear systems on the basis of the canonical matrix Lyapunov function is described. The transformation of the initial quasilinear system and the analysis of the case of "nonautonomous" uncertainties is made on the basis of some results obtained by Tikhonov [1965a, 1969b]. Our statement of the Theorem 4.3 is based on the results of the article by Chen and Zhang [2007]. The Examples 4.1 and 4.2 were considered in the article by Ming-Chung Ho, Yao-Chen Huang and Min Jiang [2007].

Chapter 5 The results of this chapter are new. They were obtained by using some of the results of the monograph by Michel and Miller [1977] and the article by Martynyuk [2000a]. Some results of the article by Xing-Gang Yan and Guang-Zhong Dai [1999] were used in the analysis of the stability of the large-scale system (5.43).

Chapter 6 This chapter contains some approaches to the solution of the problem of the stability of uncertain systems, which are based on the method of the vector Lyapunov functions (see Matrosov [2001]). In the article by Martynyuk-Chernienko [1999c] the vector approach for quasilinear uncertain systems is based on the canonical vector Lyapunov function. The criterion of the interval stability of a linear mechanical system is given here on the basis of the article by Martynyuk and Slyn'ko [2003]. The conditions for the parametric stability of a large-scale system were obtained in the article by Martynyuk and Khoroshun [2008b]. Those studies join the works by Ikeda, Ohta, and Šiljak [1989, 1991, 1994] and are based on the results of paragraph 3.6.

Chapter 7 The results of this paragraph are based on the articles by Martynyuk [2002a], Martynyuk-Chernienko [2004b], Liu Bin et al. [2005] and Lakshmikantham and Devi [1993] (see also Martynyuk [2007d], containing a large bibliography of studies of the theory of stability of impulsive systems, which have been carried out recently).

Chapter 8 The necessary definitions and preliminary information from the mathematical analysis on a time scale are given here according to the article by Bohner and Martynyuk [2007]. More detailed information in this direction is available in the monograph by Bohner and Peterson [2001]. The Theorems 8.1–8.4 are new for the given class of systems of dynamic equations. In their proof the results obtained by Martynyuk-Chernienko [2007i], Bohner and Martynyuk [2001] were used. Some results of the work by DaCunha [2004] were applied during the analysis of the Examples 8.1–8.3.

Linear interval matrix systems with continuous and discrete dynamics are considered in the article by Pastravanu and Voicu [2002]. The authors discuss the conditions for the exponential stability for this class of equations, which are adequate to the Shur or Gurwitz conditions.

Chapter 9 The basic concepts of the theory of large-scale systems under structural perturbations are given in monographs by Grujić et al. [1984].

The results of the analysis of the stability of the zero solution of the singularly perturbed system (9.1) are new in the context of the proposed approach (see Martynyuk and Miladzhanov [2011]). Some of the results of the monograph by Grujić et al. [1984] are used here along with the method of matrix-valued Lyapunov functions (see Martynyuk [1998c]).

Chapter 10 The general information from the set theory and the theory of matrix spaces is given here in accordance with the monograph by Zubov [1957b] and Lakshmikantham, Bhaskar, and Devi [2006]. The matrix-valued Lyapunov function (10.5) is introduced here for the first time.

Chapter 11 This chapter is based on the paper by Martynyuk [2002f]. The theory of equations with a causal operator is set out in the monograph by Corduneanu [2002] with a vast bibliography. The paper by Drici et al. [2006] deals with the sets of systems of equations with a causal operator.

Chapter 12 The contents of Chapter 12 are taken from Martynyuk [2009]. See also Lakshmikantham, Bhaskar and Devi [2006].

Appendix The estimates of the domains of the parametric stability are given according to the work by Martynyuk and Khoroshun [2007a]. They were obtained by using some information of their theory of operator equations (see Collatz [1966]).

Appendix

In the problem of the parametric stability of the system (1.1), the estimates of the domains of variations of parameters p and the phase vector x are important, at which there exists a solution $x(t,p)$ of the system (1.1), remaining in a certain domain of the phase space $\mathbb{R}^n$ at all $t \geq 0$ (see Ikeda, Ohta, and Šiljak [1991]). Below you will find an approach to the solution of that problem, which is based on some results of the monograph by Collatz [1966].

Let us recollect some concepts from the theory of pseudometric spaces.

Definition A.1 The space $\mathbb{R}$ is called *pseudometric*, if to any pair of elements f, g from $\mathbb{R}$ the (pseudo)distance $\rho(f,g)$ is put in correspondence, which is an element of (generally speaking, another) linear partially ordered space H above the numeric field K with a zero element Θ_H that satisfies the following conditions:

(a) $\rho(f,g) = \Theta_H$ if, and only if, $f = g$;

(b) $\rho(f,h) \leq \rho(f,g) + \rho(g,h)$ for any triplet f, g, h from $\mathbb{R}$.

The convergence in a linear partially ordered space H is introduced in a usual way, namely, a correspondence is set up between some sequences ρ_n and their limit element ρ from H; those sequences are called convergent, and then $\lim_{n\to\infty} \rho_n = \rho$ or $\rho_n \to \rho$.

The concept of convergence in a pseudometric space $\mathbb{R}$ is defined as follows.

Definition A.2 A sequence $f_n \in \mathbb{R}$ is called *convergent* to the boundary element $f \in \mathbb{R}$ if the sequence of distances $\rho(f_n, f)$ converges to the zero element Θ_H, $\rho(f_n, f) \to \Theta_H$. Here it is assumed that there exists a boundary element $f \in \mathbb{R}$.

Consider the operator equation

$$Tu = u \qquad (A.1)$$

and the iteration method of its solution

$$u_{n+1} = Tu_n, \quad n = 0, 1 \ldots . \qquad (A.2)$$

Assume that the definition domain D of the operator T lies in the full pseudometric space $\mathbb{R}$ with distances taking on values in a linear semiordered space H. Consider the following conditions:

(a) The operator T is bounded, i. e., there exists a linear continuous operator P such that for any two elements u, w from D

$$\rho(Tu, Tw) \leq P\rho(u, w).$$

(b) P is a monotone operator, i. e., for the distances ρ, ρ' from the space H, such that $\rho \leq \rho'$, the following inequality holds:

$$P\rho \leq P\rho'.$$

(c) The expression $\sum\limits_{j=0}^{\infty} P^j \rho$ makes sense, i. e., the series from iterations P^j converges for any ρ from H.

(d) The ball K consisting of the elements v satisfying the inequality

$$\rho(v, u_1) \leq (E - P)^{-1} \rho(u_0, u_1) - (u_0, u_1),$$

and the element u_0 belong to the definition domain D.

Theorem A.1 *Under the conditions (a)–(d) there exists a unique solution of the operator equation (A.1), and a sequence u_n determined by the equality (A.2) converges to that solution.*

The proof of the Theorem A.1 can be found in the monograph by Collatz [1966].

Consider the system of equations

$$f_i(x_1, \ldots, x_n, p_1, \ldots, p_m) = 0, \quad i = 1, \ldots, n, \tag{A.3}$$

with respect to the unknown $x_1, \ldots, x_n$, considering $p_1, \ldots, p_m$ to be independent variables. Assuming $x = (x_1, \ldots, x_n)^{\mathrm{T}}$, $p = (p_1, \ldots, p_m)^{\mathrm{T}}$, $f(x, p) = (f_1(x, p), \ldots, f_n(x, p))^{\mathrm{T}}$, write the system of equations (A.3) in the vector form

$$f(x, p) = 0. \tag{A.4}$$

A solution of the equation (A.4) will be understood as any continuous vector function $x = \varphi(p)$ of vector argument p, defined on some open set of the space $\mathbb{R}^m$ of variables $p_1, \ldots, p_m$, which, when substituted into the set (A.4), turns it into the identity

$$f(\varphi(p), p) = 0,$$

which holds for all points p from that set.

Let us make the following assumptions regarding the system (A.3) and the vector equation (A.4).

Assumption A.1 The system of equations (A.3) and the vector function $f(x, p)$ are such that:

(1) the functions $f_i(x_1, \ldots, x_n, p_1, \ldots, p_m)$, $i = 1, \ldots, n$, are defined and continuous on some open set Γ of the space $\mathbb{R}^n \times \mathbb{R}^m$ together with the partial derivatives

$$\frac{\partial f_i}{\partial x_j}, \quad i, j = 1, \ldots, n,$$

and

$$\frac{\partial^2 f_i}{\partial x_l \partial x_t}, \quad i, l, t = 1, \ldots, n, \quad \frac{\partial^2 f_i}{\partial x_l \partial p_s}, \quad i, l, s = 1, \ldots, n;$$

(2) for some value of the vector-parameter p^* there exists a solution $x^* \in \mathbb{R}^n$ of the vector equation (A.4), such that

$$f(x^*, p^*) = 0$$

and the point (x^*, p^*) belongs to the set Γ;

(3) $\det \left(\dfrac{\partial f(x, p)}{\partial x} \right) \neq 0$ in the point (x^*, p^*).

Since the domain of existence of an implicit function lies in the domain where the Jacobian of the equation, which determines that function, is nonzero, at first we will indicate the method of estimation of the domain in the space $\mathbb{R}^n$ in the open set Γ, where $\det \left(\frac{\partial f(x,p)}{\partial x} \right) \neq 0$. Note that since, according to the conditions of the Assumption A.1, $\det \left(\frac{\partial f(x,p)}{\partial x} \right) \neq 0$ in the point (x^*, p^*), the function $f(x, p)$ is continuous in the whole set Γ, and the point (x^*, p^*) belongs to that set, such a domain always exists.

Represent $\frac{\partial f(x,p)}{\partial x}$ in the form

$$\frac{\partial f(x, p)}{\partial x} = \frac{\partial f(x^*, p^*)}{\partial x} + B(x, p)$$
$$= \left(\frac{\partial f(x^*, p^*)}{\partial x} \right) \left[E + \left(\frac{\partial f(x^*, p^*)}{\partial x} \right)^{-1} B(x, p) \right],$$

where $B(x, p)$ is a matrix of the respective dimensions. Since the matrix $\left(\frac{\partial f(x^*, p^*)}{\partial x} \right)$ is nonsingular, the non-singularity of the matrix $\left(\frac{\partial f(x,p)}{\partial x} \right)$ is equivalent to the non-singularity of the matrix

$$\left[E + \left(\frac{\partial f(x^*, p^*)}{\partial x} \right)^{-1} B(x, p) \right],$$

which is equivalent to the condition

$$\left\| \left(\frac{\partial f(x^*, p^*)}{\partial x} \right)^{-1} B(x, p) \right\| < 1.$$

Taking into account that

$$B(x,p) = \left(\frac{\partial f(x,p)}{\partial x}\right) - \left(\frac{\partial f(x^*,p^*)}{\partial x}\right),$$

one finally obtains that for the non-singularity of the matrix $\left(\frac{\partial f(x,p)}{\partial x}\right)$ in some domain it is required that for all points (x,p) from that domain the following inequality should hold:

$$\left\| \left(\frac{\partial f(x^*,p^*)}{\partial x}\right)^{-1} \left[\frac{\partial f(x,p)}{\partial x} - \frac{\partial f(x^*,p^*)}{\partial x}\right] \right\| < 1. \qquad (A.5)$$

Let

$$x = (x_1,\ldots,x_n)^{\mathrm{T}} = (X_1^{\mathrm{T}},\ldots,X_s^{\mathrm{T}})^{\mathrm{T}},$$

where

$$X_1 = (x_1,\ldots,x_{n_1})^{\mathrm{T}},$$
$$X_2 = (x_{n_1+1},\ldots,x_{n_1+n_2})^{\mathrm{T}},$$
$$\cdots\cdots\cdots\cdots\cdots\cdots\cdots$$
$$X_s = (x_{n_1+\ldots+n_{s-1}+1}, \ldots, x_n)^{\mathrm{T}},$$
$$X_i \in \mathbb{R}^{n_i}, \quad 1 \le n_i < n, \quad i = 1,\ldots,s, \quad n_1+\ldots+n_s = n, \quad n_0 = 0,$$

and

$$p = (p_1,\ldots,p_m)^{\mathrm{T}} = (P_1^{\mathrm{T}},\ldots,P_t^{\mathrm{T}})^{\mathrm{T}},$$

where

$$P_1 = (p_1,\ldots,p_{m_1})^{\mathrm{T}},$$
$$P_2 = (p_{m_1+1}, \ldots, p_{m_1+m_2})^{\mathrm{T}},$$
$$\cdots\cdots\cdots\cdots\cdots\cdots\cdots$$
$$P_t = (p_{m_1+\ldots+m_{t-1}+1},\ldots,p_m)^{\mathrm{T}},$$
$$P_i \in \mathbb{R}^{m_i}, \quad 1 \le m_i < m, \quad i = 1,\ldots,t, \quad m_1+\ldots+m_t = m, \quad m_0 = 0.$$

Let the sought domain

$$\Pi_{r,q} = \{(x,p) \mid \quad \Omega_r: |X_i - X_i^*| \le r_i, \quad \Omega_p: |P_j - P_j^*| < q_j,$$
$$i = 1,\ldots,s, \quad j = 1,\ldots,t\}.$$

Then, using the condition (A.5), the representation of variables x, p and the Taylor expansion of the elements of the matrix $\frac{\partial f(x,p)}{\partial x}$ in the neighborhood of the point (x^*, p^*), we obtain:

$$\left\| \left(\frac{\partial f(x^*, p^*)}{\partial x} \right)^{-1} \left[\frac{\partial f(x, p)}{\partial x} - \frac{\partial f(x^*, p^*)}{\partial x} \right] \right\| \leq \left\| \left(\frac{\partial f(x^*, p^*)}{\partial x} \right)^{-1} \right\|$$

$$\times \left\| \begin{array}{ccc} \left| \dfrac{\partial f_1(x, p)}{\partial x_1} - \dfrac{\partial f_1(x^*, p^*)}{\partial x_1} \right| & \cdots & \left| \dfrac{\partial f_1(x, p)}{\partial x_n} - \dfrac{\partial f_1(x^*, p^*)}{\partial x_n} \right| \\ \cdots\cdots\cdots\cdots\cdots & & \cdots\cdots\cdots\cdots\cdots \\ \left| \dfrac{\partial f_n(x, p)}{\partial x_1} - \dfrac{\partial f_n(x^*, p^*)}{\partial x_1} \right| & \cdots & \left| \dfrac{\partial f_n(x, p)}{\partial x_n} - \dfrac{\partial f_n(x^*, p^*)}{\partial x_n} \right| \end{array} \right\| \qquad \text{(A.6)}$$

$$< \left\| \left(\frac{\partial f(x^*, p^*)}{\partial x} \right)^{-1} \right\|$$

$$\times \left(\sum_{i=1}^{s} r_i \max_{(x,p)\in\Pi_{r,q}} \|A_i(x, p)\| + \sum_{j=1}^{t} q_j \max_{(x,p)\in\Pi_{r,q}} \|D_j(x, p)\| \right),$$

where

$$A_i(x, p) = \left(\begin{array}{ccc} \left| \dfrac{\partial^2 f_1(x, p)}{\partial x_1 \partial X_i} \right| & \cdots & \left| \dfrac{\partial^2 f_1(x, p)}{\partial x_n \partial X_i} \right| \\ \cdots\cdots\cdots\cdots\cdots & & \cdots\cdots\cdots\cdots\cdots \\ \left| \dfrac{\partial^2 f_n(x, p)}{\partial x_1 \partial X_i} \right| & \cdots & \left| \dfrac{\partial^2 f_n(x, p)}{\partial x_n \partial X_i} \right| \end{array} \right),$$

$$D_j(x, p) = \left(\begin{array}{ccc} \left| \dfrac{\partial^2 f_1(x, p)}{\partial x_1 \partial P_j} \right| & \cdots & \left| \dfrac{\partial^2 f_1(x, p)}{\partial x_n \partial P_j} \right| \\ \cdots\cdots\cdots\cdots\cdots & & \cdots\cdots\cdots\cdots\cdots \\ \left| \dfrac{\partial^2 f_n(x, p)}{\partial x_1 \partial P_j} \right| & \cdots & \left| \dfrac{\partial^2 f_n(x, p)}{\partial x_n \partial P_j} \right| \end{array} \right).$$

The estimate (A.6) is obtained, taking into account that

$$\left| \frac{\partial f_k(x, p)}{\partial x_l} - \frac{\partial f_k(x^*, p^*)}{\partial x_l} \right|$$

$$= \left| \sum_{i=1}^{n} \frac{\partial^2 f_k(x, p)}{\partial x_l \partial x_i} \right|_{\substack{x=x^*+\theta(x-x^*) \\ p=p^*+\theta(p-p^*)}} (x_i - x_i^*)$$

$$+ \sum_{j=1}^{m} \frac{\partial^2 f_k(x, p)}{\partial x_l \partial p_j} \bigg|_{\substack{x=x^*+\theta(x-x^*) \\ p=p^*+\theta(p-p^*)}} (p_j - p_j^*) \bigg|$$

$$\leq \sum_{i=1}^{s} \max_{\Pi_{r,q}} \left| \frac{\partial^2 f_k(x, p)}{\partial x_l \partial X_i} \right| |X_i - X_i^*| + \sum_{j=1}^{t} \max_{\Pi_{r,q}} \left| \frac{\partial^2 f_k(x, p)}{\partial x_l \partial P_j} \right| |P_j - P_j^*|,$$

where $k, l = 1, \ldots, n$.

Thus, taking into account (A.5) and (A.6), we calculate the estimates for r_i and q_j, which determine the boundaries of the domain in which

$\det \left(\frac{\partial f(x,p)}{\partial x} \right) \neq 0$, in the form

$$\left\| \left(\frac{\partial f(x^*,p^*)}{\partial x} \right)^{-1} \right\|$$

$$\times \left(\sum_{i=1}^{s} r_i \max_{(x,p)\in\Pi_{r,q}} \|A_i(x,p)\| + \sum_{j=1}^{t} q_j \max_{(x,p)\in\Pi_{r,q}} \|D_j(x,p)\| \right) \leq 1. \qquad (A.7)$$

Note that for the estimation of this domain for matrices, the Hilbert-Schmidt norm is used.

Let $(x,p) \in \Pi_{r,q}$. Determine a domain in the parameter space, where for each parameter value there exists a solution of the equation (A.4), and a domain in the space of variables, to which that solution belongs.

For that, rewrite the equation (A.4) in a somewhat modified form. Assume

$$b_j^i = \frac{\partial}{\partial x_j} f_i(x^*,p^*), \quad i,j = 1,\ldots,n. \qquad (A.8)$$

Since the functional determinant $\det \left(\frac{\partial f(x,p)}{\partial x} \right)$ is nonzero in the point (x^*,p^*), the matrix $B = (b_j^i)$ has an inverse matrix $B^{-1} = (c_j^i)$. Now rewrite the system of equations (A.3) in the form

$$\sum_j b_j^i x_j = \sum_j b_j^i x_j - f_i(x,p), \quad i = 1,\ldots,n. \qquad (A.9)$$

Let $h_j(x,p)$ denote the right-hand part of the relation (A.9). In view of the relation (A.8), we obtain

$$\frac{\partial}{\partial x_j} h_i(x^*,p^*) = 0, \quad i,j = 1,\ldots,n. \qquad (A.10)$$

The equation (A.9) can be written in the vector form as follows:

$$Bx = h(x,p).$$

Applying the matrix B^{-1} to the above relation, we obtain an equivalent relation

$$x = g(x,p), \qquad (A.11)$$

where $g(x,p) = (g_1(x,p),\ldots,g_n(x,p)) = B^{-1}h(x,p)$. From (A.10) it follows that

$$\frac{\partial}{\partial x_j} g_i(x^*,p^*) = 0. \qquad (A.12)$$

Since the equation (A.12) is equivalent to the equation (A.4) and in the point (x^*,p^*) the relation (A.4) holds, then

$$x^* = g(x^*,p^*). \qquad (A.13)$$

Apply the method of successive approximations to the equation (A.11) which satisfies the conditions (A.12) and (A.13). Namely, set up a correspondence between each vector function $x = x^e(p)$ vector variable p, and the function $\tilde{x}^e(p)$, assuming

$$\tilde{x}^e(p) = g(x^e(p), p). \tag{A.14}$$

Then use the "operator" notation

$$\tilde{x}^e(p) = Gx^e(p). \tag{A.15}$$

Apply the Theorem A.1 to the equation (A.15). Let the domain of definition of the operator G lie in the full pseudometric space $X = \{x = (x_1, \ldots, x_n)^{\mathrm{T}} = (X_1^{\mathrm{T}}, \ldots, X_s^{\mathrm{T}})^{\mathrm{T}}\}$ with distances taking on values in the linear semiordered space $K = \{z = (z_1, \ldots, z_s)^{\mathrm{T}} : z_i \geq 0, \ i = 1, \ldots, s)\} \subseteq \mathbb{R}_+^s$, here for all x and y from X

$$\rho(x, y) = (|X_1 - Y_1|, \ \ldots, \ |X_s - Y_s|)^{\mathrm{T}}. \tag{A.16}$$

Show that the operator G is bounded, i. e., that there exists a linear continuous operator P in the space K, such that for any two elements x and y from X the following relation holds:

$$\rho(Gx, Gy) \leq P\rho(x, y).$$

Taking into account the notation (A.14)–(A.16), obtain

$$\rho(Gx, Gy) = (|\overline{g_1}(x, p) - \overline{g_1}(y, p)|, \ \ldots, \ |\overline{g_s}(x, p) - \overline{g_s}(y, p)|)^{\mathrm{T}}, \tag{A.17}$$

where $\overline{g_i}(x, p) \in \mathbb{R}^{n_i}$,

$$\overline{g_1}((X_1^{\mathrm{T}}, \ldots, X_s^{\mathrm{T}})^{\mathrm{T}}, p) = (g_1(x, p), \ldots, g_{n_1}(x, p))^{\mathrm{T}},$$
$$\overline{g_2}((X_1^{\mathrm{T}}, \ldots, X_s^{\mathrm{T}})^{\mathrm{T}}, p) = (g_{n_1+1}(x, p), \ldots, g_{n_1+n_2}(x, p))^{\mathrm{T}},$$
$$\ldots \ldots \ldots \ldots \ldots \ldots \ldots \ldots \ldots \ldots \ldots \ldots \ldots \ldots \ldots$$
$$\overline{g_n}((X_1^{\mathrm{T}}, \ldots, X_s^{\mathrm{T}})^{\mathrm{T}}, p) = (g_{n_1+\ldots+n_{s-1}+1}(x, p), \ldots, g_n(x, p))^{\mathrm{T}}.$$

Consider the k-th coordinate of this vector. On the basis of the Lagrange finite increments formula, obtain

$$\|\overline{g_k}((X_1^{\mathrm{T}}, \ldots, X_s^{\mathrm{T}}), p) - \overline{g_k}((Y_1^{\mathrm{T}}, \ldots, Y_s^{\mathrm{T}}), p)\|$$
$$= \left\| \sum_{j=1}^{s} \frac{\partial \overline{g_k}((X_1^{\mathrm{T}}, \ldots, X_s^{\mathrm{T}}), p)}{\partial X_j} \right|_{x=x+\theta(y-x)} (X_j - Y_j) \right\| \tag{A.18}$$

$$\leq \sum_{j=1}^{s} \left\| \begin{pmatrix} \left. \dfrac{\partial g_{n_1+\ldots+n_{k-1}+1}(x,p)}{\partial x_{n_1+\ldots+n_{j-1}+1}} \right|_{\substack{x=x+\\ \theta(y-x)}} & \cdots & \left. \dfrac{\partial g_{n_1+\ldots+n_{k-1}+1}(x,p)}{\partial x_{n_1+\ldots+n_j}} \right|_{\substack{x=x+\\ \theta(y-x)}} \\ \cdots\cdots\cdots\cdots\cdots\cdots\cdots\cdots & & \cdots\cdots\cdots\cdots\cdots\cdots\cdots\cdots \\ \left. \dfrac{\partial g_{n_1+\ldots+n_k}(x,p)}{\partial x_{n_1+\ldots+n_{j-1}+1}} \right|_{\substack{x=x+\\ \theta(y-x)}} & \cdots & \left. \dfrac{\partial g_{n_1+\ldots+n_k}(x,p)}{\partial x_{n_1+\ldots+n_j}} \right|_{\substack{x=x+\\ \theta(y-x)}} \end{pmatrix} \right\|$$

$$\times \|X_j - Y_j\|$$

$$\leq \sum_{j=1}^{s} \max_{(x,p)\in\Pi_{r,q}} \left\| \begin{pmatrix} \dfrac{\partial g_{n_1+\ldots+n_{k-1}+1}(x,p)}{\partial x_{n_1+\ldots+n_{j-1}+1}} & \cdots & \dfrac{\partial g_{n_1+\ldots+n_{k-1}+1}(x,p)}{\partial x_{n_1+\ldots+n_j}} \\ \cdots\cdots\cdots\cdots\cdots\cdots\cdots\cdots & & \cdots\cdots\cdots\cdots\cdots\cdots\cdots\cdots \\ \dfrac{\partial g_{n_1+\ldots+n_k}(x,p)}{\partial x_{n_1+\ldots+n_{j-1}+1}} & \cdots & \dfrac{\partial g_{n_1+\ldots+n_k}(x,p)}{\partial x_{n_1+\ldots+n_j}} \end{pmatrix} \right\|$$

$$\times \|X_j - Y_j\| = \sum_{i=1}^{s} \alpha_j^k \|X_j - Y_j\|.$$

If the application of the spectral norm results in significant computational difficulties, then it is advisable to apply the Hilbert-Schmidt norm for matrices, and the following technique.

Taking into account that

$$g(x,p) = B^{-1}h(x,p), \quad h_i(x,p) = \sum_j b_j^i x_j - f_i(x,p),$$

and applying the expansion of the elements of the matrix $\frac{\partial f(t,x)}{\partial x}$ in the neighborhood of the point (x^*, p^*) into a Taylor series, obtain

$$\left| \dfrac{\partial g_{n_1+\ldots+n_{k-1}+l}(x,p)}{\partial x_{n_1+\ldots+n_{j-1}+c}} \right|_{x=x+\theta(y-x)} \right|$$

$$= \left| \sum_{i=1}^{n} c_i^{n_1+\ldots+n_{k-1}+l} \left(b_i^{n_1+\ldots+n_{j-1}+c} - \dfrac{\partial f_i(x,p)}{\partial x_{n_1+\ldots+n_{j-1}+c}} \right) \right|_{x=x+\theta(y-x)} \right|$$

$$\leq \sum_{i=1}^{n} |c_i^{n_1+\ldots+n_{k-1}+l}| \left(\sum_{h=1}^{s} \max_{(x,p)\in\Pi_{r,q}} \left| \dfrac{\partial^2 f_i(x,p)}{\partial x_{n_1+\ldots+n_{j-1}+c}\partial X_h^{\mathrm{T}}} \right| r_h \right.$$

$$\left. + \sum_{f=1}^{t} \max_{(x,p)\in\Pi_{r,q}} \left| \dfrac{\partial^2 f_i(x,p)}{\partial x_{n_1+\ldots+n_{j-1}+c}\partial P_f^{\mathrm{T}}} \right| q_f \right)$$

$$= \beta_{n_1+\ldots+n_{j-1}+c}^{n_1+\ldots+n_{k-1}+l}, \quad k,j=1,\ldots,s, \quad l=1,\ldots,n_k, \quad c=1,\ldots,n_j.$$

Then,

$$\left| \dfrac{\partial g_{n_1+\ldots+n_{k-1}+l}(x,p)}{\partial x_{n_1+\ldots+n_{j-1}+c}} \right|_{x=x+\theta(y-x)} \right| \leq \beta_{n_1+\ldots+n_{j-1}+c}^{n_1+\ldots+n_{k-1}+l}, \tag{A.19}$$

$$k,j=1,\ldots,s, \quad l=1,\ldots,n_k, \quad c=1,\ldots,n_j,$$

$$
\left\|\left(\begin{array}{ccc}
\left.\dfrac{\partial g_{n_1+\ldots+n_{k-1}+1}(x,p)}{\partial x_{n_1+\ldots+n_{j-1}+1}}\right|_{x=x+\theta(y-x)} & \cdots & \left.\dfrac{\partial g_{n_1+\ldots+n_{k-1}+1}(x,p)}{\partial x_{n_1+\ldots+n_j}}\right|_{x=x+\theta(y-x)} \\
\cdots\cdots\cdots\cdots\cdots\cdots\cdots\cdots\cdots\cdots\cdots\cdots\cdots\cdots\cdots\cdots\cdots\cdots \\
\left.\dfrac{\partial g_{n_1+\ldots+n_k}(x,p)}{\partial x_{n_1+\ldots+n_{j-1}+1}}\right|_{x=x+\theta(y-x)} & \cdots & \left.\dfrac{\partial g_{n_1+\ldots+n_k}(x,p)}{\partial x_{n_1+\ldots+n_j}}\right|_{x=x+\theta(y-x)}
\end{array}\right)\right\|
$$

$$
\leq \left\|\left(\begin{array}{ccc}
\beta_{n_1+\ldots+n_{j-1}+1}^{n_1+\ldots+n_{k-1}+1} & \cdots & \beta_{n_1+\ldots+n_j}^{n_1+\ldots+n_{k-1}+1} \\
\cdots\cdots\cdots\cdots\cdots\cdots\cdots\cdots\cdots\cdots \\
\beta_{n_1+\ldots+n_{j-1}+1}^{n_1+\ldots+n_k} & \cdots & \beta_{n_1+\ldots+n_j}^{n_1+\ldots+n_k}
\end{array}\right)\right\| = \alpha_j^k.
$$

From the estimate (A.18), taking into account (A.19), we obtain the estimate of the k-th coordinate of the vector (A.17):

$$
|\overline{g_k}(x,p) - \overline{g_k}(y,p)| \leq \sum_{j=1}^{s} \alpha_j^k |X_j - Y_j| \tag{A.20}
$$

for all $k = 1,\ldots,s$. From (A.17), using the estimate (A.18) or (A.20), we obtain the estimate

$$
\rho(Gx, Gy) \overset{K}{\to} \leq \left(\sum_{j=1}^{s} \alpha_j^1 |X_j - Y_j|,\ \ldots,\ \sum_{j=1}^{s} \alpha_j^s |X_j - Y_j|\right)^{\mathrm{T}}
$$

$$
= \begin{pmatrix} \alpha_1^1 & \cdots & \alpha_s^1 \\ \cdots\cdots\cdots\cdots\cdots \\ \alpha_1^s & \cdots & \alpha_s^s \end{pmatrix} (|X_1 - Y_1|,\ \ldots,\ |X_s - Y_s|)^{\mathrm{T}} = P\rho(x,y),
$$

where $Pz = (\alpha_j^k)_{k,j=1}^s \cdot z$ for all z from K and the operator P is linear and continuous. The boundedness of the operator G is proved.

It is obvious that P is a monotone operator, i.e., for the distances ρ_1 and ρ_2 from the space K such that $\rho_1 \overset{K}{\to} \leq \rho_2$, the inequality $P\rho_1 \overset{K}{\to} \leq P\rho_2$ holds true.

The expression $\sum_{j=0}^{\infty} P^j \rho$ makes sense, i.e., for the iteration P^j the series $\sum_{j=0}^{\infty} P^j \rho$ converges for any ρ from K, if for the matrix of the operator P the following condition is satisfied

$$
r((\alpha_j^k)_{k,j=1}^s) < 1, \tag{A.21}
$$

where $r(A)$ is the spectral radius of the matrix A.

Using the inequality (A.21), one can determine the values $r_1^1 \leq r_1,\ldots,$ $r_s^1 \leq r_s$ and $q_1^1 \leq q_1,\ldots,q_t^1 \leq q_t$, which define the domain

$$
\Pi_{r^1,q^1} = \{(x,p) \mid\ \Omega_{r_1}\colon\ |X_i - X_i^*| < r_i^1,\ \ \Omega_{p_1}\colon\ \|P_j - P_j^*\| < q_j^1,
$$
$$
i = 1,\ldots,s,\ \ j = 1,\ldots,t\} \subseteq \Pi_{r,q},
$$

in which the conditions (A.7) and (A.21) are satisfied.

We now show that the ball D with the elements $x = (X_1^{\mathrm{T}}, \ldots, X_s^{\mathrm{T}})^{\mathrm{T}}$ of the space X, which satisfy the inequality

$$\rho(x, x_1) \overset{K}{\to} \leq (E - P)^{-1} \rho(x_0, x_1) - \rho(x_0, x_1),$$

and the element $x^* = ((X_1^*)^{\mathrm{T}}, \ldots, (X_s^*)^{\mathrm{T}})^{\mathrm{T}}$ belong to the domain of definition of the operator G. First of all, note that in the domain Π_{r^1,q^1} the existence of the operator G is determined by the inequalities

$$|X_1 - X_1^*| \leq r_1^1,$$

$$\cdots\cdots\cdots\cdots\cdots$$

$$|X_s - X_s^*| \leq r_s^1.$$

Since $\rho(x, x^*) \overset{K}{\to} \leq \rho(x, x_1) + \rho(x_1, x^*)$ and $x_0 = x^*$, for all values of x from the ball D we obtain:

$$\rho(x, x^*) \overset{K}{\to} \leq (E - P)^{-1} \rho(x_1, x^*).$$

Taking into account that $x_n = g(x_{n-1}, p)$, from (A.13) and (A.16), we find

$$\rho(x_1, x^*) = (|\overline{g_1}(x^*, p) - \overline{g_1}(x^*, p^*)|, \ldots, |\overline{g_s}(x^*, p) - \overline{g_s}(x^*, p^*)|). \quad \text{(A.22)}$$

Consider the k-th coordinate of this vector. Applying the Lagrange finite increments formula, for $p = p^* + \theta(p - p^*)$ at all $k = 1, \ldots$ we obtain

$$|\overline{g_k}(x^*, p) - \overline{g_k}(x^*, p^*)| = \left| \sum_{j=1}^{t} \frac{\partial \overline{g_k}(x^*, p)}{\partial P_j^{\mathrm{T}}} (P_j - P_j^*) \right|$$

$$\leq \sum_{j=1}^{t} \left\| \begin{pmatrix} \dfrac{\partial g_{n_1+\ldots+n_{k-1}+1}(x^*, p)}{\partial p_{m_1+\ldots+m_{j-1}+1}} & \cdots & \dfrac{\partial g_{n_1+\cdots+n_{k-1}+1}(x^*, p)}{\partial p_{m_1+\cdots+m_j}} \\ \cdots\cdots\cdots\cdots\cdots\cdots\cdots\cdots\cdots\cdots\cdots\cdots \\ \dfrac{\partial g_{n_1+\ldots+n_k}(x^*, p)}{\partial p_{m_1+\ldots+m_{j-1}+1}} & \cdots & \dfrac{\partial g_{n_1+\ldots+n_k}(x^*, p)}{\partial p_{m_1+\ldots+m_j}} \end{pmatrix} \right\| \quad \text{(A.23)}$$

$$\times |P_j - P_j^*|.$$

Continuing the estimate (A.23), we obtain

$$|\overline{g_k}(x^*, p) - \overline{g_k}(x^*, p^*)|$$

$$\leq \sum_{j=1}^{t} \max_{p \in \Omega_{p1}} \left\| \begin{pmatrix} \dfrac{\partial g_{n_1+\ldots+n_{k-1}+1}(x^*, p)}{\partial p_{m_1+\ldots+m_{j-1}+1}} & \cdots & \dfrac{\partial g_{n_1+\ldots+n_{k-1}+1}(x^*, p)}{\partial p_{m_1+\ldots+m_j}} \\ \cdots\cdots\cdots\cdots\cdots\cdots\cdots\cdots\cdots\cdots\cdots\cdots \\ \dfrac{\partial g_{n_1+\ldots+n_k}(x^*, p)}{\partial p_{m_1+\ldots+m_{j-1}+1}} & \cdots & \dfrac{\partial g_{n_1+\ldots+n_k}(x^*, p)}{\partial p_{m_1+\ldots+m_j}} \end{pmatrix} \right\|$$

$$\times |P_j - P_j^*| = \sum_{j=1}^{t} \Delta_j^k |P_j - P_j^*|.$$

$$\text{(A.24)}$$

If the application of the spectral norm results in significant technical difficulties, one can use the Hilbert-Schmidt norm and the following technique.

Taking into account that

$$g(x,p) = B^{-1}h(x,p), \quad h_i(x,p) = \sum_j b_j^i x_j - f_i(x,p),$$

and applying the expansion of the elements of the matrix $\frac{\partial f(x^*,p)}{\partial p}$ into a Taylor series in the neighborhood of the point p^*, we obtain

$$\left| \frac{\partial g_{n_1+\ldots+n_{k-1}+l}(x^*,p)}{\partial p_{m_1+\ldots+m_{j-1}+c}} \right|_{p=p^*+\theta(p-p^*)}$$

$$= \left| \sum_{i=1}^n c_i^{n_1+\ldots+n_{k-1}+l} \frac{\partial f_i(x^*,p)}{\partial p_{m_1+\ldots+m_{j-1}+c}} \right|_{p=p^*+\theta(p-p^*)}$$

$$\le \sum_{i=1}^n |c_i^{n_1+\ldots+n_{k-1}+l}| \max_{p \in \Omega_{P_1}} \left| \frac{\partial f_i(x^*,p)}{\partial p_{m_1+\ldots+m_{j-1}+c}} \right| = \gamma_{m_1+\ldots+m_{j-1}+c}^{n_1+\ldots+n_{k-1}+l},$$

where $k = 1,\ldots,s$, $j = 1,\ldots,t$, $l = 1,\ldots,n_k$, $c = 1,\ldots,m_j$.

Hence,

$$\left| \frac{\partial g_{n_1+\ldots+n_{k-1}+l}(x^*,p)}{\partial p_{m_1+\ldots+m_{j-1}+c}} \right|_{p=p^*+\theta(p-p^*)} \le \gamma_{m_1+\ldots+m_{j-1}+c}^{n_1+\ldots+n_{k-1}+l}, \tag{A.25}$$

$$k = 1,\ldots,s, \quad j = 1,\ldots,t, \quad l = 1,\ldots,n_k, \quad c = 1,\ldots,m_j.$$

From (A.22), on the basis of (A.23) and (A.25), obtain the estimate

$$\rho(x_1,x^*) \overset{K}{\to} \le \left(\sum_{j=1}^T \Delta_j^1 |P_j - P_j^*|, \ldots, \sum_{j=1}^t \Delta_j^s |P_j - P_j^*| \right)^T$$

$$= \begin{pmatrix} \Delta_1^1 & \cdots & \Delta_t^1 \\ \cdots\cdots\cdots\cdots \\ \Delta_1^s & \cdots & \Delta_t^s \end{pmatrix} (|P_1 - P_1^*|, \ldots, |P_t - P_t^*|)^T$$

$$\overset{K}{\to} \le \begin{pmatrix} \Delta_1^1 & \cdots & \Delta_t^1 \\ \cdots\cdots\cdots\cdots \\ \Delta_1^s & \cdots & \Delta_t^s \end{pmatrix} (q_1^1, \ldots, q_t^1)^T,$$

where

$$\Delta_j^k = \left\| \begin{pmatrix} \gamma_{m_1+\ldots+m_{j-1}+1}^{n_1+\ldots+n_{k-1}+1} & \cdots & \gamma_{m_1+\ldots+m_j}^{n_1+\ldots+n_{k-1}+1} \\ \cdots\cdots\cdots\cdots\cdots\cdots\cdots\cdots \\ \gamma_{m_1+\ldots+m_{j-1}+1}^{n_1+\ldots+n_k} & \cdots & \gamma_{m_1+\ldots+m_j}^{n_1+\ldots+n_k} \end{pmatrix} \right\|.$$

$$k = 1,\ldots,s, \quad j = 1,\ldots,t.$$

We obtain

$$\rho(x_1,x^*) \overset{K}{\to} \le \begin{pmatrix} \Delta_1^1 & \cdots & \Delta_t^1 \\ \cdots\cdots\cdots\cdots \\ \Delta_1^s & \cdots & \Delta_t^s \end{pmatrix} (q_1^1, \ldots, q_t^1)^T. \tag{A.26}$$

Note that a similar estimate can be obtained from the formulas (A.22) and (A.24), where Δ_j^k is determined from the formula (A.24).

Then from (A.23), taking into account (A.26), for all values of x from the ball D we obtain the estimate

$$\rho(x, x^*) \overset{K}{\to} \leq (E - P)^{-1} \begin{pmatrix} \Delta_1^1 & \cdots & \Delta_t^1 \\ \cdots\cdots\cdots\cdots \\ \Delta_1^s & \cdots & \Delta_t^s \end{pmatrix} (q_1^1, \ldots, q_t^1)^{\mathrm{T}}.$$

Under the condition

$$(E - P)^{-1} \begin{pmatrix} \Delta_1^1 & \cdots & \Delta_t^1 \\ \cdots\cdots\cdots\cdots \\ \Delta_1^s & \cdots & \Delta_t^s \end{pmatrix} (q_1^1, \ldots, q_t^1)^{\mathrm{T}} \overset{K}{\to} \leq (r_1^1, \ldots, r_s^1)^{\mathrm{T}} \qquad (A.27)$$

the ball D belongs to the domain of definition of the operator G. It is obvious that the element x^* belongs to that domain.

Thus, in the domain determined from the relations (A.7), (A.21), and (A.27), all the conditions of the Theorem A.1 are satisfied. Then, using the above relations, with the norm corresponding to the problem situation, one can determine the domain in the space of parameters for which the equation (A.4) has a unique and continuous solution and the domain in the space of variables to which this solution belongs.

Bibliography

Aliev, F.A. and Larin, V.B.
Optimization of Linear Control Systems. Analytical Methods and Computational Algorithms. Amsterdam: Gordon and Breach Publishers, 1998.

Aubin, J.P. and Cellina, A.
Differential Inclusions. New York: Springer Verlag, 1984.

Aumann, R.J.
Integrals of set-valued functions. *J. Math. Anal. Appl.* **12** (1965) 1–12.

Bin, L., Xinzhi, L., and Xiaoxin L.
Robust global exponential stability of uncertain impulsive systems. *Math. Acta. Sci.* **25B** (2005) 161–169.

Bittanti, S., Laub, A.J. and Willems, J.C.
The Riccati Equation. Berlin: Springer Verlag, 1991.

Bohner M. and Martynyuk A.A.
Elements of the theory of stability of A.M.Lyapunov for dynamic equations on a time scale. *Prikl. Mekh.* **49**(9) (2007) 3–27.

Bohner, M. and Peterson, A.
Dynamic Equations on Time Scales: An Introduction with Applications. Boston: Birkhauser, 2001.

Boyd, S., Ghaoui, L.E., Feron, E., and Balakrishnan, V.
Linear Matrix Inequalities in System and Control Theory. Philadelphia: SIAM, 1994.

Burton T.A. and Hatvani L.
Stability theorems for nonautonomous functional differential equations. *Tohoku Math. J.* **41** (1989) 65–104.

Cao, D.Q. and Shu, Z.Z.
Robust stability bounds for multi-degree of freedom linear systems with structured perturbations. *Dynamics and Stability of Systems* **9**(1) (1999) 79–87.

Česari, L.
Asymptotic Behavior and Stability Problems in Ordinary Differential Equations. Berlin: Springer Verlag, 1959.

Chen, G. and Dong, X.
From Chaos to Order: Methodologies, Perspectives and Applications. Singapore: World Scientific, 1998.

287

Chen, Y.-H.
  Performance analysis of controlled uncertain systems. *Dynamics Control* **6** (1996) 131–142.

Chen, Y.-H. and Leitmann, G.
  Robustness of uncertain systems in the absence of matching assumptions. *Int. J. Control* **454** (1987) 1527–1542.

Chen, Y.H. and Chen, J.S.
  Robust composite control for singularly perturbed systems with time-varying uncertainties. *J. Dynamic Systems, Measurement Control.* **117** (1995) 445–452.

Chen, F. and Zhang, W.
  LMI criteria for robust chaos synchronization of a class of chaotic systems. *Nonlinear Anal.* **67** (2007) 3384–3393.

Chetaev N.G.
  Stability of Motion. Works on Analytical Mechanics. Moscow: Publ.house of USSR Academy of Science, 1962. 535 p.

Collatz L.
  Functional Analysis and Numerical Mathematics. New York: Academic Press, 1966.

Corduneanu C.
  Functional Equations with Causal Operators. London: Taylor & Francis, 2002.

Corless, M.
[a]   Guaranteed rates of exponential convergence for uncertain systems. *J. Optimiz. Theory and Appl.* **64** (1990) 471–484.
[b]   Control of uncertain nonlinear systems. *ASME J. Dynam. Systems, Measurement, Control.* **115** (1993) 362–372.

Corless, M. and Leitmann, G.
[a]   Exponential convergence for uncertain systems with component-wise bounded controllers. In: *Robust Control via Variable Structure and Lyapunov Techniques*, (Eds.: F. Garofalo and L. Glielmo), Lecture Notes in Control and Information Sciences, Vol. 217, Berlin: Springer Verlag, 1996, p. 175–196.
[b]   *Deterministic Control of Uncertain System via a Constructive use of Lyapunov Stability Theory.* Berlin: Springer Verlag, 1989.

DaCunha, J.J.
  Lyapunov Stability and Floquet Theory for Nonautonomous Linear Dynamic Systems on Time Scales. Waco: Texas, 2004 (PhD thesis).

Deimling, K.
  Multivalued Differential Equations. New York: Walter de Gruyter, 1992.

Demidovich B.P.
  Lectures on the Mathematical Stability Theory. Moscow: Nauka, 1967.

Djordjević, M.Z.
  Stability analysis of interconnected systems with possibly unstable subsystems. *Syst. Contr. Lett.* **3** (1983) 165–169.

Drici, Z., McRae, A. and Devi J.V.
  Stability results for set differential equations with causal maps. *Dynamic Systems Appl.* **15** (2006) 451–464.

Freedman, H.I
Deterministic Mathematical Models in Population Ecology. Edmonton: HIFR Consulting, 1987. 254 p.

Galperin, E.A., and Skowronski, J.M.
Geometry of V-functions and the Lyapunov stability theory. *Proc. 24th Conf. Decision and Control.* 1985, p. 302–303.

Garofalo, F. and Glielmo, L. (Eds.)
Robust Control via Variable Structure and Lyapunov Techniques. Berlin: Springer Verlag, 1996.

Grujich L.T., Martynyuk A.A., and Ribbens-Pavella M.
Stability of Large-Scale Systems under Structural and Singular Perturbations. Kiev: Nauk. Dumka, 1984.

Gutman, S.
Uncertain dynamical systems — A Lyapunov min-max approach. *IEEE Trans. Automat. Contr.* **AC-24** (1979) 437–443.

Hahn, W.
Stability of Motion. Berlin: Springer Verlag, 1967.

Hilger, S.
Analysis on measure chains — A unified approach to continuous and discrete calculus. *Results Math.* **18** (1990) 18–56.

Hukuhara, M.
Sur l'application semicontinue dont la valeur est un pact convexe. *Funck. Ekvac.* (10) (1967) 48–66.

Ikeda, M., Ohta, Y. and Šiljak, D.D.
Parametric stability. *Proc. Univer. Genova,* The Ohio State University Joint Conference. Boston: Birkhäuser, 1991, p. 1–20.

Ikeda, M. and Šiljak, D.D.
Generalized decompositions of dynamic systems and vector Lyapunov functions. *IEEE Trans. Automat. Contr.* **AC-26**(5) (1981) 1118–1125.

Khlaeo-om, P. and Kuntanapreeda, S.
A stability condition for neural network control of uncertain systems. *ESANN 2005 Proc.,* Bruges (Belgium), 2005, p. 187–192.

Khoroshun A.S.
The global parametric quadratic stabilization of nonlinear systems with uncertainties. *Prikl. Mekh.* **44**(6) (2008) 126–133.

Krasovskiy N.N.
Some Problems of Stability of Motion. Moscow: Fizmatgiz, 1959.

Ladyzhenskaya O.A.
Boundary-Value Problems of Mathematical Physics. Moscow: Nauka, 1973.

Lakshmikantham, V., Bainov, D.D., and Simeonov, P.S.
Theory of Impulsive Differential Equations. Singapore: World Scientific, 1989.

Lakshmikantham, V., Bhaskar, T.G. and Devi, V.
Theory of Set Differential Equations. Cambridge: Cambridge Scientific Publishers, 2006.

Lakshmikantham, V. and Devi, V.

  Strict stability criteria for impulsive differential systems. *Nonlinear Anal.* **21**(10) (1993) 785–794.

Lakshmikantham, V. and Leela, S.

[a]   Fuzzy differential systems and the new concept of stability. *Nonlinear Dynamics and Systems Theory* **1**(2) (2001) 111–119.

[b]   Controlled uncertain dynamic systems and stability of moving invariant sets. *Problems of Nonlinear Analysis in Engineering Systems*, Ed. 2 (1995) 9–13.

Lakshmikantham, V., Leela, S., and Martynyuk, A.A.

[a]   *Stability of Motion: Comparison Method.* Kiev: Naukova Dumka, 1991.

[b]   *Stability Analysis of Nonlinear Systems.* New York: Marcel Dekker, 1988.

Lakshmikantham, V., Leela, S., and Sivasundaram S.

  Lyapunov functions on product spaces and stability theory of delay differential equations. *J. Math. Anal. Appl.* **154**(2) (1991) 391–402.

Lakshmikantham, V., Matrosov, V.M., and Sivasundaram, S.

  Vector Lyapunov Functions and Stability Analysis of Nonlinear Systems. Amsterdam: Kluwer Academic Publishers, 1991.

Lakshmikantham, V. and Shahzad, N.

  Stability of moving invariant sets and uncertain systems. *Nonlinear Studies* **3**(1) (1996) 31–34.

Lakshmikantham, V. and Vatsala, A.S.

  Stability of moving invariant sets. In: *Advances in Nonlinear Dynamics* (Eds.: S. Sivasundaram and A.A. Martynyuk). Amsterdam: Gordon and Breach Scientific Publishers, 1997, p. 79–83.

LaSalle, J.P.

  The Stability of Dynamical Systems. Philadelphia: SIAM, 1976.

Leela, S.

  Uncertain dynamic systems on time scales. *Mem. Diff. Equns. Math. Phys.* **12** (1997) 142–148.

Leela, S. and Shahzad, N.

  On stability of moving conditionally invariant sets. *Nonlinear Analysis* **27**(7) (1996) 797–800.

Leitmann, G.

[a]   Deterministic control of uncertain systems via a constructive use of Lyapunov stability theory. In: *System Modeling and Optimization.* (Eds.: H.J. Sebastian and K. Tammer), Lecture Notes in Control and Information Sciences, **143**, Berlin: Springer Verlag, 1990, p. 38–55.

[b]   One approach to the control of uncertain dynamical systems. *Appl. Math. Comput.* **70** (1995) 261–272.

Letov A.M.

  Stability of Nonlinear Controlled Systems. Moscow: Fizmatgiz, 1964.

Lorenz, E.N.
Deterministic nonperiodic flow. *J. Atmos. Sci.* **20** (1963) 130–141.

Lyapunov A.M.
The general problem of the stability of motion. Leningrad–Moscow: ONTI, 1935.

Martynyuk, A.A.
[a]   Hierarchic matrix Lyapunov functions and the stability of solutions of uncertain systems. *Proceedings of the Institute of mathematics of National Academy of Sciences of Belarus* **4** (2000) 109–114.
[b]   Block-diagonal matrix Lyapunov function and the stability of uncertain impulsive systems. *Prikl. Mekh.* **40**(3) (2004) 322–327.
[c]   *Liapunov's Matrix Functions Method with Applications.* New York: Marcel Dekker, Inc., 1998.
[d]   On the stability of a set of trajectories of nonlinear dynamics. *Dokl. Acad. Nauk* **414**(3) (2007) 385–389.
[e]   *Stability of Motion: The Role of Multicomponent Liapunov's Functions.* London: Cambridge Scientific Publishers, 2007.
[f]   Comparison principle for a set of equations with a robust causal operator. *Doklady Mathematics* **80**(1) (2009) 1–4.

Martynyuk A.A. and Khoroshun A.S.
[a]   To the theory of parametric stability. *Dokl. NAS of Ukraine* **7** (2007) 59–65.
[b]   On the parametric asymptotic stability of large-scale systems. *Prikl. Mekh.* **44**(5) (2008) 104–114.

Martynyuk, A.A. and Martynyuk-Chernienko, Yu.A.
Stability analysis of quasilinear uncertain dynamical systems *Nonlinear Analysis* doi:10.1016/j.nonrwa.2007.01.019.

Martynyuk, A.A. and Miladzhanov, V.G.
Large-Scale Dynamical Systems: A Systematic Approach to Stability Analysis. Kiev: Institute of Mechanics NAS of Ukraine, Manuscript, (2009). (Russian)

Martynyuk, A.A. and Obolenskii, A.Yu.
Stability of autonomous Wazewski systems. *Differents. Uravn.* **16** (1980) 1392–1407.

Martynyuk, A.A. and Slyn'ko, V.I.
Choosing the parameters of a mechanical system with interval stability. *Int. Appl. Mech.* **39**(9) (2003) 1089–1092.

Martynyuk-Chernienko, Yu.A.
[a]   To the theory of stability of uncertain systems. *Dokl. NAS of Ukraine* **1** (1998) 28–31.
[b]   On the uniform asymptotic stability of solutions of an uncertain system with respect to an invariant set. *Dokl. Acad. Nauk* **364**(2) (1999) 163–166.
[c]   On the stability of motion of systems with uncertain parameter values. *Prikl. Mekh.* **35**(2) (1999) 101–104.
[d]   On the stability of solutions of a quasilinear uncertain system. *Ukr. Mathem. J.* **51**(4) (1999) 458–465.
[e]   On instability of solutions of an uncertain system with respect to a given moving set. *Dokl. NAS of Ukraine* **3** (2000) 23–29.
[f]   Exponential convergence of motions of an uncertain system. *Dokl. Acad. Nauk* **385**(3) (2002) 305–308.

[g]   *The conditions for the stability of motion of nonlinear systems with uncertain parameter values.* Thesis. Kiev: Institute of mechanics of Academy of Science of Ukraine 2001. [in Ukrainian]

[h]   To the theory of stability of the motion of an impulsive system with uncertain parameter values. *Dokl. Acad. Nauk* **395**(2) (2004) 160–163.

[i]   On the stability of dynamic systems on time scales. *Dokl. Acad. Nauk* **413**(1) (2007) 1–5.

[j]   A new generalization of direct Lyapunov method for uncertain dynamical systems. *Nonlinear Dynamics Systems Theory* **3**(2) (2003) 191–202.

[k]   Stability analysis of uncertain systems via matrix-valued Liapunov functions. *Nonlinear Analysis* **63** (2005) 388–404.

[l]   Exponential convergence and instability of solutions of uncertain dynamical systems. *Nonlinear Analysis* **66** (2007) 1318–1328.

Martynyuk-Chernienko Yu.A. and Slynko V.I.

[a]   On the stability of conditionally invariant sets of dynamic systems with uncertain parameter values. *Dokl. Acad. Nauk* **385**(3) (2002) 305–308.

[b]   On the synthesis of controls of an uncertain system with a conditionally invariant set. *Prikl. Mekh.* **39**(11) (2003) 125–136.

Matrosov V.M.

The Method of Vector Lyapunov Functions: Analysis of Dynamic Properties of Nonlinear Systems. Moscow: Fizmatgiz, 2001.

Matrosov V.M., Vassilyev S.N., Karatuyev V.G., et al.

Derivation Algorithms for Theorems of the Method of Vector Lyapunov Functions. Novosibirsk: Nauka, 1981.

McRae, F.A. and Devi, V.

Impulsive set differential equations with delay. *Applicable Analysis* **84**(4) (2005) 329–341.

Melnikov G.I.

Some questions of the direct Lyapunov method. *Dokl. Acad. Nauk* **110**(3) (1956) 326–329.

Michel, A.N. and Miller, R.K.

Qualitative Analysis of Large Scale Dynamical Systems. New York: Academic Press. 1977.

Ming-Chung, Ho, Yao-Chen, Hung and Min Jiang, I.

On the synchronization of uncertain chaotic systems. *Chaos, Solitons Fractals* **33** (2007) 540–546.

Neimark Yu.I., Fufaev N.A.

The Dynamics of Nonholonomic Systems. Moscow: Nauka, 1967.

Ohta, Y. and Šiljak, D.D.

Parametric quadratic stabilizability of uncertain nonlinear systems. *Systems and Control Lett.* **22** (1994) 437–444.

Pastravanu, O. and Voicu, D.

Internal matrix systems — Flow invariance and componentwise asymptotic stability. *Differential Integral Equations* **15**(11) (2002) 1377–1394.

Pucci, P. and Serrin, J.
Remarks on Lyapunov stability. *Differential Integral Equations* **8**(6) (1995) 1265–1278.

Rotea, M.A. and Khargonekar, P.P.
Stabilization of uncertain systems with norm bounded uncertainty — A control Lyapunov function approach. *SIAM J. Control Optimization* **27** (1989) 1462–1476.

Samoilenko A.M., and Perestyuk N.A.
Impulsive Differential Equations. Singapore: World Scientific, 1995.

Šiljak, D.D.
[a] *Large-Scale Dynamic Systems: Stability and Structure.* New York: North Holland, 1978. 416 p.
[b] Parameter space methods for robust control design: a guided tour. *IEEE Trans.* **AC-34** (1989) 674–688.

Šiljak, D.D. and Šiljak, M.D.
Nonnegativity of uncertain polynomials. *Math. Problems Engineering* **4** (1998) 135–163.

Silva, G. and Dzul, F.A.
Parametric absolute stability of a class singularly perturbed systems. *Proc. 37th IEEE Conf. Decision & Control* (Tampa, FL, USA, 1998), p. 1422–1427.

Skowronski, J.M.
Parameter and state identification in non-linearizable uncertain systems. *Int. J. Non-Linear Mechanics* **19**(5) (1984) 421–429.

Tikhonov A.A.
[a] On the stability of motion under continuously acting perturbations. *Vestnik LGU.* **1**(1) (1965) 95–101.
[b] To the problem of the stability of motion under continuously acting perturbations. *Vestnik LGU* **4**(19) (1969) 116–122.

Wada, T., Ikeda, M., Ohta, Y. and Šiljak, D.D.
[a] Parametric absolute stability of Lur'e systems. *IEEE Trans. Automat. Contr.* **43** (1998) 1649–1653.
[b] Parametric absolute stability of multivariable Lur'e systems. *Automatica* **36** (2000) 1365–1372.

Xing-Gang, Yan, and Guang-Zhong, Dai
Stability analysis and estimation of the parametric robust space of a nonlinear composite system. *IMA J. Math. Contr. Inform.* **16** (1999) 353–362.

Yakubovich, V.A.
Absolute stability of nonlinear regulating systems in critical cases. 1. *Automatica Telemekhanika* **24** (1963) 293–303.

Yoshizawa, T.
Stability Theory by Liapunov's Second Method. Tokyo: Math. Soc. Japan, 1966.

Zadeh, L.A.
Fuzzy sets. *Informat. Contr.* **8**(12) (1965) 338–353.

Zubov, V.I.
[a] Exponential asymptotic stability. *Russ. Acad. Sci. Dokl. Math.* **48**(2) (1994) 328–331.
[b] *Methods of A.M.Lyapunov and its Applications.* Leningrad: Izdat. 1957.

Index

Milton Keynes UK
Ingram Content Group UK Ltd.
UKHW021619071024
449327UK00020BA/1116